ISO/IEC 导则 第 1 部分
JTC 1 技术工作程序合集

(第 15 版,2019)

ISO/IEC 著

国家标准化管理委员会
国家市场监督管理总局标准创新管理司 译

中国标准出版社

北 京

IEC Central Office
3, rue de Varembé
CH-1211 Geneva 20
Switzerland
Tel: +41 22 919 02 11
Fax: +41 22 919 03 00
info@iec.ch
www.iec.ch

ISO Copyright Office
Case postale 401
CH-1214 Vernier Geneva
Switzerland
Tel: + 41 22 749 01 11
Fax: +41 22 749 09 47
copyright@iso.org
www.iso.org

THIS PUBLCATION IS COPYRIGHT PROTECTED

Copyright © 2019 ISO/IEC, Geneva Switzerland

All rights reserved. It is permitted to download this electronic file, to make a copy and to print out the content for the purpose of preparing ISO and IEC documents only. You may not copy or "mirror" the file, or any part of it, for any other purpose without permission in writing from the publishers.

北京市版权局著作权合同登记号：图字 01-2019-6269

图书在版编目（CIP）数据

ISO/IEC 导则.第 1 部分.JTC 1 技术工作程序合集：第 15 版：2019：汉文、英文/ISO/IEC 著.—北京：中国标准出版社，2020.7
书名原文：ISO/IEC Directives, Part 1 Consolidated JTC 1 Supplement 2019—Procedures specific to JTC 1
ISBN 978-7-5066-9576-3

Ⅰ.①Ⅰ…　Ⅱ.①Ⅰ…　Ⅲ.①电子工业—质量管理体系—国际标准—2019—汉、英　Ⅳ.①TM-65

中国版本图书馆 CIP 数据核字（2020）第 051173 号

中国标准出版社出版发行
北京市朝阳区和平里西街甲 2 号（100029）
北京市西城区三里河北街 16 号（100045）
网址：www.spc.net.cn
总编室：（010）68533533　发行中心：（010）51780238
读者服务部：（010）68523946
中国标准出版社秦皇岛印刷厂印刷
各地新华书店经销
＊
开本 787×1092　1/16　印张 23.25　字数 423 千字
2020 年 7 月第一版　2020 年 7 月第一次印刷
＊
定价：99.00 元

如有印装差错　由本社发行中心调换
版权专有　侵权必究
举报电话：（010）68510107

审定和翻译委员会

主任委员 田世宏

副主任委员 崔 钢

主　　审 李玉冰　　肖 寒　　白德美

审稿人员（按姓氏音序排序）

　　　　　　高丽稳　　黄 立　　李东方
　　　　　　李鹏程　　刘永东　　田 武
　　　　　　余云涛　　余 化　　张 亮

翻译人员（按姓氏音序排序）

　　　　　　丁 宁　　田 武　　张 戈

译者的话

国际标准是全球治理体系和经贸合作发展的重要技术基础。随着我国对外开放程度的不断扩大，经济快速发展，科技水平持续进步，对我国标准化工作者参与国际标准化活动的能力提出了更高的要求，迫切需要我国专家和相关人员熟悉和掌握国际标准制修订工作程序和规则。为此，国家标准化管理委员会组织翻译了《ISO/IEC 导则》的重要出版物。

《ISO/IEC 导则》是国际标准化组织（ISO）和国际电工委员会（IEC）技术工作的重要文件，是 ISO/IEC 各成员体参与国际标准化技术活动的指导性文件。随着国际标准化事业的不断发展，ISO/IEC 在认真分析和总结实践经验的基础上，对技术工作程序不断进行修改和完善。2019 年版《ISO/IEC 导则》分三本书出版：第 1 本为《ISO/IEC 导则 第 1 部分 ISO 技术工作程序合集（第 10 版，2019）》；第 2 本为《ISO/IEC 导则 IEC 技术工作程序补充部分（第 13 版，2019）》；第 3 本为《ISO/IEC 导则 第 1 部分 JTC 1 技术工作程序合集（第 15 版，2019）》。

《ISO/IEC 导则 第 1 部分 ISO 技术工作程序合集（第 10 版，2019）》由《ISO/IEC 导则》的第 1 部分和 ISO 专用程序组成。书中提供了完整的 ISO 技术工作程序，供 ISO 委员会遵守，主要涉及通过技术委员会及其附属机构的活动制定和维护国际标准。

《ISO/IEC 导则 IEC 技术工作程序补充部分（第 13 版，2019）》，包括了对 ISO/IEC 导则的修改和补充内容，提供了 IEC 委员会应遵循的完整程序规则，经 IEC 标准化管理局批准，已在 IEC 内部实施。

《ISO/IEC 导则　第 1 部分　JTC 1 技术工作程序合集（第 15 版，2019）》包含了《ISO/IEC 导则》的第 1 部分和 JTC 1 补充部分给出的程序规则，纳入了自 2018 年第 14 版出版以来 ISO 和 IEC 一致同意修改的所有内容。带修订模式（灰色底纹）的第 15 版可用于查阅修改的细节，英文版本中已删除的内容未做翻译，为方便读者查阅，保留了英文原文（含删除线）。因第 14 版的附录 JC 修改为第 15 版的附录 L，附录 JC 英文全部内容以带删除线形式呈现，为了避免重复，未将其收入本书。

上述三本书中文版的出版，有助于我国标准化工作者不断跟踪和研究导则的变化内容，进而更加实质性地参与 ISO/IEC 国际标准化技术工作。为方便读者对照参考，书中附有相应的英文版。同时，希望通过本书普及国际标准化组织的相关知识，加强我国专家国际标准化能力建设，不断提升我国参与国际标准化活动的整体水平。

在本书编译过程中，我们得到了国内外标准化专家和标准化出版机构专家的大力支持和帮助，在此谨向他们表示衷心的感谢。

由于时间仓促，书中疏漏在所难免，恳请广大读者和标准化同行提出宝贵意见。

译　者

2020 年 3 月

目 录

0 引言（《JTC 1 技术工作程序合集》）...1

前 言...3

1 技术工作的组织构架和职责..7

   1.1 技术管理局的任务...7
   1.2 技术管理局咨询组...7
   1.3 联合技术工作...8
   1.4 首席执行官的职责...8
   1.5 技术委员会的设立...9
   1.6 分委员会的建立...10
   1.7 参加技术委员会和分委员会工作...11
   1.8 技术委员会和分委员会主席...13
   1.9 技术委员会和分委员会秘书处...14
   1.10 项目委员会...16
   1.11 编辑委员会...17
   1.12 工作组...17
   1.13 委员会中具有咨询职能的小组...20
   1.14 临时工作组...20
   1.15 技术委员会间的联络...21
   1.16 ISO 与 IEC 之间的联络...21
   1.17 与其他组织的联络...22

2 国际标准的制定 26

2.1 项目管理方法 26
2.2 预研阶段 30
2.3 提案阶段 31
2.4 准备阶段 33
2.5 委员会阶段 34
2.6 征询意见阶段 36
2.7 批准阶段 38
2.8 出版阶段 39
2.9 可交付使用的成果的维护 40
2.10 更正和修正 43
2.11 维护机构 45
2.12 注册机构 45
2.13 版权 45
2.14 专利项目的引用（见附录I） 46

3 其他可交付使用的文件的制定 46

3.1 技术规范 46
3.2 可公开提供的规范（PAS） 47
3.3 技术报告 47

4 会议 48

4.1 概述 48
4.2 召开会议的程序 49
4.3 会议语言 50
4.4 会议的取消 51

5 申诉 51

5.1 概述 51
5.2 对分委员会决定的申诉 51

5.3 对技术委员会决定的申诉 52

5.4 对技术管理局决定的申诉 52

5.5 申诉期间工作的运行 52

附录 A （规范性）指南 53

附录 B （规范性）ISO/IEC 联络和分工程序 55

附录 C （规范性）对制定标准的提案的论证 59

附录 D （规范性）秘书处资源和秘书资格 64

附录 E （规范性）语言使用通则 66

附录 F （规范性）项目制定的选择方案 70

附录 G （规范性）维护机构 82

附录 H （规范性）注册机构 83

附录 I （规范性）ITU-T/ITU-R/ISO/IEC 共用专利政策实施指南 84

附录 J （规范性）技术委员会和分委员会的制定范围 101

附录 K （规范性）项目委员会 103

附录 L （规范性）管理体系标准提案 105

附录 M （规范性）制定行业专用管理标准和管理体系标准（MSS）的政策 127

附录 JA （规范性）投票 132

附录 JB （规范性）ITU-T 与 ISO/IEC JTC 1 的合作 137

附录 JD （规范性）阶段编码的矩阵表示 139

附录 JE （规范性）图形符号标准化程序 141

附录 JF （规范性）注册机构（RA）政策 146

参考文献 152

0 引言（《JTC 1 技术工作程序合集》）

0.1 关于《JTC 1 技术工作程序合集》

《ISO/IEC 导则》的第 1 部分规定了制定国际标准和其他出版物应遵循的基本程序。本《JTC 1 技术工作程序合集》包含 JTC 1 专用程序。

《ISO/IEC 导则》的第 1 部分和本《JTC 1 技术工作程序合集》规定了 ISO/IEC JTC 1 应遵循的程序规则。此外，可能还需参照其他指导文件，如 JTC 1 的常备文件。JTC 1 的专用表单可见 JTC 1 模板文件夹，网址 http://isotc.iso.org/livelink/livelink?func=ll&objId=8913214&objAction=browse&sort=name。

0.2 《JTC 1 技术工作程序合集》与《ISO/IEC 导则》的第 1 部分的关系

本《JTC 1 技术工作程序合集》包括了 2019 年发布的第 15 版 ISO/IEC 导则的第 1 部分以及 JTC 1 专用程序。

0.3 JTC 1 技术工作程序合集的结构

《JTC 1 技术工作程序合集》的条款结构遵循《ISO/IEC 导则》的第 1 部分的规定。

0.4 获取《JTC 1 技术工作程序合集》

可访问 https://www.iso.org/directives-and-policies.html 获取《ISO/IEC 导则》的第 1 部分和第 2 部分、《ISO 技术工作程序合集》《JTC 1 技术工作程序合集》及其他相关文件。可访问 www.jtc1.org 获取 JTC 1 常备文件。可访问 http://www.iec.ch/members_experts/refdocs/iec/isoiecdir iecsup%7bed10.0%7den.pdf 获取《IEC 技术工作程序补充部分》。

0.5 《JTC 1 技术工作程序合集》的联系信息

有关《JTC 1 技术工作程序合集》的建议或问题，请联系：

国际标准化组织（ISO） 地址：Chemin de Blandonnet 8 CP 401-1214 Vernier, Geneva Switzerland 电话：+41227490111 传真：+41227333430 电邮：central@iso.org	国际电工委员会（IEC） 地址：3, rue de Varembé P.O.Box 131 CH-1211 GENEVA 20 Switzerland 电话：++41229190211 传真：+41229190300 电邮：info@iec.ch	ISO/IEC 第一联合技术委员会（JTC 1）秘书处 地址：美国国家标准学会（c/o ANSI）转交 25 West 43rd Street, 4th Floor New York, NY 10036 USA 电话：+12126424932 传真：+12128402298 电邮：lrajchel@ansi.org

前 言

《ISO/IEC 导则》分为两部分出版：

—— 第1部分：技术工作程序

—— 第2部分：ISO和IEC文件的结构和起草原则与规则

另外，国际标准化组织（ISO）、国际电工委员会（IEC）和ISO/IEC第一联合技术委员会（JTC 1）还分别出版了对第1部分的独立补充部分，包括各自的专用程序。

本部分阐述了ISO和IEC在开展技术工作时应遵循的程序：主要涉及通过技术委员会及其附属机构的活动制定和维护国际标准。

ISO、IEC和ISO/IEC JTC 1还在各自网站（http：//www.iso.org/directives；http：//www.iec.ch/ members_experts/refdocs/ 和 http：//www.jtc1.org）上向那些关注技术文件制定的人员提供其他指南和工具。

本文件为第15版，纳入了自2018年第14版出版以来ISO技术管理局和IEC标准化管理局一致同意修改的所有内容。《ISO/IEC 导则》的专用程序，分别以《ISO 技术工作程序》（亦称《ISO 技术工作程序合集》）、《IEC 技术工作程序补充部分》或 ISO/IEC《JTC 1 技术工作程序》（亦称《JTC 1 技术工作程序合集》）的形式给出。这些补充部分应与本文件结合使用。

对前一版本修改的条款如下：前言、1.7.3、1.12.2、1.13.2、1.13.6、1.17.2、3.2.1、3.2.2、3.2.4、附录B、附录D、附录E、附录F、附录I和附录M。带修订模式的第15版可用于查阅修改的细节。考虑到国际标准具有成本效益和适时性的特点，ISO 和 IEC 制定本程序，并且得到广泛认可和普遍应用。为了实现这些目标，本程序的制定基于以下原则。

a） 现行技术和项目管理

在这些程序框架内，通过应用现行技术（如信息技术工具）和项目管理方法，提高工作速度，方便专家和秘书处完成任务。

b） 协商一致

协商一致原则旨在解决实质性的反对意见，这是制定能获得许可并被广泛采用的国际标准的一条基本原则和必要条件。虽然加速开展技术工作是必要的，但在标准进入批准阶段之前要为讨论、协商和解决重大技术争议问题留出足够的时间。

关于"协商一致"原则的详细信息，见 2.5.6。

c) 纪律

国家成员体必须保证遵守时间要求的纪律，以避免过长和不确定的"停滞时间"。同样，为了避免重复讨论，国家成员体有责任保证其技术观点是在国家层面考虑所有相关利益方的基础上提出的，并且应在技术工作初期明确提出观点而非在最后（批准）阶段提出。国家成员体还必须认识到不要把实质性意见提交会议上讨论，因其他会议代表没有机会就该意见提前在本国进行充分磋商，因而无法迅速达成协商一致。

d) 成本效益

本程序考虑了运行总成本。"总成本"的概念包括各国家成员体的直接支出、日内瓦办公机构的支出（主要由国家成员体交纳的会费资助）、差旅费以及国家和国际工作组和委员会专家的工时费。

e) 表决和决定的通用原则

委员会秘书处应确保委员会无论是在全体会议上还是以通信方式作出的决定都应有记录，在 ISO 以委员会决议的形式体现。

在 JTC 1 中，对于信函投票或在委员会会议期间，除非 ISO/IEC 导则的第 1 部分另有规定，要求 P 成员简单多数同意批准。

在 JTC 1 中，委员会领导层应确保在会议期间审议在委员会会议前以书写方式进行的投票。

在 JTC 1 中，如果文件需要投票（NP、CD 或任何随后阶段），禁止在会议期间进行正式讨论，或通过正式委员会表明途径表明国家体立场。

在 JTC 1 中，所有弃权票不计。

在 JTC 1 中，不允许进行委托投票。

f) ISO 国际标准的全球相关性

在 JTC 1 中，推行 ISO 结对程序。ISO 期望其每一项国际标准都能代表全球共识，并符合全球市场需求。为了实现这一目标，ISO 认识到，须采取特殊措施，以确保在 ISO 的技术工作中考虑到发展中国家的需要。其中一项措施是将本《JTC 1 技术工作程序合集》关于"结对"的具体规定纳入《ISO/IEC 导则》，即国家成员体为提升能力结成伙伴关系。（见 1.7、1.9.2、1.9.3、1.9.4 和《ISO 技术工作程序合集》的附录 ST。）

虽然这些规定应限于技术工作，但"结对"可能体现在多个层面上，特别是帮助结对伙伴形成建立标准化体系、合格评定和 IT 基础设施的能力，从而使其最终能够独立承担工作。

g) 本文件使用的术语

注1：为方便起见，本文件在适用处均采用了下表中的术语，表示 ISO 和 IEC 中相似或相同的概念。

本文件术语	ISO	IEC
国家成员体 National Body	成员体 Member Body（MB）	国家委员会 National Committee（NC）
技术管理局 Technical Management Board（TMB）	ISO 技术管理局 Technical Management Board（ISO/TMB）	标准化管理局 Standardization Management Board（SMB）
首席执行官 Chief Executive Officer（CEO）	秘书长 Secretary-General	秘书长 General Secretary
CEO 办公室 Office of the CEO	中央秘书处 Central Secretariat（CS）	中央办公室 Central Office（CO）
理事局 Council Board	理事会 Council	理事局 Council Board（CB）
咨询组 Advisory Group	技术咨询组 Technical Advisory Group（TAG）	咨询委员会 Advisory Committee
秘书（委员会或分委员会） Secretary（of a Committee or Subcommittee）	委员会主管 Committee Manager	秘书 Secretary
关于其他概念，请参照 ISO/IEC 指南 2。		

在 JTC 1 中，国家成员体指 JTC 1 成员的国家成员体。在 JTC 1 中，"首席执行官办公室"指信息技术任务组（ITTF）。在本《JTC 1 技术工作程序合集》中，"技术管理局"等术语指的是 ISO 和 IEC 机构。例如，在使用术语"首席执行官（CEO）"时，宜认为其包括 ISO 秘书长和 IEC 秘书长。

在 JTC 1 中，使用了委员会主管一词。此称谓更改仅适用于 TC 和 SC 层面。秘书仍然适用于工作组层面。

除上述提及的条款外，已对本《JTC 1 技术工作程序合集》(2019) 前一版本的以下条款进行了实质性修改：已从 1.13.2、1.14、2.1.3.3、2.1.6、2.1.9、2.4、2.5 和附录 JC 中删除了对 JTC 1 专用程序的引用。为与前言、2.5.3、2.7.3、2.9.2.1、2.9.5、附录 L 和附录 M 保持一致，本版本新增了一些内容。

有关修改细节，请查阅 2019 版修订模式。

注 2：此外，本文件中还使用了下列缩略语：

JTAB	联合技术咨询局
JCG	联合协调小组
JPC	联合项目委员会
JTC	联合技术委员会
JWG	联合工作组
TC	技术委员会
SC	分委员会
PC	项目委员会
WG	工作组
PWI	预研工作项目
NP	新工作项目提案
WD	工作草案
CD	委员会草案
DIS	国际标准草案（ISO）
CDV	投票用委员会草案（IEC）
FDIS	最终国际标准草案
PAS	可公开提供的规范
TS	技术规范
TR	技术报告
AMD	修正
COR	勘误表
DCOR	技术勘误草案
DR	缺陷报告
SD	常备文件
HoD	代表团团长
NWI	新工作项目

1 技术工作的组织构架和职责

1.1 技术管理局的任务

各组织的技术管理局负责技术工作的全面管理,具体如下:

a) 设立技术委员会。

b) 任命技术委员会主席。

c) 指派或重新指派技术委员会秘书处和分委员会秘书处(在某些情况下)。(在任何情况下,应由 JTC 1 决定指派分委员会秘书处。)

d) 批准技术委员会的名称、范围和工作计划。

e) 批准由技术委员会提出的设立和解散分委员会的决定。

f) 如有必要,安排技术工作特定项目的优先顺序。

g) 协调技术工作,包括对多个技术委员会相关的主题的标准制定进行指派、分工或协调。为协助此项工作,技术管理局可在相关领域设立专家咨询小组,就其基础、行业及跨行业协调工作提供咨询,并对相关的计划及新工作需求提出建议。

h) 在 CEO 办公室的协助下,监控技术工作的进展情况,并采取适当的行动。

i) 审查对新技术领域的工作的需求并进行规划。

j) 维护 ISO/IEC 导则和其他技术工作规则。

k) 考虑国家成员体提出的原则性事项及对新工作项目提案、委员会草案、征询意见草案或最终国际标准草案决定的申诉。

注1:对新工作项目提案、委员会草案、征询意见草案及最终国际标准草案的术语解释见第2章。

注2:关于 ISO 技术管理局的任务和职责的详细信息,见 ISO/TMB(技术管理局)的职权范围——https://www.iso.org/committee/4882545.html;关于 IEC,见 http://www.iec.ch/dyn/www/f?p=103:47:0::::FSP_ORG_ID,FSP_LANG_ID:3228,25。

1.2 技术管理局咨询组

1.2.1 就 1.1 g)而言,具有咨询职能的小组可由:

a) 其中一个技术管理局设立;

b) 两个技术管理局联合设立。

注:在 IEC 中,这类小组被称为咨询委员会。

1.2.2 成立这类小组的提案中应包括拟成立小组的职责和组成建议,切记为了保证其有效运作,在确保充分代表相关方的同时,要尽可能限制小组的规模。例如,可以决定其成员仅由相关技术委员会的主席和秘书组成。无论如何,技术管理局都应决定拟实施的准则,并指定小组的成员。

这类小组对其职责、组成或工作方法（适当时）提出的任何变更建议都应提交技术管理局审批。

1.2.3 指派给这类小组的任务可包括对出版物（特别是国际标准、技术规范、可公开提供的规范及技术报告）的起草或协调提出建议，但不应包括起草这些文件，除非由技术管理局特别授权。

1.2.4 任何准备出版的文件都应按照附录 A 的程序制定。

1.2.5 咨询组的工作结果应以建议的形式提交到技术管理局。这种建议可包括为起草出版物成立工作组（见 1.12）或联合工作组（见 1.12.6）的提案。如有这类工作组，应在相关的技术委员会内开展工作。

1.2.6 咨询组内部文件只应分发给其成员，并抄送给 CEO 办公室。

1.2.7 规定任务一旦完成，或如果随后决定其工作可由正常的联络机构完成，咨询组应解散（见 1.16）。

1.3 联合技术工作

1.3.1 联合技术咨询局（JTAB）

联合技术咨询局的任务是避免或消除 ISO 和 IEC 技术工作中可能或实际重复的工作，并在其中一个组织认为需要做出联合计划时采取措施。联合技术咨询局只处理现行程序下较低层面不能解决的问题（见附录 B）。这些问题除了技术工作外，可能还涉及策划和程序等内容。

联合技术咨询局的决定要通知 ISO 和 IEC 两个组织立即实施，这种决定至少在 3 年内不得申诉。

1.3.2 联合技术委员会（JTC）和联合项目委员会（JPC）

1.3.2.1 根据 ISO 技术管理局和 IEC 标准化管理局的共同决定或联合技术咨询局的决定可设立联合技术委员会和联合项目委员会。

1.3.2.2 应由两个组织协商同意，由其中一个组织承担联合项目委员会的管理责任。按照一个成员/国家一票的原则参与其工作。

若在同一国家的两个国家成员体都选择参加一个联合项目委员会的工作，应确定一个成员体承担该委员会的管理责任。承担管理责任的国家成员体有责任协调在其国家的活动，包括文件分发、评论和投票。

此外，还要遵循项目委员会的正常程序（见 1.10）。

1.4 首席执行官的职责

各组织的首席执行官主要负责在其组织内部实施《ISO/IEC 导则》和其他技术工作规则。为此，CEO 办公室负责各技术委员会、理事局和技术管理局间的衔接。

未经 ISO 或 IEC 首席执行官、ISO/IEC 联合技术咨询局或技术管理局的授权，不得制定与本文件不一致的程序。

在 JTC1 中，"首席执行官"由信息技术任务组（ITTF）代表。

1.5 技术委员会的设立

1.5.1 技术委员会由技术管理局设立和解散。

1.5.2 经与相关技术委员会协商后，技术管理局可将现有的分委员会转换成新的技术委员会。

1.5.3 针对新技术活动领域设立新技术委员会的提案可由下列组织提出：
—— 国家成员体；
—— 技术委员会或分委员会；
—— 项目委员会；
—— 政策委员会；
—— 技术管理局；
—— 首席执行官；
—— 有相关组织支持的负责管理认证体系的机构；
—— 具有国家成员体资格的其他国际组织。

1.5.4 提案应使用规定的表格，可从 www.iso.org/forms 和 http：//www.iec.ch/standardsdev/resources/docpreparation/forms_templates/ 下载电子版（通常为 word 文档），提案的内容包括：

a） 提案方；
b） 提案的主题；
c） 提案的范围和初始工作计划；
d） 提案的论证；
e） 如果适用，在其他机构开展的类似工作的调查报告；
f） 认为有必要与其他机构的联络。

有关新工作提案的详细内容，见附录 C。该表格应提交给 CEO 办公室。

1.5.5 CEO 办公室应确保提案是按 ISO 和 IEC 的要求（见附录 C）正确编制的，并应提供足够信息以支持国家成员体做出明确决定。CEO 办公室还应评估提案与现行工作的关系，可以征询有关方面意见，包括技术管理局或主持现行有关工作的委员会。必要时，可设立一个临时工作组对该提案进行审查。

CEO 办公室审查后，可以决定将提案退给提案方，以便在分发投票前做进一步修改。在这种情况下，提案方应按建议进行修改或提供不做修改的理由。如提案方不进行修改并且要求对原提交的提案分发投票，技术管理局将决定采取适当的措施。这可能包括停止该提案直到对提案进行修改或对收到的建议进行投票。

在所有情况下，CEO 办公室都可以将评议意见和建议填入提案表格中。

有关提案论证的详细内容，见附录 C。

积极鼓励提案方在编制提案过程中与其他国家成员体进行非正式沟通。

1.5.6 CEO 办公室应将提案分发给各组织（ISO 或 IEC）的所有国家成员体，询问他们是否：

　　a）支持设立新的技术委员会，并提供一份陈述理由的说明（"理由陈述"）；

　　b）打算积极参与新技术委员会的工作（见 1.7.1）。

该提案还应提交给另一组织（IEC 或 ISO）征询意见并达成协议（见附录 B）。

应在提案分发后的 12 周内用适当的表格对提案做出答复。若国家成员体未提供上述 1.5.6 a）提及的说明，其赞成票或反对票均不予以登记和考虑。

JTC 1 的提案答复表格已由电子投票系统取代。未使用电子投票系统的答复将不予统计。

1.5.7 技术管理局对答复进行评估，并且决定设立一个新的技术委员会，基于：

　　1）参加投票表决的国家成员体中 2/3 多数赞成此项提案；

　　2）至少有 5 个投赞成票的国家成员体表达了其积极参加活动的意向。

并指派秘书处（见 1.9.1），或者，根据上述相同的验收标准，把工作指派给现有技术委员会。

1.5.8 应按其成立的顺序对技术委员会进行编号，如果某个技术委员会解散，其编号不得指派给另一个技术委员会。

1.5.9 在做出设立新的技术委员会决定之后应尽快安排必要的联络（见 1.15 ~ 1.17）。

1.5.10 新技术委员会在成立之后应就其名称和范围尽快形成一致意见，最好采用通信方式。范围是明确界定技术委员会工作界限的声明。规定技术委员会的范围应以"……的标准化"或"……领域的标准化"句子开始，措辞应尽可能简明。

有关业务范围的建议见附录 J。达成一致意见的名称和范围应由首席执行官提交技术管理局批准。

1.5.11 技术管理局或技术委员会可以提出修改技术委员会名称和范围的建议，修改的措辞应由技术委员会确定并由技术管理局批准。

1.6 分委员会的建立

1.6.1 设立和解散分委员会由上级技术委员会参加投票的 P 成员的 2/3 多数决定，并需技术管理局批准。只有当某一国家成员体表示愿意承担秘书处工作的条件下方可设立分委员会。

1.6.2 分委员会设立时，应该包括至少 5 个表示愿意积极参加（见 1.7.1）分委员会工作的上级技术委员会成员。

1.6.3 一个技术委员会的分委员会应按顺序设立编号，如果某个分委员会已经解散，其编号不得分配给另一个分委员会，除非解散分委员会是技术委员会进行全面调整工作的一部分。

1.6.4 分委员会的名称及范围应由上级技术委员会决定，并应包括在已确定的上级技术委员会范围内。

1.6.5 上级技术委员会秘书处应采用适当表格将其设立分委员会的决定通知 CEO 办公室，CEO 办公室再将表格提交技术管理局批准。

1.6.6 设立新分委员会的决定得到批准后，应尽快安排与其他机构建立必要的联络关系（见 1.15 ~ 1.17）。

1.7 参加技术委员会和分委员会工作

在 JTC 1 中，推行用于结对的 ISO 程序。人们认识到，发展中国家的成员体往往缺乏资源，无法参加可能对其国民经济有重要影响的所有委员会。因此，邀请发展中国家成员体与经验更为丰富的发达国家 P 成员建立 P 成员结对关系。在这种安排下，发达国家 P 成员确保将结对 P 成员的意见传达给归口 ISO 委员会并加以审议。因此，结对 P 成员也应具有 P 成员的地位（见注），并由中央秘书处登记为结对 P 成员。

> **注：** 有关成员体决定实施结对的最有效方法。例如，这可能包括 P 成员赞助结对成员体的专家参加委员会会议或在工作组中担任专家，也可能包括 P 成员征询结对成员体对特定议程项目/文件的意见，确保结对国家成员体以文字形式将其立场传达给委员会秘书处。

所有结对安排的细节应通知有关委员会的秘书处和主席，委员会成员和 CEO 办公室也应得到相应的通知，以确保最大透明度。

在具体的委员会中，某个发达国家的 P 成员只能与一个发展中国家的 P 成员结对。

结对 P 成员应针对委员会的所有表决事宜自行提交投票。

有关结对的更多信息，请见《ISO 技术工作程序合集》附录 ST "结对指南"。

根据 ISO 章程和议事规则，通讯成员和订户成员不具备 P 成员资格。ISO 的通讯成员可以注册为委员会观察员，但无权提出意见。

1.7.1 所有的国家成员体都有权参加技术委员会和分委员会工作。

在 JTC 1 中，每个国家不允许超过一个国家成员体（ISO 成员体或 IEC 国家委员会成员体）成为 JTC 1 的成员，同样，每个国家也只允许一个国家体成为 JTC 1 分委员会的成员。

为了获得最高的工作效率，维护必要的工作纪律，各国家成员体都应针对每个技术委员会和分委员会的情况向 CEO 办公室明确表示它是否打算：

—— 积极参加其工作，有义务对技术委员会或分委员会内正式提交投票的所有问题、新工作项目提案、征询意见草案和最终国际标准草案进行投票以及对会议做出贡献（P 成员）；

—— 或者，以观察员身份跟进工作，因此接收委员会文件并有权提出评议意见和参加会议（O 成员）。

在 JTC 1 中，选择成为委员会 P 成员的 JTC 1 国家成员体还有义务对该委员会负责的所有系统复审项目进行投票。

国家成员体可以选择既不作为某一技术委员会的 P 成员也不作为 O 成员。在这种情况下，该国家成员体对该技术委员会工作既无以上规定的权利也不承担上述义务，但所有国家成员体无论其在技术委员会或分委员会中的身份如何，都有权对国际标准的征询意见草案（见 2.6）和最终草案（见 2.7）进行投票。

在 JTC 1 中，一国一票。

国家成员体有责任组织国内各相关方，考虑其利益，有效、及时地提交国家层面的意见。

1.7.2 分委员会的成员资格对任何国家成员体开放，无论他们在上级技术委员会中的身份如何。

在分委员会成立之际，应为技术委员会成员提供机会明确他们成为分委员会 P 成员或 O 成员的意向。

技术委员会成员不意味着自动成为分委员会成员，因此国家成员体应正式告知其成为分委员会身份的意向。

1.7.3 国家成员体可在任何时间开始、结束或改变其在技术委员会或分委员会中的成员身份。对于 IEC，须通知 CEO 办公室或相关委员会秘书处；对于 ISO，须按照 1.7.4 和 1.7.5 的要求在全球目录系统中直接输入。

1.7.4 如果某技术委员会或分委员会的 P 成员出现以下情况，该技术委员会或分委员会秘书处应通知 CEO 办公室：

—— 长期不积极，并且连续两次既未直接参加，也未以通信方式为会议做出贡献，未任命专家参加技术工作；

—— 在 IEC，对技术委员会或分委员会内正式提交投票的问题没有投票（见 1.7.1）；

—— 在 ISO，在超过一个日历年内对技术委员会或分委员会内正式提交进行委员会内部投票（CIB）的 20% 以上（并且至少 2 个）的问题没有投票（见 1.7.1）。

在 JTC 1 中，应遵守 ISO 政策。

一旦收到这种通知，首席执行官应提醒国家成员体有义务积极参加技术委员会和分委员会活动。如果该团体对此提示没有做出令人满意的回复，而且连续出现不符合 P 成员要求的行为，该国家成员体的身份应自动降为 O 成员。这样改变身份的国家成员体在 12 个月之后可以向首席执行官表明其希望重新成为该技术委员会 P 成员的愿望，在这种情况下，可以批准其成为 P 成员。

注： 本条款不适用于指南的制定。

1.7.5 如果某技术委员会或分委员会的 P 成员没有参加对相关委员会起草的征询意见草案和最终国际标准草案投票表决，或者没有对本委员会负责的可交付使用的文件进行系统复审投票，首席执行官应提醒该国家成员体履行投票义务。如果对该提醒没有做出满意的回应，则该国家成员体应自动降为 O 成员。这样改变身份的国家成员体在 12 个月后可以向首席执行官表明其重新成为该委员会 P 成员的愿望，在这种情况下，可以批准其为 P 成员。

注： 本条款不适用于指南的制定。

在 JTC1 中，应遵循 ISO 的系统复审政策。

1.8 技术委员会和分委员会主席

1.8.1 任命

技术委员会主席应由本技术委员会秘书处提名，由技术管理局批准，任期最长为 6 年，或酌情缩短。允许延长任期，累计最长为 9 年。

分委员会主席应由该分委员会秘书处提名，并由相关技术委员会批准，任期最长为 6 年，或酌情缩短。允许延长任期，累计最长为 9 年。批准任命和延期的准则是相关技术委员会 P 成员中 2/3 多数赞成。

技术委员会或分委员会秘书处可在现有主席任期结束前一年内提交新主席的提名。提前一年任命的主席应指定为该委员会的"候任主席"。这一做法旨在为候任主席担任委员会主席之前提供学习机会。

1.8.2 职责

技术委员会主席负责全面管理其技术委员会的工作，其中包括各分委员会和工作组的工作。技术委员会或分委员会主席应：

a) 仅以国际身份工作，放弃他／她的本国观点，因此，他／她不能在其委员会内同时作为国家成员体的代表。

b） 指导技术委员会或分委员会秘书履行其职责。

c） 召集会议，以对委员会草案达成一致意见（见 2.5）。

d） 保证在会议上充分归纳所表达的各种观点，以便所有与会者都能理解。

e） 保证在会议上明确阐述所有决定，并由秘书提供书面文件，以便会议期间确认。

f） 在征询意见阶段作出适当的决定（见 2.6）。

g） 就技术委员会的重要事项通过技术委员会秘书处向技术管理局提供咨询意见。为此，他/她应通过分委员会秘书处得到任意一个分委员会主席的报告。

h） 确保技术管理局的政策和战略决定在委员会内的实施。

i） 确保制定和持续维护战略业务计划，战略业务计划涵盖技术委员会和所有向技术委员会报告工作的组织，包括所有分委员会的活动。

j） 确保委员会战略业务计划得到正确一贯的实施并应用于技术委员会或分委员会的工作计划中。

k） 在对委员会的决定有投诉的情况下给予协助。

如果主席因不可预见的原因不能出席会议，与会代表可选举本次会议主席。

分委员会主席应按要求出席上级委员会的会议，并可参加讨论，但无投票权。在特殊情况下，如果主席不能出席，应委派秘书（或 ISO 和 IEC 中的一名代表）代表分委员会。如果分委员会代表不能出席，则应提交书面报告。

1.9 技术委员会和分委员会秘书处

1.9.1 指派

技术委员会秘书处工作应由技术管理局指派给国家成员体承担。

分委员会秘书处应由上级技术委员会指派给国家成员体承担。但是，如有两个或多个国家成员体提出申请承担同一分委员会秘书处工作，技术管理局应对分委员会秘书处的指派做出决定。

在任何情况下，应由 JTC 1 决定指派分委员会秘书处。

无论是技术委员会还是分委员会，其秘书处应指派给符合下列条件的国家成员体，即该国家成员体应：

a） 表明其积极参与技术委员会或分委员会工作的意愿；

b） 同意履行秘书处的职责，保证有足够的资源用于支持秘书处工作（见 D.2）。

一旦将技术委员会或分委员会秘书处指派给某一国家成员体，该国家成员体则应任命一位合格人员担任秘书（见 D.1 和 D.3）。

1.9.2 职责

已被指派承担秘书处工作的国家成员体应确保对相应的技术委员会或分委员会提

供技术或行政服务。

秘书处负责监控、报告并确保工作积极进展，应竭力早日圆满完成工作。这些任务应尽可能以通信方式进行。

秘书处有责任确保 ISO/IEC 导则和技术管理局的决定得以执行。

秘书处仅以国际身份工作，放弃其本国观点。

秘书处应确保按时完成下列各项工作：

a) 工作文件

1) 准备委员会草案，安排草案的分发，并处理收到的意见。

2) 准备征询意见草案和作为最终国家标准草案分发或作为国际标准出版的文本。

3) 必要时，在能够并愿意负责语言版本的其他国家成员体的协助下，确保英语和法语文本的等效（另见 1.11 和对《ISO/IEC 导则》的独立补充部分）。

在 JTC 1 中，只要求用英文起草文本，除非有例外情况。

b) 项目管理

1) 协助确定每个项目的优先次序和目标日期。

2) 向 CEO 办公室通报所有工作组和维护小组召集人和项目负责人的姓名等。

3) 积极提出出版备选的可交付使用的文件，或由于严重超时和／或缺乏足够的支持取消项目的建议。

c) 会议（见第 4 章）

1) 确立会议议事日程，并安排分发。

2) 安排分发会议议事日程中列出的所有文件，包括工作组报告，并指明会议期间需要讨论的所有其他文件（见 E.5）。

3) 起草会议的决定（亦称"决议"），确保批准工作组建议的决定，包含被赞同的具体要素：

—— 以书面形式提供决定，供会议期间确认（见 E.5）；

—— 在会议结束后 48 小时内将决定公布在委员会的电子文件夹中。

4) 编写会议记录，在会议后 4 周内分发。

5) 编写报告提交技术管理局（TC 秘书处），IEC 会后 4 周内提交，或提交至上级委员会（SC 秘书处）。在 JTC 1 中，也可见常备文件 19 "会议"。

d) 建议

向主席、项目负责人和召集人就项目进展相关的程序提供咨询意见。

在任何情况下，各秘书处应与其技术委员会或分委员会的主席保持密切联系。

技术委员会秘书处应就其活动，包括其分委员会和工作组的活动，与 CEO 办公室和技术委员会成员保持密切联系。

分委员会秘书处应与上级技术委员会秘书处保持密切联系，并在必要时与 CEO 办公室保持密切联系。委员会还应就其活动，包括其工作组的活动，与分委员会成员保持联系。

技术委员会或分委员会秘书处应与 CEO 办公室一起更新委员会成员身份的记录。

1.9.3 技术委员会秘书处的变更

如果某一国家成员体希望放弃技术委员会秘书处工作，该国家成员体应立即通知 CEO 办公室，至少提前 12 个月发出通知。由技术管理局做出将该秘书处移交另一国家成员体的决定。

如果某一技术委员会秘书处始终未能履行本程序规定的职责，首席执行官或国家成员体可将此事提交技术管理局，技术管理局可对秘书处的指派情况进行评价，以便将秘书处工作移交另一个国家成员体。

1.9.4 分委员会秘书处的变更

如果某一国家成员体希望放弃分委员会秘书处工作，该国家成员体应立即通知上级技术委员会秘书处，至少提前 12 个月发出通知。

如果某一分技术委员会的秘书处始终未能履行本程序中规定的职责，首席执行官或国家成员体可以将此事提交上级技术委员会处理，可以通过 P 成员的多数票同意决定重新指派分委员会秘书处。

无论是上述哪种情况。技术委员会秘书处都应发出通知征询分委员会其他 P 成员承担秘书处的意愿。

如有两个或多个国家成员体愿意承担同一分委员会秘书处工作，或是由于技术委员会结构的原因，该秘书处的重新指派与技术委员会秘书处的重新指派密切相关，则由技术管理局对分委员会秘书处的重新指派做出决定。如果只有一个国家成员体愿意承担秘书处工作，则由上级技术委员会自行指派。

在任何情况下，应由 JTC 1 决定重新指派分委员会秘书处。

1.10 项目委员会

项目委员会由技术管理局设立，制定不属于现有技术委员会范围内的标准。

注：此类标准有一个参考编号但可以分为多个部分。

附录 K 中规定了项目委员会的程序。

希望转化为技术委员会的项目委员会应遵循设立新技术委员会的程序（见 1.5）。

1.11 编辑委员会

建议技术委员会成立一个或多个编辑委员会，以便更新和编辑委员会草案、征询意见草案和最终国际标准草案，确保这些文件符合《ISO/IEC 导则　第 2 部分》的有关要求（见 2.6.6）。

这类委员会至少应包括：
—— 一位以英语为母语并具有一定法语知识的技术专家；
—— 一位以法语为母语并具有一定英语知识的技术专家；
—— 项目负责人（见 2.1.8）。

项目负责人和 / 或秘书可直接负责一种相关语言的文本。

JTC 1 的工作语言为英语，虽然某些文件可能需要法语知识。除非正在编制法语文本，否则不需要法语技术专家。

为更新和编辑已经以通信方式通过的草案，以便进一步处理，若相关技术委员会和分委员会秘书处提出要求，编辑委员会应举行会议。

编辑委员会应配备处理和提供电子文本的手段（见 2.6.6）。

JTC 1 采用替代过程。

任命的项目编辑负责编辑和更新工作草案、委员会草案、征询意见草案、最终国际标准草案和 TR、TS 修订文件，并确保其符合《ISO/IEC 导则》的第 2 部分（也可见 2.6.6）的相关要求。

对于制定中的每个标准或其他文件，宜尽早确定项目编辑。项目编辑由分委员会任命，并应遵循项目组的指令。

项目编辑负责在技术工作的各个阶段（即直到出版）维护文件。可交付的最终文本的前言应表明负责可交付文件的 JTC 1 分委员会。

出版后，项目编辑宜保留一份包含所有批准的更正和修订的更新文件，以便适当时尽快发布修订版。所有修订和更正应在修订版的前言中列出。

JTC 1 或其下属组织可设立编辑组，协助项目编辑确保按照《ISO/IEC 导则》的第 2 部分的规定，对草案进行编辑。编辑组在 JTC 1 秘书处或成立该编辑组的分委员会领导下工作。

项目编辑应仅以国际身份工作，放弃其所在国家的观点。

项目编辑的变更由技术委员会负责，而不是由 JTC 1 国家成员体（或联络组织）负责。

1.12 工作组

1.12.1 技术委员会或分委员会可根据委员会决定为完成专项任务设立工作组（见 2.4）。

工作组应通过上级技术委员会指定的召集人向上级技术委员会或分委员会报告工作。

工作组召集人应由委员会任命，最长任期为 3 年，在任期期满后上级委员会的下一次全会上终止其任期。这种任命应经相关国家成员体（或联络组织）确认。召集人可被再次任命，最长任期为 3 年。任期次数不限。

召集人的变更均由委员会负责，而不是由国家成员体（或联络组织）负责。

需要时秘书处可支持召集人工作。在 JTC 1 中，提名工作组秘书须由提名人所在国家成员体确认。

工作组是由 P 成员、上级委员会的 A 类联络组织和 C 类联络组织分别指定的限定数量的专家组成，旨在承担指派给工作组的特定任务。专家以个人身份工作，不作为指派他们的 P 成员或 A 类联络组织（见 1.17）的官方代表，那些由 C 类联络组织（见 1.17）指派的专家除外。但建议他们与指派他们的 P 成员或组织保持密切联系，以便将工作组的工作进展情况和各种意见尽量在早期阶段通报给 P 成员或组织。

建议对工作组的规模予以合理限制。为此，技术委员会或分委员会可决定每个 P 成员和联络组织指派专家的最大数。

一旦做出成立工作组的决定，应正式通知 P 成员及 A 类联络组织和 C 类联络组织，以便任命专家。应按其成立的顺序对工作组编号。

当委员会在其会议上决定成立工作组时，应立即指定召集人或代理召集人，召集人或代理召集人应在 12 周内安排召开第 1 次工作组会议。委员会会议结束后应立即将上述信息通报给技术委员会 P 成员、A 类联络组织和 C 类联络组织，并邀请他们在 6 周内任命专家。可酌情将其他项目指派给现有的工作组。

在 JTC 1 中，上级组织应将工作组的管理责任指派给召集人，必要时可配备秘书处给予支持。秘书处应为 JTC 1 国家成员体或 JTC 1 国家成员体认可的组织。JTC 1 国家成员体应以书面形式确认其同意该安排后，任命才可生效。所有被提名的工作组召集人任期均为 3 年，任期 3 年后在上级组织下一次全体会议结束。召集人可以连任 3 年。

1.12.2 工作组的组成在 ISO 全球目录系统（GD）或 IEC 专家管理系统（EMS）中界定。未在 ISO GD 或 IEC EMS 中注册的工作组的专家不得参加其工作。召集人可邀请相关人员旁听会议。

1.12.3 持续不积极的专家是指没有通过参加工作组会议或以通信方式为会议做出贡献的专家。CEO 办公室可以根据技术委员会或分委员会秘书的请求，征求 P 成员意见将其从工作组中清除。

1.12.4 其任务一旦完成（通常是在上一项目征询意见阶段结束时）（见 2.6），工作组应由委员会决定解散，项目负责人继续承担顾问工作，直至出版阶段完成（见 2.8）。

1.12.5 工作组的内部文件及其报告应依据 ISO/IEC 导则独立补充部分中规定的程序发行。

1.12.6 在特殊情况下可设立联合工作组（JWG），承担多个 ISO 和 / 或 IEC 技术委员会或分委员会相关的专项任务。收到联合工作组请求的委员会应及时答复。

注：有关 ISO 和 IEC 委员会的联合工作组的具体规则，除下列规定外还见附录 B。

设立联合工作组的决定应附有涉及下列事项的委员会双方协议：

—— 负责管理该项目的委员会 / 组织；

—— 联合工作组的召集人，应由其中一个委员会的 P 成员提名，可选择由其他委员会任命共同召集人；

—— 联合工作组的成员资格（成员资格可对希望参加的所有 P 成员、A 类和 C 类联络组织开放，如果同意，可以限定每个委员会派出同等数量的代表）。

负责管理该项目的委员会 / 组织应：

—— 在其工作计划中记录该项目；

—— 在 JTC 1 中，召集所有隶属于 JWG 的委员会的专家；

—— 负责处理评议意见（通常反馈给联合工作组），并确保该项目在其各个阶段的评议意见和投票得到适当的汇总和处理（见 2.5、2.6 和 2.7），并向委员会领导提供所有评议意见；

—— 根据 2.5、2.6 和 2.7 规定的程序准备用于委员会阶段、征询意见阶段和批准阶段的草案；

—— 在 JTC 1 中，将所有相关文件（会议记录、工作草案、委员会草案、征询意见和批准阶段文件）发送给其他委员会秘书处，供其在各自委员会分发；以及或

—— 负责该出版物的维护。

批准准则依据牵头管理的委员会使用的导则。如果牵头委员会是 JTC1 委员会，JTC1 补充部分合集也适用。

对于提案阶段（NP）：

—— 对于 ISO/ISO JWG，只需进行一次 NP 投票：如果 NP 已在某一委员会中启动或批准，则不能在其他 TC 中再次投票，但是，可为 ISO/IEC JWG 启动两次（2）NP（每个组织一次（1））；

—— 在较后阶段建立联合工作组，在这种情况下将由相关技术委员会确定其牵头管理；

— 一旦同意设立联合工作组，负责牵头管理的委员会将其牵头事宜和参加其工作的委员会分别通告 ISO CEO 办公室和 IEC 中央办公室；

— 向其他技术委员会发出征集参加联合工作组的专家的通知。

对于准备阶段（WD）：

— 联合工作组与其他工作组的功能一样：需要协商一致进入委员会草案（CD）。

对于委员会阶段（CD）：

— 分发委员会草案（CD），供各委员会审阅和评论；

— 按《ISO/IEC 导则》的第 1 部分的规定，委员会草案（CD）最终要求所有委员会的协商一致。

对于国际标准草案（DIS）和最终国际标准草案（FDIS）投票：

— 要求国家成员体征求所有相关的国家对口委员会意见以确定立场。在征求意见函的封面页有提请国家成员体（NSB）注意的说明；

— 对于 ISO/IEC 联合工作组（JWG），进行两次 DIS/FDIS 投票，每个组织各进行一次。对于 ISO 和 ISO/IEC JTC 1 JWG，仅进行一次 DIS/FDIS 投票。

"前言"中明确参与制定可交付使用的文件的所有委员会。

1.13 委员会中具有咨询职能的小组

1.13.1 技术委员会或分委员会可成立具有咨询职能的小组，帮助委员会主席和秘书完成与协调、策划及指导委员会工作或其他具有咨询特性的具体任务。

1.13.2 成立咨询组的提案中应包括关于咨询组的组成建议和职责范围，包括成员资格条件。为保证其有效工作，在尽可能限制其规模的同时要切记充分代表利益相关方的要求。咨询组成员应为委员会官员和其他个人，由国家成员体和/或相关 A 类联络组织提名。上级委员会应在咨询组成立和提名前批准其最终组成和职责范围。

对于主席的咨询组，应审议平等参与条款。

1.13.3 指派给这类小组的任务可包括提出有关起草或协调出版物（特别是国际标准、技术规范、可公开提供的规范及技术报告）的提案，但不包括这些文件的起草。

1.13.4 咨询组应以建议的形式将其成果提交设立该小组的机构。建议中可包括为准备出版物设立工作组（见 1.12）或设立联合工作组（见 1.12.6）的提案。

1.13.5 咨询组内部文件仅发给本组成员，并抄送给相关委员会秘书处及 CEO 办公室。

1.13.6 专项任务一旦完成并得到上级委员会批准，咨询组即宣布解散。

1.14 临时工作组

技术委员会或分委员会可成立临时工作组，研究某一特定问题。该小组应在本次会议或最迟在下一次会议上向其上级委员会报告其所研究的问题。

在 JTC 1 中，工作组也可设立临时工作组。然而，由于 O 成员不能参加工作组，所以也不能参加工作组的临时工作组。

临时工作组成员应从出席上级委员会会议的代表中选出，如有必要，可由委员会指派的专家补充。上级委员会还应指定一名召集人。

临时工作组应在提交报告的会议上自动解散。

1.15 技术委员会间的联络

在 JTC 1 中，见常备文件 15 "联络"中有关附加要求的内容。

1.15.1 每个组织内部工作在相关领域的技术委员会和／或分委员会间应建立并保持联络关系。需要时，还要与负责标准化基础工作（如术语和图形符号）的技术委员会建立联络关系，联络中还应包括基础文件（包括新工作项目提案和工作草案）的交换。

在 JTC 1 中，委员会可通过决议确定建立内部联络。收到内部联络请求的委员会不能拒绝此类请求，并且不需通过决议确认其接受。

1.15.2 保持上述这种联络关系是相关技术委员会秘书处的职责。技术委员会秘书处可将这项任务指派给分委员会秘书处。

1.15.3 技术委员会或分委员会可指定一名或若干名联络组织代表跟进另一个与其建立联络关系的技术委员会或其所属的一个或多个分委员会的工作，应将这些联络组织代表的有关信息通知相关委员会秘书处，该秘书处应将所有相关的文件传递给联络组织代表及其所属的技术委员会或分委员会秘书处。被任命的联络组织代表应向其所属秘书处提交工作进展报告。

1.15.4 这些联络组织代表应有权参加指派跟进其工作的技术委员会或分委员会会议，但他们不应有投票权。他们可以参加会议讨论，包括技术委员会权限内的相关事宜和基于从所属委员会收集的反馈信息提交书面意见。他们还可参加该技术委员会或分委员会的工作组会议，但是只能就自己权限内的相关事宜发表自己委员会的观点，不能参加其工作组的活动。

1.16 ISO 与 IEC 之间的联络

1.16.1 安排 ISO 和 IEC 技术委员会和分委员会之间适当的联络十分重要。ISO 和 IEC 技术委员会和分委员会之间建立联络关系的通信渠道是通过 CEO 办公室。就任一组织研究的新课题而言，每当某一组织考虑一个新的或修订的工作计划并且可能与另一个组织相关时，首席执行官就力求使两个组织间达成协议，从而使此项工作不出现重叠或重复等现象（见附录 B）。

1.16.2 由 ISO 或 IEC 指定的联络组织代表应有权参加其跟进的另一个组织的技术委员会或分委员会的讨论，可以提交书面意见，但他们不应有投票权。

1.17 与其他组织的联络

1.17.1 适用于所有联络类别的通用要求

为确保有效性，联络应双向运行并有合适的互惠安排。

在工作初期阶段就应考虑对联络的需求。

无论版权归联络组织所有，还是归其他方所有，联络组织应遵循依据《ISO/IEC 导则》的有关版权的政策（见 2.13）。应将版权政策声明提供给联络组织并要求联络组织就其可接受性提出明确说明。联络组织无权就提交文件收取费用。

联络组织应按需要自愿参与 ISO 或 IEC 的技术工作。

联络组织应在其限定的相关技术和工业领域内具有足够的代表资格。

联络组织应遵循 ISO/IEC 程序，包括知识产权（IPR）（见 2.13）。

联络组织应接受 2.14 有关专利权的要求。

技术委员会和分委员会应对他们所有联络协议定期进行复审，至少每两年复审一次或在委员会的每次会议上进行复审。

在 JTC 1 及其小组中，联络关系应每年复审一次。

1.17.2 联络类别

在 JTC 1 中，见常备文件 15 "联络"。

1.17.2.1 在技术委员会／分委员会层面（A 类和 B 类联络）

在技术委员会／分委员会层面的联络类别如下：

—— A 类：在技术委员会或分委员会解决问题中，对技术委员会或分委员会工作做出有效贡献的组织。这类组织收到所有相关文件，并被邀请参加会议，他们可以提名专家参加 WG（见 1.12.1）；

—— B 类：已表明希望了解技术委员会或分委员会工作的组织。这类组织可收到技术委员会或分委员会的工作报告。

注：B 类联络供政府间组织使用。

A 类和 B 类联络的建立程序：

—— 希望建立 A 类和 B 类联络的组织应向 CEO 办公室递交申请，并向技术委员会或分委员会官员和 IEC CO 技术官员和 ISO CS 技术项目主管递交申请副本，申请包括下列内容：

—— 组织为非盈利组织；

—— 组织将向全球或较大范围区域成员开放；

—— 组织的活动和成员资格能够说明其具备起草国际标准的能力和专业技能，并且在推动标准在相关技术委员会或分委员会领域实施具备权威性（只与A类联络相关）；

—— 主要联系人姓名。

注：在提交申请之前，组织将一直与技术委员会或分委员会官员保持联系，这种情况下，技术委员会或分委员会官员应确保组织了解1.17.1规定的版权义务，并且同意ISO/IEC程序，包括知识产权和专利权。

—— CEO办公室将对满足合格标准予以确认，然后与申请组织总部所在地的国家成员体进行商议；

—— 如果申请组织总部所在地国家成员体反对，申请事宜将移交至技术管理委员会予以裁决；

—— 如果申请组织总部所在地国家成员体未反对，申请事宜将提交至技术委员会或分委员会秘书，并要求分发投票；

—— 批准A类和B类联络须P成员的2/3多数赞成。

1.17.2.2 在工作组层面（C类联络）

在工作组层面（以及JTC 1中的项目层面）的联络类别为：

—— C类：积极参与工作组活动并作出技术贡献的组织，包括制造商协会、商业协会、产业联盟、用户群以及行业协会和科学学会。联络组织应是具有个人、公司或国家成员资格的多国组织（在其目标和标准制定活动中），性质上可以是长期的或临时的。

1.17.3 资格

1.17.3.1 在技术委员会/分委员会层面（A类和B类）

某一组织申请作为ISO技术委员会/分委员会的联络组织时，CEO办公室将与该组织所在的国家成员体进行核查。如果该国家成员体认为该联络申请不符合资格准则，应将该问题提交技术管理局（TMB）确定资格。

CEO办公室还要确保该组织符合下列资格准则：

—— 不以营利为目的；

—— 是合法实体，CEO办公室将要求其提供其章程；

—— 其成员资格对全球或广泛区域成员开放；

—— 通过其活动和成员资格证明其有能力和专门知识参与国际标准的制定或有权推进国际标准的实施；

—— 有利益相关方参与和达成协调一致的决策过程，以便形成利益相关方提供的建议（见 ISO 联络组织指南——凝聚利益相关方并达成协商一致。http：//www.iso.org/iso/guidance_liaison-organizations.pdf）。

1.17.3.2　在工作组层面（C 类联络）

当某组织申请与某一工作组建立联络关系，CEO 办公室将与该组织所在的国家成员体进行核查，并且要确保该组织符合下列资格准则：

—— 不以营利为目的；

—— 通过其活动和成员资格证明其有能力和专业知识参与国际标准的制定或有权利推进国际标准的实施；

—— 有利益相关方参与和达成协调一致的决策过程，以便形成利益相关方提供的建议（见 ISO 联络组织指南——凝聚利益相关方并达成协商一致。http：//www.iso.org/iso/guidance_liaison-organizations.pdf）。

1.17.4　受理（A、B 和 C 类联络）

建立 A、B 和 C 类联络组织的协议需要参加投票 P 成员的 2/3 多数赞成。

鼓励委员会在一项工作项目研制的初期就寻求所有相关方的参与。如果 C 类联络组织在某一特定工作项目研制阶段晚期提交申请，相关的 P 成员将考虑该组织加入后的价值。

在 JTC 1 中，当收到相应 JTC 1 下属组织（及分委员会或直接向 JTC 1 报告的工作组）的推荐后，JTC1 向 ITTF 提出 C 类联络建议。由下属组织提交的每项联络申请应包含 JTC 1 下属组织和申请联络组织的协议（内容涉及双方的预期利益和责任声明）。

在 JTC 1 中，如果继续积极参与工作组或项目的工作，并且有相关的国家成员体参与，ITTF 应重新确认该组织的联络身份。如果 JTC 1 首先提出了联络请求，并且认为 C 类联络适用，则 JTC 1 可要求相应的下属组织提出申请，并适用上述程序。

1.17.5　权利与义务

1.17.5.1　在技术委员会/分委员会层面（A 和 B 类联络）

技术委员会和分委员会如有可能，应寻求具有联络身份的组织对每份相关文件的充分且正式的支持。应同样对待来自联络组织的任何评议意见和来自国家成员体的评议意见。联络组织未提供充分支持不应认为就是反对。若认为这些不支持（异议）是反对，委员会可参考 2.5.6 寻求进一步指导。

1.17.5.2　在工作组层面（C 类联络）

C 联络组织有权作为正式成员参加工作组、维护组或项目组（见 1.12.1），但不能作为项目负责人或召集人。C 类联络组织专家是任命他们的组织的正式代表。他们只可应邀参加委员会全体会议。如果他们受委员会邀请参会，只可作为观察员。

在 JTC 1 中，对于受到指派跟进分委员会或工作组工作的代表，有权参加分委员会或工作组的会议，但无投票权。该类代表可在会议上就其组织能力范围内的事项进行讨论，包括提交书面意见。

在 JTC 1 中，JTC1 将努力消除接触与参与 JTC1 活动及其工作的障碍，尤其针对残疾人和老年用户。

1.17.6 项目委员会转换为技术委员会或分委员会时带转联络关系

若项目委员会转化为技术委员会或分委员会，新的技术委员会或分委员会应通过一项决议确认带转 A 类和 B 类联络组织。这一决议的批准要求参加投票 P 成员的 2/3 多数赞成。联络类别见表 1。

表 1 联络类别

类别	A	B	C
目的	为委员会的工作做出有效的贡献	了解委员会的工作情况	为工作组起草标准做出技术贡献
资格	——非盈利 ——合法实体 ——会员制（全球或广泛区域） ——相关能力和专门知识 ——利益相关方参与和达成协商一致的决策过程 （详见 1.17.3.1）	仅政府间组织 ——非盈利 ——合法实体 ——会员制（全球或广泛区域） ——相关能力和专门知识 ——利益相关方参与和达成协商一致的决策过程 （详见 1.17.3.1）	——非盈利 ——相关能力和专门知识 ——利益相关方参与和达成协商一致的决策过程 （详见 1.17.3.1）
层面	TC/SC	TC/SC	工作组
参与	参加 TC/SC 会议，获取文件，可任命专家参加工作组，这些专家可担任召集人或项目负责人	仅了解工作（获取文件）	作为工作组成员充分参与（但不能 成为召集人或项目负责人）
权利与义务	没有投票权，但可以发表意见（其评论与成员体的评论相同对待）。可以提出新的工作项目（见 2.3.2）	没有投票权，但可以发表意见（评论与成员体的评论相同对待）。 不能提出新的工作项目	如果委员会明确邀请，专家可出席委员会会议，但只能以观察员身份出席。 不能提出新的工作项目

1.17.7 与 ITU-T 的 A 类联络

在 JTC1 中，与 ITU-T 保持独特的 A 类联络。见附录 JB 和 JTC 1 常备文件 3 "ITU-T 和 ISO/IEC JTC 1 合作指南"。

1.17.7.1 与 ITU-T 的联络

所有对 ITU-T 的贡献都宜遵守 ITU-T 建议 A.1 和 A.2，以及 ITU-T 的其他要求。特别是：

- 如果存在取代之前的贡献的情况，宜识别每项贡献；
- 每项贡献应只针对一个研究小组。但是，也宜识别其他相关研究组。

1.17.7.2 与 ITU-T 的协同关系

常备文件 3 "ITU-T 和 ISO/IEC JTC 1 合作指南"中规定了与 ITU-T 合作的两种模式，即协作交流和协作团队。JTC 1 SC 可与相应的 ITU-T 研究组达成协议，酌情建立这两种合作模式中的任何一种。JTC 1 应在与 ITU-T 合作时做出合理的决定，并根据具体情况对每个提案项目进行评估。

针对每个提案，JTC 1 应至少考虑到以下方面：

1. 考虑到稀缺的技术资源；
2. 考虑到 JTC 1 的范围；
3. 最大限度地提高标准制定过程的效率；
4. 促进标准尽快实施；
5. 考虑可能出现的重复标准的影响；
6. 确认与 ITU-T 在该提案有关的特定技术领域的合作工作。

当计划从新工作项目开始进行合作时，合作项目的理由（比如，承认 JTC 1 SC 中缺少的专业知识也存在于相应工作范围内的 ITU-T 研究组中）和职责范围应包括在 NP 文件中，以确保合作提案在 JTC 1 内都可被知晓。

如果在 JTC 1 项目开始后考虑合作，则可以将增加的项目合作视为对 SC 工作计划的修改，并通过默认投票按照《JTC 1 技术工作程序合集》的规定进行处理（见 2.1.5.7 和 JA 1.4）。在默认投票时，应附上合作项目的理由和职责范围。

常备文件 3 "ITU-T 和 ISO/IEC JTC 1 合作指南"中规定了这两种合作模式的运作程序。这些程序主要涉及 JTC 1 和 ITU-T 同步的批准活动，旨在补充而非修改 JTC 1 的批准要求。

2 国际标准的制定

2.1 项目管理方法

2.1.1 概述

技术委员会或分委员会的主要职责是制定并维护国际标准，同时，也要积极鼓励技术委员会和分委员会考虑出版第 3 章中所描述的中间出版物。

应按照下面描述的项目管理方法制定国际标准。

2.1.2 战略业务计划

每个技术委员会都应为其专业领域活动制定战略业务计划：

a) 考虑制定工作计划的商业环境；

b) 指明工作计划中正在拓宽的领域、已经完成的领域、接近完成或稳步发展的领域，以及没有进展宜撤销的领域（见2.1.9）；

c) 评价需要修订的工作（见ISO/IEC导则独立补充部分）；

d) 提出对新需求的展望。

战略业务计划应由技术委员会正式商定，并纳入到技术委员会报告中以供技术管理局定期复审和批准。

2.1.3 项目阶段

2.1.3.1 表2说明了技术工作项目各阶段的顺序，并列出了与项目各阶段有关的文件名称。有关技术规范、技术报告和可公开提供的规范的制定在第3章中予以描述。在JTC 1中，JTC 1 PAS（可公开提供的规范）转换过程与ISO和IEC中产生PAS可提交文件的过程不同（见附录F）。

表2 项目阶段及有关文件

项目阶段	有关文件	
	名称	缩写
预研阶段	预研工作项目 [a]	PWI
提案阶段	新工作项目提案 [a]	NP
准备阶段	工作草案 [a]	WD
委员会阶段	委员会草案 [a]	CD
征询意见阶段	征询意见草案 [b]	ISO/DIS IEC/CDV
批准阶段	最终国际标准草案 [c]	FDIS
出版阶段	国际标准	ISO、IEC 或 ISO/IEC

[a] 如附录F所述，这些阶段可省略。
[b] ISO为国际标准草案，IEC为投票用委员会草案。对于JTC 1，征询意见草案为DIS。
[c] 可省略（见2.6.4）。

2.1.3.2 F.1用图示说明了国际标准出版前的步骤。

2.1.3.3 ISO/IEC导则ISO补充部分和IEC补充部分给出了项目阶段矩阵图，并用数字标出相关的子阶段。在JTC 1中，使用附录JD。

2.1.4 项目描述和接受

项目是推进完成一个新的、修正或修订国际标准的任何工作。一个项目可细分成若干子项目（见2.1.5.4）。

只有根据相关程序〔见2.3中关于新工作项目提案及ISO/IEC导则独立补充部分中

关于现行国际标准的复审和维护]接受了某个项目的提案，此项目方可开展工作。

2.1.5 工作计划

2.1.5.1 技术委员会或分委员会的工作计划包括所有指派给该技术委员会和分委员会的项目，其中包括对已出版标准的维护。

2.1.5.2 在制定工作计划过程中，每个技术委员会或分委员会都应考虑行业规划的需求，以及外部技术委员会对国际标准的需求，即其他技术委员会、技术管理局咨询组、政策级委员会及 ISO 和 IEC 以外的组织对国际标准的需求（见 2.1.2）。

2.1.5.3 项目应在技术委员会通过的范围内，项目的选择应依据 ISO 和 IEC 的政策目标和资源进行仔细推敲（见附录 C）。

2.1.5.4 对工作计划中的每个项目都应给出编号（见《ISO/IEC 导则 IEC 技术工作程序》：有关 IEC 的文件编号），并应将这个编号保留在工作计划中直至该项目完成，或直至同意撤销该项目为止。如果技术委员会或分委员会认为有必要把一个项目分成若干子项目，则可给每个子项目一个分编号。子项目应被完全涵盖在原项目的范围之内，否则应提交一个新工作项目提案。

在 JTC 1 中，为了避免在授权项目细分或现行工作的微小改进方面出现不当延误，如果变更不超出原项目的范围，分委员会可在其 P 成员投票批准的情况下继续进行此类工作。但是，变更必须提交 JTC 1 批准，如果 JTC 1 不批准，则必须停止工作。

2.1.5.5 适用时，工作计划应注明承担每个项目的分委员会或工作组。

2.1.5.6 新技术委员会一致同意的工作计划应提交技术管理局批准。

2.1.5.7 在 JTC 1，在全体会议之后，分委员会应向 JTC 1 秘书处单独提交一份改后的工作方案，包括所有项目细分情况和现行工作的微小改进，但不包括新工作提案。此文件应采用默认投票程序（见附录 JA1.4 "投票"）。

2.1.6 目标日期

对于工作计划中的每个项目，技术委员会或分委员会应确定完成下列每个阶段的目标日期：

—— 完成工作草案初稿（如果新工作项目提案方只提供工作文件大纲，见 2.3）；
—— 分发委员会草案初稿；
—— 分发征询意见草案；
—— 分发最终国际标准草案（经 CEO 办公室同意）；
—— 出版国际标准（经 CEO 办公室同意）。

注：在 JTC 1 中，如果不包含技术变更，应略过最终国际标准草案（FDIS）。

这些目标日期应适应尽可能短的制定时间，以便快速制定国际标准，并应向 CEO

办公室报告，由 CEO 办公室向各国家成员体发送相关信息。关于目标日期的确定，请见 ISO/IEC 导则的独立补充部分。

在确定目标日期时，应考虑项目之间的关系。应优先考虑作为其他国际标准实施基础的国际标准制定项目。对于技术管理局认可的对国际贸易有重要影响的项目应给予最优先考虑。

技术管理局还可指示相关技术委员会或分委员会秘书处向 CEO 办公室提交最新的有效草案作为技术规范出版（见 3.1）。

应对所有的目标日期经常进行复审，必要时应进行修改，并在工作计划表中清楚标明。修改后的目标日期应通知技术管理局。对于工作计划表中 5 年以上尚未达到批准阶段的所有工作项目，技术管理局将予以撤销（见 2.7）。

2.1.6.1 概述

在 JTC 1 中，当新项目提案（无论是新的出版物还是对现行出版物的修订）获批时，委员会秘书处应向 ISO 中央秘书处提交项目成果，并标明所选标准的完成时间，如下所示，[所有目标日期从 许可项目 AWI（批准的工作项目）通过之日算起，阶段 10.99]：

注：各阶段的截止日期应根据具体情况确定。

SDT 18 标准制定完成时间——18 个月后出版

SDT 24 标准制定完成时间——24 个月后出版

SDT 36 标准制定完成时间——36 个月后出版

SDT 48 标准制定完成时间——48 个月后出版

注：选择 18 个月完成的项目，如果在项目注册后 13 个月内成功完成了 DIS 投票，则符合 ISO/CS 提出的"直接出版过程"。此过程将出版物的完成时间缩短了大约三分之一。

委员会秘书处应注意在项目规划期间进行风险评估，以便提前发现潜在问题并适当确定目标日期。委员会秘书处应经常复审目标日期，以便在每次委员会会议上对其进行确认或修改。此类审查还应设法确认项目仍与市场相关，如果发现不再需要这些项目，或者完成日期可能太晚，从而导致市场上采用替代解决方案，则应取消这些项目或将其转化为另外的出版物（见 2.1.6.2）。

注：标准起草过程中循环测试所花费的时间不应计入总体起草时间。在循环测试期间，一旦 ISO/CS 收到秘书处的请求，则应暂停标准起草时间。

2.1.6.2 项目的自动撤销（及重新启动）

在 JTC 1 中，如果超过 DIS（40.00 阶段）或出版（60.60 阶段）的目标日期，委员会应在 6 个月内决定采取某项下列行动：

a) 准备阶段或委员会阶段的项目：如果技术内容成熟并可接受，提交 DIS；
b) 征询意见阶段的项目：如果技术内容成熟并可接受，二次提交 DIS 或 FDIS；
c) 出版 TS：如果技术内容可接受，但未来不大可能达到国际标准的成熟程度；
d) 出版 TR：如果认为技术内容不可作为 TS 或未来的国际标准出版，但与公众相关；
e) 向 JTC 1 和 ISO/TMB 以及 IEC/SMB 提交延期申请：如果无法达成一致意见，但与利益相关方利益关系明显。在 JTC 1 中，委员会可在整个项目期限内获得最多 9 个月的延期，但建议出版中间可交付文件（如 TS）；
f) 撤销工作项目：如果委员会无法找到解决方案。

如果在 6 个月期限结束时，没有采取上述行动，ITTF 将自动撤销该项目。只有在 ISO 技术管理局和 IEC 标准化管理局批准后，才能恢复这些项目。

2.1.7　项目管理

技术委员会或分委员会秘书处负责管理该技术委员会或分委员会工作计划中的所有项目，包括对已达成一致同意的目标日期的进展情况进行监控。

如果没有在目标日期内完成工作（见 2.1.6），并对此项工作没有给予充分的支持（即不再符合 2.3.5 中规定的新工作接受要求），负责管理的委员会应撤销这个工作项目。

2.1.8　项目负责人

对于每个项目的起草，技术委员会或分委员会应在考虑新工作项目提案方提名（见 2.3.4）的基础上指定项目负责人（工作组召集人、指派的专家，如果适当时，也可包括秘书）。应确定项目负责人能够获得开展起草工作的适宜资源。项目负责人应仅从国际立场出发，放弃其国家的观点。当对提案阶段到出版阶段（见 2.5 ~ 2.8）产生的技术问题提出要求时，项目负责人应做好顾问角色的准备。

秘书处应将项目负责人的姓名、地址及相关项目的编号报送 CEO 办公室。

在 JTC1 中，不设项目负责人。工作组由召集人领导，并设项目编辑。

2.1.9　进度控制

分委员会和工作组应定期向技术委员会报告项目的进展情况（见《ISO/IEC 导则》的 ISO 和 IEC 技术工作程序），它们的秘书处之间的会议将有助于控制进度。

CEO 办公室负责监控所有工作的进度，并定期向技术管理局汇报有关情况。为此，CEO 办公室应收到在《ISO/IEC 导则》的 ISO 和 IEC 各自技术工作程序中指明的文件副本。

2.2　预研阶段

2.2.1　通过其 P 成员的简单多数表决，技术委员会或分委员会可将尚不完全成熟、不能进入下一阶段且不能确定目标日期的预研工作项目（例如涉及新兴技术的项目）纳入工作计划中。

这类项目中可包括战略业务计划中列出的那些项目，特别是在 2.1.2 d）"对新需求的展望"中所列的项目。

2.2.2 所有预研工作项目都应列入工作计划中。

2.2.3 对于所有的预研工作项目，有关委员会都应定期复审。委员会应对每个预研工作项目的市场相关性和所需要的资源进行评价。

任何预研工作项目若在 TC/SC 规定的截至期满时（针对 IEC）、在 3 年内还未进入提案阶段（针对 ISO），将自动从工作计划中删除。

2.2.4 本阶段可用于预研新工作项目提案（见 2.3）并制定初始草案。

2.2.5 所有这类项目在进入准备阶段前，都应经过 2.3 中规定的程序的批准。

2.3 提案阶段

起草管理体系可交付文件时，见附录 L。

2.3.1 新工作项目提案（NP）涉及下列内容：

—— 新标准；

—— 现行标准的新增部分；

—— 技术规范（见 3.1）。

在 JTC 1 中，以下情况无需进入 NP 阶段（2.3）：

—— 对现行标准或技术规范的修订或修正；

—— 将 TS 转换为 IS。

然而，委员会应通过包含以下内容的决议并：

1) 目标日期；

2) 确认没有超出范围；

3) 是否任命了项目编辑。

委员会还应征集专家。

将 TS 转换为 IS 需 2/3 多数赞成。

如果修订或修正导致超出范围，则应发起 NP 投票。

2.3.2 现有技术委员会或分委员会范围内新工作项目提案可由下列相关组织提出：

—— 国家成员体；

—— 技术委员会或分委员会秘书处；

—— 另一个技术委员会或分委员会；

—— A 类联络组织；

注：在 JTC 1 中，仅有 JTC 1 A 类联络组织。

——技术管理局或其中某个咨询组；

——首席执行官。

2.3.3 如果涉及到 ISO 和 IEC 两个组织的技术委员会，首席执行官应安排必要的协调（见附录 B）。

2.3.4 每个新工作项目提案均应使用适当表格提交，且经过充分论证并恰当编制成文件（见附录 C）。

新工作项目提案方应：

——尽量提供工作草案初稿供讨论，或至少应提供该工作草案大纲。

——提名一名项目负责人。在 JTC 1 中，不设项目负责人。在 JTC1 中，如果这项新工作项目提案指派给现有的工作组，提案方应提名"项目编辑"。

——在 JTC 1 中，填写有关无障碍问题的详细信息，填写位置位于表 4 "新工作项目建议书"中"其他"字段下"与建议书相关的补充信息"栏中。

对于属于现有委员会范围内的建议，该表应提交给 CEO 办公室或相关的委员会秘书处。

CEO 办公室或相关委员会主席和秘书处应确保该提案按照 ISO 和 IEC 要求（见附录 C）正常制定，并且提供充分信息支持国家成员体做出合理决定。

CEO 办公室或相关委员会主席和秘书处还应评定提案与现行工作的关系，他们可以咨询利益相关方，包括技术管理局或开展现行工作的委员会。必要时，可以成立临时工作组审查提案。对提案的任何评审均不应超过 2 周时间。

在所有情况下，CEO 办公室或相关委员会主席和秘书处都可在提案表中加入评议意见和建议。

项目委员会的新工作项目提案见附录 K。

已完成的表格的副本应分发给技术委员会或分委员会成员，供 P 成员进行书面投票和供 O 成员和联络组织成员参考。

应在表格中标明建议出版的日期。

可以用通信方式对一项新工作项目提案做出决定。

应在 12 周内反馈投票结果。

委员会可以根据具体情况通过决议来决定对新工作项目投票期限缩短至 8 周。

填写投票表格时，国家成员体应提供论证其反对票决定的陈述（见论证说明）。如果未提供这种说明，该国家成员体投的反对票将不予登记和考虑。

2.3.5 满足以下要求，项目即获得许可：

a）参加投票的技术委员会或分委员会 P 成员的 2/3 多数票批准工作项目，统计时不包括弃权票；

b) 在 16 个或以下 P 成员的委员会至少有 4 个 P 成员，在 17 个或以上的 P 成员的委员会至少有 5 个 P 成员，承诺积极参与项目制定工作，也就是，在准备阶段通过提名技术专家和对工作草案提出评议意见作出有效贡献。

在统计时，只考虑已经批准将该工作项目纳入其工作计划的 P 成员。如果一个国家成员体承诺积极参与，但在附有赞成票的表中未指定专家，那么在确定本次投票是否满足批准准则时不予以登记和考虑。

对于 JTC 1，在 NP 语境下，如果 JTC 1 的 P 成员对新工作项目提案投了反对票，但没有提供清晰的理由陈述，委员会秘书处应将该反对票退回 JTC 1 国家成员体，并给予 2 周的延迟投票时间。

如果 JTC 1 的国家成员体在 2 周内没有回应，则不计入该投票。

秘书处不得对理由陈述作出是否有价值的判断，如有疑问，应向 JTC 1 国家成员体提出。

如果 JTC 1 的国家成员体没有在表格中填写提名专家，应在投票结果产生后两周内提名专家。如果没有遵守规定，该 JTC 1 国家成员体的参与将不被计算在内，从而影响上述 b) 的批准要求。

各委员会可提高对指定专家的最低要求。

在可以证明只有极少数 P 成员具有产业和/或技术知识的前提下，技术委员会可以请求技术管理局允许开始时可少于 4 位或 5 位提名专家。

在 JTC 1 中，可运用附加投票规则；见附录 JA.1 和 JA.2。

2.3.6 某个新工作项目提案一旦获得许可，它将以适当的顺序作为一个新项目纳入相关技术委员会或分委员会的工作计划中。在适当表格中应注明一致同意的目标日期（见 2.1.6）。

投票结束后 4 周内向 ISO 中央秘书处（用表 6）和 IEC 中央办公室（用表 RVN）报告表决结果。

2.3.7 工作项目被纳入工作计划后，提案阶段即告结束。

2.4 准备阶段

2.4.1 准备阶段包括：依据《ISO/IEC 导则 第 2 部分》的要求准备工作草案（WD）。

2.4.2 如果一个新工作项目获得许可，项目负责人应与 P 成员在批准阶段提名的专家一起工作［见 2.3.5 a)］。在 JTC 1 中，不设项目负责人。

2.4.3 秘书处可在会议上或以通信方式向技术委员会或分委员会提出成立工作组的建议。通常情况下，工作组的召集人担任项目负责人职务。

这样的工作组应由技术委员会或分委员会负责成立，技术委员会或分委员会应确定工作组的任务，并确定向技术委员会或分委员会提交草案的目标日期（见 1.12）。工作组召集人应保证所开展的工作在已投票的工作项目的范围内。

2.4.4 在答复设立工作组的提议时，表示同意积极参与［见 2.3.5 a）］的每个 P 成员应确认其技术专家。其他 P 成员或 A 类、C 类联络组织也可提名专家。

2.4.5 项目负责人负责项目的制定工作，通常负责召集和主持工作组所有会议。项目负责人可邀请工作组的一名成员担任秘书。

在 JTC1 中，当没有项目负责人时，宜确定项目编辑（见 2.1.8）。针对一项标准，工作组起草一份或多份工作草案。通常，在工作草案推进到委员会阶段之前，会经过几次修订。在决定工作草案的内容时，召集人宜注意确保达成一致意见。

2.4.6 为避免项目制定在后期拖延，应尽一切努力准备英语、法语两种语言文本。

如果要制定三种语言（英语、法语、俄语）的标准，应包括俄语文本在内。

在 JTC 1 中，除非有例外情况，仅要求准备英文文本。

2.4.7 关于本阶段的时间限制，见 2.1.6。

2.4.8 当工作草案作为委员会草案（CD）初稿分发给技术委员会或分委员会成员时，CEO 办公室负责登记，准备阶段即告结束。委员会还可决定将最终工作草案作为 PAS（见 3.2）出版，以满足市场的特定需求。

2.5 委员会阶段

2.5.1 委员会阶段是考虑国家成员体意见的主要阶段，旨在在技术内容上达成一致。因此国家成员体应认真研究委员会草案文本并在本阶段提交所有相关评议意见。

在 JTC 1 中，图形符号应提交给负责规范图形符号的相关 ISO 委员会和/或 IEC 委员会（如适用）(见附录 JE）。

2.5.2 一旦得到委员会草案，应立即将其分发给技术委员会或分委员会的 P 成员及 O 成员供其考虑，并明确注明提交答复的最迟日期。在 JTC 1 中，也要求联络组织提交答复。

经技术委员会或分委员会同意，应为国家成员体提供 8 周、12 周或 16 周的时间对委员会草案进行评论。

在 JTC 1 中，默认的 CD/CDAM/DTS/DTR 分发时间为 8 周。

应根据给定的说明提交评议意见，以便进行汇总。

会前，国家成员体应向其代表充分介绍国家的立场。

2.5.3 答复日期终止后的 4 周内，秘书处应准备好意见汇总，并将其分发给技术委员会或分委员会的全体 P 成员及 O 成员。在准备意见汇总的过程中，秘书处应与技术委员会或分委员会主席协商，必要时与项目负责人协商，提出项目处理意见：

  a）在下次会议上讨论委员会草案及评议意见；或

  b）或者，分发修改后的委员会草案供考虑；

  c）或者，登记征询意见阶段用委员会草案（见2.6）。

  在b）和c）情况下，秘书处应在评议意见汇总报告中说明针对收到的每条评议意见采取的措施。在将修改的委员会草案（CD）提交给委员会考虑［在b）情况下］之前将这些信息提供给所有P成员，或者在向CEO办公室提交最终草案进行征询意见阶段登记时［在c）情况下］同时提供。必要时通过分发修改的评议意见汇总报告方式提供。

  各委员须对收到的所有意见作出答复。如果在分发之日起的8周内，2个或更多的P成员不同意秘书处在b）或c）情况下提出的建议，委员会草案应在会议上讨论（见4.2.1.3）。

  ==在JTC 1中，起草CD修订本、处理评议意见以及建议进入下一步处理程序职责可分配给WG、临时工作组或向上级委员会传达报告的项目编辑。==

2.5.4 如果经过会议研究未对委员会草案达成一致意见，应综合会议上的各项决定形成另一草案，在12周内予以分发考虑。经委员会或分委员会同意，应为国家成员体提供8周、12周或16周的时间对该草案及后续的文本进行评论。

  ==在JTC 1中，默认的CD/CDAM/DTS/DTR分发时间为8周。==

2.5.5 应一直研究每一版草案文本，直到技术委员会或分委员会中的P成员达成协商一致，或做出撤销或推迟这个项目的决定。

2.5.6 应在协商一致的原则基础上做出分发征询意见草案（2.6.1）的决定。

  技术委员会或分委员会主席有责任与秘书进行协商，必要时与项目负责人进行协商，根据ISO/IEC指南2：2004给出的"协商一致"的定义，判断该草案是否得到足够的支持。

  协商一致： 总体同意，其特点在于利益相关方的任何重要一方对重大问题不坚持反对立场，并具有寻求考虑所有相关方的意见和协调任何冲突的过程。

  注： 协商一致不意味着一致同意。

  以下内容适用于"协商一致"的定义：

  在达成共识的过程中，在文件的制定中不同观点得以表达和处理。然而，"持续反对"是在委员会、工作组或其他小组（例如任务组、咨询组等）的会议上表达的意见，是由相关利益的一个重要部分维持的、不符合委员会的协商一致意见。"相关利益"的概念将视委员会的动态而有所不同，因此应由委员会领导层逐案确定。"持续反对"的概念不适用于成员体就CD、DIS或FDIS进行的投票，因为这些都受适用的投票规则的制约。

表达"持续反对"意见者有权发表意见，并建议在宣布"持续反对"时采取以下办法：

—— 领导层应首先评估反对是否可被视为"持续反对"，即是否得到相关利益的一个重要部分的支持。如果情况并非如此，领导层将登记反对意见（例如以会议纪要、记录等方式），并继续领导有关文件的工作。

—— 如果领导层确定存在持续的反对意见，则需努力真诚地解决这一问题。然而，持续的反对并不等同于否决。处理持续存在的反对意见的义务并不意味着有义务成功地解决这些问题。

评估是否达成共识的责任完全在于领导层。这包括评估是否存在持续的反对意见，或任何持续的反对意见能否在不损害关于文件其余部分现有协商一致意见的情况下得到解决。在这种情况下，领导人将登记反对意见并继续开展工作。

持反对意见的当事方可使用第 5 章详述的上诉机制。

如对协商一致的意见有疑问，经技术委员会或分委员会表决的 P 成员 2/3 多数赞成，可视为足以接受该委员会草案登记为征询意见草案；不过，应尽一切努力解决反对票问题。

在 JTC 1 中，当计票时，不计弃权票，也不计没有附上技术原因的反对票。

负责委员会草案的技术委员会和分委员会秘书处应确保征询意见草案充分体现在会议上或以通信方式做出的所有决定。

2.5.7 如果某一技术委员会或分委员会已达成协商一致，其秘书处应在最长 16 周内将草案的最终文本用一种适于分发给国家成员体征询意见的电子表格（2.6.1）提交给 CEO 办公室（如果是分委员会，应抄送给技术委员会秘书处一份）。

2.5.8 关于本阶段的时间限制，见 2.1.6。

2.5.9 当所有的技术问题得到解决，委员会草案作为征询意见草案分发，并由 CEO 办公室登记后，委员会阶段即告结束。登记前，凡是不符合《ISO/IEC 导则 第 2 部分》的文本应退回秘书处进行修改。

2.5.10 如果技术问题不能在适当的时间范围内得到完全解决，技术委员会或分委员会可以考虑在该文件成为国际标准的协议达成之前，以技术规范（见 3.1）形式作为一种中间出版物出版。

2.6 征询意见阶段

2.6.1 在征询意见阶段，征询意见草案（ISO 的 DIS 和 IEC 的 CDV）应由 CEO 办公室分发给所有国家成员体进行为期 12 周的投票。在 JTC 1 中，征询意见草案为 DIS 或

DAM。DIS 或 DAM 应在 8 周的翻译期后进行 12 周的投票。

关于语言使用方面的政策，见附录 E。在 JTC 1 中，除非有例外情况，仅要求用英文文本准备草案。

CEO 办公室应将其接收投票的截止日期通知国家成员体。

投票终止时，首席执行官应在 4 周内将投票结果及所收到的意见交给技术委员会或分委员会主席和秘书处，以便迅速采取下一步行动。

2.6.2 国家成员体提交的投票应是明确的：赞成、反对或弃权。

赞成票可以附有编辑性或少量的技术性意见，其条件是技术委员会或分委员会的秘书及项目负责人与主席进行磋商，决定如何处理这些意见。

如果某国家成员体发现某一征询意见草案是不可接受的，应投反对票并说明技术理由。该国家成员体可以注明在具体的技术性修改意见获得许可的条件下将反对票改为赞成票，但不得投以接受修改意见为条件的赞成票。

在 JTC 1 中，如果 JTC 1 国家成员体在没有提交理由的情况下投了反对票，那么投票将不计算在内。

在 JTC 1 中，如果 JTC 1 国家成员体投了反对票，但提交的技术理由不明确，委员会主管应在投票结束后 2 周内与技术项目主管联系。

在 JTC 1 中，对评议意见类型（技术性、编辑性或总体性）没有限制，JTC 1 国家成员体可在投票时提交；但是，如果对征询意见草案投反对票，则要求 JTC 1 国家成员体说明其技术性原因。

2.6.3 如果满足下列条件，征询意见草案则可通过：

a） 参加投票的技术委员会或分委员会 P 成员 2/3 多数赞成；

b） 反对票不超过投票总数的 1/4。

计票时，弃权票及未附有技术理由的反对票不计算在内。

将正式投票期过后收到的评议意见提交给技术委员会或分委员会秘书处，待下次国际标准复审时进行研究。

在 JTC 1，运用附加的投票规则；见附录 JA.1。

2.6.4 技术委员会或分委员会主席收到投票结果及意见后，应协同其秘书处或项目负责人，并与 CEO 办公室磋商采取下列行动之一：

a） 如符合 2.6.3 的批准准则且无任何技术改动，直接予以出版（见 2.8）；

b） 如符合 2.6.3 的批准准则但须包括技术改动，则将经修改的征询意见草案登记为国际标准的最终草案；

c） 如果没有满足 2.6.3 的批准准则；则

1) 分发修改后的征询意见草案进行投票（见 2.6.1）；

注：分发修改后的征询意见草案并进行为期 8 周的投票。如有一个或多个相关委员会 P 成员提出要求，可延长至 12 周。在 JTC 1 中，修改的征询意见草案的分发期限可以延长至 12 周。

2) 或者，分发修改后的委员会草案进行评论；

3) 或者，下次会议讨论和评议征询意见草案。在 JTC 1 中，可以通过电话会议或网络会议举行评议意见决议会议。

2.6.5 投票期结束后的 12 周内，技术委员会或分委员会秘书处应准备一份完整报告，并由 CEO 办公室分发到各国家成员体。报告应：

a) 说明投票结果；

b) 陈述技术委员会或分委员会主席的决定；

c) 复制所收到的评议意见内容；

d) 包括技术委员会或分委员会秘书处对所提交的每条评议意见的看法。

应尽最大努力解决反对票问题。

如果自报告分发之日起 8 周内，2 个或更多的 P 成员表示不赞成主席所做出的决定 2.6.4 c) 项中的 1) 或 2.6.4 c) 项中的 2)，该草案则应在会议上进行讨论（见 4.2.1.3）。

各委员会应对收到的所有意见作出答复。

2.6.6 如果主席决定该草案进入批准阶段（见 2.7）或出版阶段（见 2.8），技术委员会或分委员会秘书处应在投票期结束后的最长 16 周内，在编辑委员会的帮助下准备一份最终文本并将其送至 CEO 办公室，以便准备和分发最终国际标准草案。

秘书处应向 CEO 办公室提供可修改的电子格式文本和允许确认修改形式的电子格式文本。

不符合《ISO/IEC 导则 第 2 部分》规定的文本应退回秘书处并在登记前要求进行修改。

2.6.7 关于本阶段的时间限制，见 2.1.6。

2.6.8 当 CEO 办公室登记征询意见草案文本，作为最终国际标准草案分发或作为国际标准出版时，即 2.6.4 a) 和 b) 的情况，征询意见阶段即告结束。

2.7 批准阶段

2.7.1 在批准阶段，CEO 办公室应在 12 周内（ISO）将最终国际标准草案（FDIS）分发给所有国家成员体进行为期 8 周（IEC6 周）的投票。

CEO 办公室应将其接收投票的截止日期通知国家成员体。

2.7.2 国家成员体提交的投票应明确：赞成、反对或弃权。

国家成员体可就任何关于 FDIS 的投票提出意见。

如果国家成员体认为最终国际标准草案是不可接受的，应投反对票并陈述技术理由，但不能投以接受修改意见为条件的赞成票。

在 JTC 1 中，国家成员体无理由投反对票的，投票不计。

在 JTC 1 中，国家成员体投反对票并提交评议意见，但意见明显不属于技术内容的，秘书处应在投票截止前 2 周内联系技术项目主管。

2.7.3 如果满足下列要求，分发进行投票的最终国际标准草案则获得通过：

a） 参加投票的技术委员会或分委员会 P 成员的 2/3 多数赞成；并且

b） 反对票不超过总数的 1/4。计票时，弃权票和未附有技术理由的反对票不包括在内。

在 JTC 1 中，运用附加的投票规则，见附录 JA.1。

2.7.4 在投票结束前，技术委员会或分委员会秘书处有责任提醒 CEO 办公室注意到草案起草中可能引入的错误。本阶段，不再接受进一步的编辑性或技术性修改意见。

2.7.5 收到的所有意见将被保留，供下次复审用，并在投票表格上记录为"注明供日后考虑"。但 CEO 办公室和秘书处可设法解决明显的编辑错误。不允许对批准的 FDIS 进行技术更改。

CEO 办公室应在投票期结束后的 2 周内，向所有的国家成员体分发报告，在报告中公布投票结果并指明国家成员体正式通过将其发布为国际标准，还是正式否决了该最终国际标准草案。

2.7.6 如果依据 2.7.3 的要求通过了最终国际标准草案，该草案可进入出版阶段（见 2.8）。

2.7.7 如果依据 2.7.3 的要求最终国际标准草案未获通过，则应将此文件退回相关的技术委员会或分委员会，委员会依据支持反对票的技术理由重新考虑。

委员会可以决定：

—— 以委员会草案、征询意见草案或在 ISO 和 JTC1 中最终国际标准草案的形式再次提交修改后的草案；

—— 出版技术规范（见 3.1）；

—— 取消项目。

2.7.8 分发投票报告，说明 FDIS 已被批准为国际标准（见 2.7.5），或作为技术规范（见 3.1.1.2）出版，或将文件退回委员会，批准阶段即告结束。

2.8 出版阶段

2.8.1 CEO 办公室应在 6 周内更正技术委员会和分委员会秘书处标明的所有错误，并且印刷和分发国际标准。

在 JTC 1 中，ITTF 应更正技术委员会或分委员会秘书处标明的所有错误，并在 8 周内印刷和分发国际标准。

2.8.2 国际标准出版，出版阶段即告结束。

2.9 可交付使用的成果的维护

2.9.1 概述

在 ISO/IEC 导则独立补充部分给出了可交付使用的文件的维护程序。

在 JTC 1 中，应适用以下程序。

国际标准的勘误附加程序见 JTC 1 常备文件 21 "国际标准的勘误"。

通过 PAS 转换过程批准、由 PAS 提交人维护并由 JTC 1 管理的文件的附加维护要求，见常备文件 9 "PAS 转换过程"。

技术规范和技术报告维护的附加要求见常备文件 6 "技术规范和技术报告"。

2.9.2 系统复审和委员会复审

2.9.2.1 通用原则

在 JTC 1 中，ISO 和 IEC 为 JTC 1 联合发布的所有国际标准和技术规范应进行系统复审，以确定是否根据表 S1 确认、修订/修正、稳定或撤销标准和规范。

在系统评审间，委员会可在任何时候通过决议启动标准修订或修正。

有关现行标准修订或修正启动程序，见 2.3.1。

对于不影响技术内容的少量更改，如更新和编辑性更正，可采用称为"少量修订"的简短程序。少量修订仅由委员会（通过决议）在批准和出版阶段（见 2.7 和 2.8）提出。归口技术委员会作出决议且咨询过归口技术项目主管后，应分发修订后的可交付使用的文件的最终草案，以进行为期 8 周的 FDIS 投票，如果为维也纳协定文件，投票期为 12 周。在下一版可交付使用的文件的前言中，应指出进行了少量修订，并列出所做的更新和编辑性更正。

表 S1 系统复审时间表

可交付使用的文件	系统复审前的最长存续时间	出版物的最多确认次数	最长有效时间
国际标准	5 年	没有限制	没有限制
技术规范（见 3.1.3）	3 年	建议一次	建议 6 年
技术报告（见 3.3.3）	未规定	未规定	没有限制

虽然没有规定对技术报告进行系统复审（见表 S1），但仍要求归口委员会 5 年内进行一次复审。

通常在以下情况下启动系统性复审：

—— 根据归口委员会秘书处提议，通常由于文件出版或最后一次确认后，达到了规定期限；

—— ITTF 的默认操作；

—— 应一个或多个国家成员体的要求；

—— 应委员会主管对 ITTF 的要求。

系统性复审的时间通常基于出版年份；如果文件已被确认，则基于最后一次确认的年份。但不需要等到最长时间才启动。

如果相关 SC 不再存在，则可将维护该标准的责任归口到 JTC 1 国家成员体或 JTC 1 A 类联络组织。

有关现行标准的修订或修正过程，见 2.10.3 和 2.3.1。

2.9.2.2 系统复审要求

在 JTC 1 中，系统复审投票期为 20 周。

在启动国家成员体的系统性复审投票之前，委员会通过其 P 成员在全体会议上投票或以信函投票的方式，提出关于标准和技术规范的处置建议，与系统性复审投票一起列入。ITTF 通过系统复审投票程序提交选票供国家成员体批准。

系统复审投票结束后，秘书处应将反映投票结果的建议分发给技术委员会或分委员会成员。在系统复审投票结束后的 6 个月内，委员会应就是否修订、确认、稳定或撤销标准做出最终决定，随后秘书处应将委员会的决定提交 ITTF。

注：系统复审投票由 ITTF 以电子方式进行管理。委员会的 P 成员有义务对该委员会归口管理的所有可交付使用的文件进行系统复审投票。邀请 ISO 和 IEC 的所有 P 成员对此类系统复审做出回应。复审还包括收集下列信息：国家成员体是否需要修改可交付使用的文件以适合其国情。委员会需要关注这些修改，以确定是否需要考虑提高标准的全球相关性。

2.9.2.3 系统复审投票结果的说明

通常情况下，有关 JTC 1 中系统复审之后所采取的适当措施的决定，应由对具体措施进行投票的 ISO 和 IEC P 成员按照简单多数的方式表决。然而，在某些情况下，进一步分析结果可能发现更为合理的解释。委员会对措施进行决策，并告知 ITTF。

注：由于答复的多样性、执行程度和评论的相对重要性，在说明投票结果时，不可能为所有情况提供具体的规则。

在提出措施时，应适当考虑所需确认项的最大数量和相关可交付使用的文件的最长有效期（见表 S1）。

如果证实使用了标准或技术规范，将继续使用这些标准或技术规范，并且无需技术变更，则可确认可交付使用的文件。准则如下：

—— 无论是否修改，已采用标准或技术规范，或至少在 5 个国家采用（如果不满足此标准，应撤销可交付使用的文件）；并且

—— 回复投票确认的 P 成员达到简单多数。

如果系统复审的投票结果不能确定确认、修订或撤销，或是不能基于收到的答复做决定，秘书处应要求在规定的时间内（例如在 8 周内）批准建议的措施。

2.9.3 修订或修正

在 JTC 1 中，委员会可在任何时候通过决议，对可交付使用的文件进行修订或修正。

如果证实已经使用了某个文件，而且将继续使用该文件，但需要进行技术更改，则可建议对可交付使用的文件进行修订或修正。准则如下：

—— 无论是否修改，已采用标准或技术规范，或至少在 5 个国家采用（如果不满足此标准，应撤销可交付使用的文件）；并且

—— 认为有必要进行修订或修正的 ISO 和 IEC P 成员达到简单多数。

在这种情况下，可登记为获批工作项目（阶段 10.99）。

应召集专家，然而，不要求委员会活跃 P 成员的最小数目。如果在委员会批准后未立即启动修订或修正活动，建议首先将该项目登记为预研工作项目，并在确认后登记该标准。当最终建议登记在阶段 10.99 时，应参考先前系统复审结果和委员会通过的决议（现行标准的修订或修正过程见 2.3.1）。

如果决定对国际标准或技术规范进行修订或修正，该标准或技术规范将成为新项目，并应列入委员会的工作方案。修订或修正的步骤与准备新可交付使用的文件的步骤相同（见 JTC 1 补充文件合集 2.3 至 2.8），包括确定完成相关阶段的目标日期。

对于不影响技术内容的少量更改，如更新和编辑性更正，可采用称为"少量修订"的简短程序。少量修订仅由委员会（通过决议）在批准和出版阶段（见 2.7 和 2.8）提出。在委员会做出决议并与 ITTF 磋商后，应分发修订后的可交付使用的文件的最终草案，供委员会成员进行为期 8 周的投票。在下一版可交付使用的文件的前言中，应指出进行了少量修订，并列出所做的更新和编辑性更正。

2.9.3.1 转换为国际标准（仅指技术规范）

除了确认、修订或修正以及撤销这三个基本选项外，对技术规范进行系统复审的情况还有第四个选项，就是将其转换为国际标准。

进行国际标准转换时，应根据制定国际标准的正式程序酌情提交一份更新的文本（见 2.3.1）。

转换过程通常从 DIS 投票开始。如果认为需要做出重大变更：需要在 DIS 投票前在委员会中进行全面评议，则应提交文件的修订版以供评议，并作为 CD 进行投票。

2.9.4 撤销

在 JTC 1 中，如果提议撤销某项国际标准或技术规范，ITTF 应将委员会的建议通

知各国家成员体。ITTF 应发起系统复审投票，以便国家成员体对该建议进行投票。

2.9.5 稳定

在 JTC1 中，标准可以是稳定的。稳定的标准是持续有效且成熟的；并且一旦确定标准置于稳定状态，则不需要任何形式的进一步维护。但这些标准对于有较长预期寿命的现用产品和设备服务继续有效。

若要确定标准处于稳定状态，则在对标准进行最后一次修改后，应至少通过一个 5 年的复审周期，然后归口委员会才能建议其置于稳定状态。

在对标准进行系统复审时，委员会可建议将其归口管理的标准置于稳定状态。在任何情况下，建议都应附有理由说明，并经过默认的系统性复审投票，对标准置于稳定状态的建议进行确认。

一旦确定标准处于"稳定"状态，ITTF 会将其记录在稳定状态标准的主列表中。该记录还包括首次列入的日期和上述提议的理由。稳定标准表示该标准在 ISO 目录列表中处于稳定状态。

置于稳定状态的标准无需进行系统复审时，归口委员会应定期（间隔不能超过 10 年）要求其 P 成员审查委员会的稳定标准，以确定标准的稳定状态是否具有效用。

当委员会或 JTC 1 P 成员或分委会 P 成员意识到置于稳定状态的标准：

- 不再使用；或者
- 已被取代；或者
- 不安全。

将对该标准是否修订或撤销进行默认投票（见 JA.1.3）。

请注意，技术规范和技术报告不能置于稳定状态。

2.9.6 已撤销标准的恢复

在 JTC 1 中，如果在撤销国际标准后，委员会确定仍然需要该标准，则可建议恢复该标准。恢复已撤销标准的投票应作为 DIS 或 FDIS 由 ISO 和 IEC P 成员投票。应运用 2.6 和 2.7 的投票程序。经批准后，该标准应以新版本出版，并注明新的出版日期。在前言中应说明该标准恢复了上一版本。

2.10 更正和修正

2.10.1 概述

已出版的国际标准可通过下列出版物进行后续修改：

—— 技术勘误表（仅在 IEC）；

在 JTC 1，也可选择出版技术勘误表。

—— 更正版。

—— 修改单；

—— 修订版（作为 2.9 维护程序的一部分）

对于 JTC 1，在出版阶段，ITTF 应与委员会和分委员会秘书处协商，考虑组织的财务影响和国际标准的用户利益，决定出版修订版或是包括修订的新版。

[**注**：如果预计将会经常增加 IS 的条款，在开始制定时，则应将这些附加条款作为系列标准的一部分（见《ISO/IEC 导则　第 2 部分》5.5.1）]

注：如有修订，将出版国际标准的新版本。

2.10.2 更正

在 JTC1 中，"更正"是通用词汇，包括技术勘误表和更正版本。在 JTC1 中，仍然可以选择发布技术勘误表或更正版本。

更正只是为了纠正在起草或出版过程中无意间引入并可能导致不正确或不安全地应用该出版物的错误或歧义。

更正不是发布自出版以来已过时的更新信息。

有疑点的错误应引起相关技术委员会或分委员会秘书处的注意。在秘书处和主席的确认，必要时与项目负责人及技术委员会或分委员会中的 P 成员进行磋商，秘书处应向 CEO 办公室提交有关更正的提案并说明其必要性。

考虑到对组织的财政影响和出版物用户的利益，在与技术委员会和分委员会秘书处磋商后，首席执行官应决定是否出版技术勘误表（仅在 IEC）和（或）现行版本的更正版（见 2.10.4）。委员会秘书处将向委员会成员通报结果。

在 JTC 1 中，也可选择出版技术勘误表。

一般情况下，凡超过 3 年的出版物，将不会出版其更正版。

对于 JTC 1，ITTF 应与委员会或分委员会秘书处协商，考虑组织的财务影响和国际标准的用户利益，决定出版技术勘误表、或对现有版本进行更正或更新（亦见 2.10.4）。一般情况下，对发布了 3 年以上的国际标准，将不会出版其技术勘误表。在 JTC 1 DID，JTC1 常备文件 21 "国际标准的更正"给出了制定和发布技术勘误表的程序。

2.10.3 修改单

修改单是对现行国际标准中原已达成协调一致的技术条款的改动和补充。修改单被视为部分修订，该国际标准的其余部分不开放评论。

修改单通常作为一份单独的文件出版，受影响的国际标准的版本仍在使用中。修改单的起草和出版程序应遵循 2.3（ISO 和 JTC 1）或见 IEC 补充部分中的审查和维护程序以及修正草案（DAM）中 2.4、2.5、2.6、最终修正草案（FDAM）中的 2.7 和 2.8

所述。JTC 1 使用和 ISO 相同的程序。

在 JTC 1，针对 CD/CDAM/DTS/DTR 的默认投票期为 8 周。

在批准阶段（见 2.7），考虑到对组织的财政影响和国际标准用户利益，首席执行官应与技术委员会和分委员会秘书处协商，决定是否出版修改单，或出版包括修正在内的新版国际标准（见 2.10.4）。

> 注：如果预见到可能对国际标准条款要进行经常性补充，那么在开始起草时，应考虑将这些补充条款作为系列的一部分（见《ISO/IEC 导则》的第 2 部分）。

2.10.4 避免繁多的修改

用以更正现行国际标准的技术勘误表（仅在 IEC）或修改单不得多于 2 个单行本，制定第 3 个这样的文件时应出版国际标准的新版本。

在 JTC 1 中，也可选择出版技术勘误表。

2.11 维护机构

技术委员会或分委员会出版了要求不断修改的标准时，该技术委员会或分委员会可确定成立维护机构。有关指定维护机构的规则，见附录 G。

2.12 注册机构

技术委员会或分委员会制定了含有注册条款的标准时，需要成立一个注册机构。有关指定注册机构的规则，见附录 H。

对于 JTC 1，也可见附录 JF"注册机构政策"。

2.13 版权

所有草案、国际标准及其他出版物的版权属于各自以 CEO 办公室为代表的 ISO、IEC 或 ISO 和 IEC 共同所有。

以国际标准为例，其内容可能来自许多渠道，包括国家现行标准、科学或贸易期刊上出版的文章、自主的研发工作或对商业化产品的描述等。这些来源可能拥有一个或多个版权。

在 ISO 和 IEC，有一种理解是提交并成为 ISO、IEC 或 ISO 和 IEC 出版物中一部分的原始材料可以在 ISO 和/或 IEC 系统内（根据情况），作为建立协商一致过程的一部分拷贝和分发。这并不影响原始版权所有者在其他地方对原始文本拥有的权利。如果材料已经受版权保护，ISO 和/或 IEC 应获得复制和分发材料的授权。这通常不需要书面协议，或至多是一个简单的书面同意声明。如果贡献者希望就此向 ISO 和/或 IEC 提交的材料签署正式版权协议，这样的请求必须分别通知 ISO 中央秘书处或 IEC 中央办公室。

需要注意的是，ISO 和 IEC 各自的成员都有权采用 ISO 和 IEC 标准并作为他们的国家标准再出版。确实存在或可能存在类似的采用形式（如区域标准化组织）。

在 JTC 1 中，DIS/FDIS、国际标准、DAM/FDAM、修改单、技术勘误表、技术规范和技术报告的版权属于 ISO 和 IEC。

对有要求的标准，应当出版注册表。注册表的版权属于 ISO 和 IEC，ISO 和 IEC 可向 JTC 1 注册机构授予版权，但前提是该机构具备运作能力。

2.14 专利项目的引用（见附录 I）

2.14.1 在例外情况下，如有技术理由证明引用专利项目是合理的，原则上不反对制定含专利权所覆盖的专利项目（基于发明专利、实用新型和其他类似的法定权利，包括前面所提及的出版物中所采用的专利项目）条款的国际标准，即使在这些标准的条款中没有其他符合性方法可供选择。下列规则应适用。

2.14.2 如有技术理由证明起草文件时使用包括专利权所涵盖的项目条款是合理的，则应符合下列程序：

a） 文件的提案方应提请委员会注意其已知的并考虑涵盖其提案的任何项目的专利权，参与起草文件的任何相关方都应提请委员会注意在文件制定的各阶段发现的专利权问题；

b） 如果根据技术理由接受了该提案，起草方应要求已确定的专利权人声明其愿意在合理和非歧视的条款和条件下与全世界的申请人谈判其权利内的全球许可。这种协商任务应由相关团体完成，并在 ISO 和 / 或 IEC 外进行。专利权人声明的记录应放入 ISO 中央秘书处或视情况放入 IEC 中央办公室登记簿，并在相应文件的引言中提及。如果专利权人不提供这种声明，在没有得到 ISO 理事会或 IEC 理事局授权的情况下，相关委员会不得在文件中包括专利权所涵盖的项目；

c） 除非得到相关理事局授权，否则在收到已确定的专利权持有者发表声明前不得出版该文件。

2.14.3 文件出版后，如果发现不能在合理和非歧视的条款和条件下获取文件所包含的专利权涵盖项目的许可，应将文件退回相关委员会再进一步考虑。

3 其他可交付使用的文件的制定

3.1 技术规范

在 JTC 1 中，常备文件 6 "技术规范和技术报告"给出了 JTC 1 制定技术规范的特定要求。

技术规范可在下列情况和条件下起草和出版。

3.1.1 当所讨论的项目正在研究过程中，或由于其他理由，现在不可能但将来可能达成出版国际标准协议，技术委员会或分委员会可以根据 2.3 中所规定的程序决定出版技术规范是否适宜。制定这种技术规范的程序应遵循 2.4 和 2.5 中的规定。将结果文件以技术规范形式出版的决定要求参加投票的技术委员会或分委员会 P 成员的 2/3 多数赞成。技术规范属规范性文件。

当最终国际标准草案不能得到通过批准阶段所要求的支持率（见 2.7），或对协商一致有疑义的情况下，技术委员会或分委员会可通过参加投票的 P 成员的 2/3 多数赞成来决定该文件以技术规范形式出版。

3.1.2 如果技术委员会或分委员会 P 成员已同意出版技术规范，技术委员会或分委员会秘书处应在 16 周内以电子版形式将技术规范草案提交 CEO 办公室出版。在不与现有的国际标准相冲突的条件下，提供不同技术解决方案的相互竞争的技术规范是可能的。

3.1.3 技术规范应在出版后 3 年内由技术委员会或分委员会进行复审，目的是重新检查促使该技术规范出版的原因，以及是否有可能就出版国际标准来取代该技术规范达成必要的协议。在 IEC 中，复审日期应以技术规范公布前已商定的稳定日期（复审日期）为基础。IEC 专用程序不适用于 JTC 1。

3.2 可公开提供的规范（PAS）

本部分不适用于 JTC 1。

在 JTC 1 中，JTC 1 PAS（可公开提供的技术规范）的转换过程与 ISO 和 IEC 不同（见附录 F）。

3.2.1 PAS 可以是在正式国际标准制定之前出版的中间性规范，或在 IEC 中可以是一种与外部组织合作出版的带有"双标识"的出版物。PAS 不是完全符合标准要求的文件。

3.2.2 PAS 的提案只能由 A 类联络组织或 C 类联络组织（见 1.17）提交，或由委员会的 P 成员提交。

3.2.3 提交的 PAS 文本经过确认、检查，与现行国际标准没有冲突并经相关委员会参加投票的 P 成员简单多数批准后方可出版。只要不与现有的国际标准相冲突，提供不同技术解决方案的 PAS 就是可能存在的。

3.2.4 PAS 的初始有效期为 3 年。有效期可以延长，最长可延长 3 年，期满后，不论是否需要修改，都应转换为另一类型的规范性文件，或撤销。

3.3 技术报告

在 JTC 1 中，常备文件"技术规范和技术报告"给出了 JTC 1 制定技术报告的特定要求。

3.3.1 如果技术委员会或分委员会已收集到数据，这些数据不同于正常作为国际标准出版的数据（可以包括诸如国家成员体从所开展的调查中获取的数据，其他国际组织在工作中所获取的数据，或与国家成员体某一专门学科标准有关的"最新技术"数据），技术委员会或分委员会可以通过投票 P 成员的简单多数，决定是否要求首席执行官以技术报告的形式出版这些数据。该文件从本质上说完全是信息性的，而不应有规范性的内容。该文件应明确说明其与主题规范性要素的关系，这些规范性要素正在或将在有关主题的国际标准中处理。如有必要，首席执行官应与技术管理局进行协商，决定是否将本文件作为技术报告出版。

3.3.2 如果技术委员会或分委员会 P 成员同意出版技术报告，技术委员会或分委员会秘书处应在 16 周内以电子版形式将报告草案提交首席执行官出版。

3.3.3 建议由承担工作的委员会对技术报告进行定期复审，以保证其有效性。撤销技术报告的决定由承担工作的技术委员会或分委员会作出。

技术报告不进行系统性复审。

4 会议

4.1 概述

提醒国家成员体不允许向代表/专家收取任何种类的参会费用，也不允许为技术委员会、分委员会、工作组、维护和项目小组的任何会议收取特定旅馆住宿费或房费。基本会议设施应完全由国家成员体和/或自愿赞助者提供资金。关于更多信息，请参阅 IEC 会议指南（http：/www.iec.ch/ members_ Experts/refdocs/iec/IEC_Meeting_Guide 2012.pdf）和 ISO 附录 SF。对于 JTC 1，见 JTC 1 常备文件 19 "会议"。

4.1.1 技术委员会或分委员会宜使用现代电子手段开展工作，只有需要讨论委员会草案（CD）或其他以通信方式无法解决的重要事宜时，才应召开技术委员会或分委员会会议。在 JTC 1 中，另请见常备文件 19 "会议"。

4.1.2 技术委员会秘书处宜与 CEO 办公室协商并考虑其工作计划，预先拟订委员会及其分委员会（如果可能，包括其工作组）至少 2 年的会议计划。

在 JTC 1 中，JTC 1 会议应由 JTC 1 秘书处组织召开，通常间隔 6 个月，会议持续时间应足以解决所有议程事宜。

4.1.3 在制定会议计划时，应考虑相关专业技术委员会或分委员会一起召开会议的益处，以增进交流和减少参加若干个技术委员会或分委员会会议代表的参会负担。

在 JTC 1 中，若干工作组一起开会也有此优势。

4.1.4 在制定会议计划时，应考虑技术委员会或分委员会会议结束后在同一地点立即召开编辑委员会会议,将有利于加快草案的制定工作。

4.2 召开会议的程序

4.2.1 技术委员会和分委员会会议

对于 JTC 1,另见常备文件 19 "会议" 有关实体会议和电子会议的策划内容。

4.2.1.1 召开会议的日期和地点应由相关技术委员会或分委员会主席和秘书处、首席执行官及承办会议的国家成员体共用商定。如果分委员会召开会议,分委员会秘书处首先应与上级技术委员会秘书处协商,确保会议的协调(见 4.1.3)。

4.2.1.2 希望承办某一会议的国家成员体应与首席执行官及相关技术委员会或分委员会秘书处联系。

该国家成员体首先应保证其所在国对以参加会议为目的的技术委员会或分委员会 P 成员代表的入境不受限制。

在 JTC 1 中,在授权代表参加会议时,P 成员国和 O 成员国的国家成员体应在 ISO 会议申请系统或 IEC 会议注册系统(MRS)中进行注册(视情况而定)。

国家成员体承办国可以通过 ISO 会议申请系统或 IEC 会议注册系统(MRS)中注册代表名单,以适当安排会议。主办国的国家成员体应向有需要的参与国的国家成员体代表直接发送邀请函。

建议承办方核实并提供关于会议设施接入方式的信息。根据 4.2.1.3,应分发一份说明会议后勤情况的文件。除提供地点和交通信息外,还应提供会议设施可接入性的详细情况。

在规划过程中,应要求通知具体的可接入性要求。承办方应尽最大努力满足这些要求。

在 JTC 1 中,除非与委员会秘书处商定了其他安排,承办会议的国家成员体负责提供会议秘书处支持和服务。

在 JTC 1 中,委员会秘书处应将所有授权参会代表通知承办方,以便承办方对会议作出适当安排。

4.2.1.3 秘书处应作好日程安排,保证会议日程和会务信息在会议召开前 16 周由 CEO 办公室(IEC)分发或由秘书处(ISO)分发并抄送 CEO 办公室。在 JTC 1 及其分委员会中,成员应在会议前 8 周内,向委员会秘书处提交关于议程或增加新工作项目提案的意见或建议。秘书处应立即分发这些意见或建议,以便代表充分准备。

注:所有新工作项目提案应经通信方式批准(委员会内部投票——CIB),见 2.3.4。

只有在召开会议前至少 6 周已经得到意见汇总的委员会草案才能纳入会议日程,并可在会上进行讨论。

所有其他工作文件，包括拟在会上进行讨论的草案评议意见汇总，应在会前至少 6 周内分发。

议程应明确说明开始和估计结束时间。

如果会议超过预计结束时间，主席应确保 P 成员愿意做出表决决定。但是，如果 P 成员离开，他们可以要求主席不采取任何进一步的投票决定。

在 JTC 1 中，会议预期完成时间后以及任何 P 成员离开后所做的任何决定，应在会议之后通过函件方式确认。

注： 参会者在预定出行计划时，宜考虑预期会议时间。

在 JTC 1 及其分委员会中，只有在会议召开至少 4 周前可获得评议意见汇总的委员会草案才可纳入议程，并可在会议上进行讨论。

4.2.2　工作组会议

对于 JTC 1，常备文件 19"会议"中给出了有关工作组会议的特定要求。

4.2.2.1　工作组应尽可能使用现代电子手段开展工作（如电子邮件、远程电话会议）。当需举行会议时，工作组会议的召集人应在会前至少 6 周内将通知交给其成员和上级技术委员会秘书处。

在 JTC 1 中，关于策划工作组实体会议或电子会议的要求，也可见常备文件 19"会议"，特别是发布会议通知和待讨论文件的要求。

会议安排应由会议召集人和会议承办国工作组的成员共用承担，后者负责所有实际工作的安排。

在 JTC 1 中，由于工作组可能包括大量的参会代表，会议日期和地点应由上级组织的秘书处和承办会议的 JTC 1 国家成员体商定。

4.2.2.2　如果工作组会议与上级技术委员会工作会议联合召开，会议召集人应与上级技术委员会秘书处协调安排。特别应确保工作组成员能够收到分发给上级技术委员会会议代表的所有一般性会议资料。

4.2.2.3　工作组（或 IEC 中的 PT/MT/AC）负责人或相关委员会秘书应将在各自国家举行的工作组（或 IEC PT/MT/AC）会议通知国家成员体秘书处。

4.3　会议语言

虽然会议官方语言是英语、法语和俄语，但会议默认语言为英语。

俄罗斯联邦国家成员体提供所有俄语的口译或笔译服务。

适当时，主席和秘书处应依据 ISO 或 IEC 通用规则以参加会议人员能够接受的方式妥善解决会议中使用语言的问题（见附录 E）。

在 JTC 1 或其下属组织的会议上，如果会议代表鉴于特殊情况希望用其他语言发言，会议主席或召集人可以在会议上授权，但前提是具备翻译手段。

4.4 会议的取消

会议一旦召开，应尽力避免取消或推迟会议。然而，如果不能在4.2.1.3所要求的时间内准备好会议议程及基本文件，首席执行官则有权取消会议。

5 申诉

5.1 概述

5.1.1 国家成员体有权在12周内（ISO）和8周内（IEC）对有问题的决定提出申诉：

a) 就分委员会的决定向母技术委员会提出申诉；

b) 就技术委员会的决定向技术管理局提出申诉；

c) 就技术管理局的决定向理事会提出申诉。对任何申诉案例，理事会的决定为最终裁决。

5.1.2 技术委员会或分委员会的P成员可以申诉技术委员会或分委员会的任何作为或不作为，如果P成员认为这样的作为或不作为：

a) 不符合：

—— 章程和程序规则；

—— ISO/IEC导则；或

b) 不利于国际贸易和商务，或诸如安全、卫生或环境等公共因素。

5.1.3 申诉的问题可以是技术性的或管理性的。

对涉及新工作项目提案、委员会草案、征询意见草案及最终国际标准草案的决定进行的申诉，仅在下述情况下予以考虑：

—— 涉及原则问题；

—— 草案的内容有损于ISO或IEC的声誉。

5.1.4 应完整记录所有申诉，以支持P成员的关注。

5.2 对分委员会决定的申诉

5.2.1 申诉文件应由P成员提交给上级技术委员会秘书处，并抄送首席执行官。

5.2.2 上级技术委员会秘书处一收到申诉文件，就应将申诉信息通知其所有P成员并立即采取行动，以通信方式或召开会议方式对申诉内容进行研究并做出决定。在此过程中应与首席执行官进行磋商。

5.2.3 如果技术委员会支持其分委员会，那么提出申诉的P成员可：

—— 或者接受技术委员会的决定；

—— 或者，对决定提出申诉。

5.3 对技术委员会决定的申诉

5.3.1 对技术委员会决定的申诉可存在 2 种情况：

—— 起因于上述 5.2.3 的申诉，或是

—— 对技术委员会的原始决定申诉。

5.3.2 在任何情况下，申诉文件应提交首席执行官，并抄送技术委员会主席和秘书处。

5.3.3 首席执行官进行适宜的协商后，应在收到申诉文件后的 4 周内，将申诉及其意见提交技术管理局。

5.3.4 技术管理局将决定是否对申诉作进一步处理。如果决定予以处理，技术管理局主席应组织一个调解小组。

调解小组应在 12 周内听取申诉意见并努力尽快解决意见分歧。协调小组应在 12 周内提出最终报告。如果调解小组没有成功地解决意见分歧，应向首席执行官报告，并就如何解决这一问题提出建议。

5.3.5 收到调解小组的报告后，首席执行官应通知技术管理局，由技术管理局做出决定。

5.4 对技术管理局决定的申诉

对技术管理局决定的申诉应连同案例各阶段的全部文件一并提交给首席执行官。

首席执行官应在收到申诉文件后 4 周内，将申诉文件及其对申诉的评议意见一并提交理事会成员。

理事会应在 12 周内做出决定。

5.5 申诉期间工作的运行

当申诉针对正在进行的工作时，该项工作应继续，直至进入并包括批准阶段（见 2.7）。

附录 A
（规范性）
指南

A.1 引言

除了技术委员会制定的国际标准、技术规范、可公开提供的规范及技术报告外，ISO 和 IEC 还出版与国际标准化有关的指南。这些指南应根据《ISO/IEC 导则 第 2 部分》起草。

指南不应由技术委员会和分委员会制定，但可以由 ISO 政策制定委员会、IEC 咨询委员会或战略组、向 ISO 技术管理局报告的 ISO 工作组或者 ISO/IEC 联合协调组制定。这些机构以下称为"项目负责委员会或工作组"。

制定和出版指南的程序如下。

A.2 提案阶段

ISO 和/或 IEC 技术管理局批准新指南或指南修订的提案，并决定项目负责委员会或工作组的组成和秘书处。

ISO 和/或 IEC 技术管理局一旦批准某项目，该项目负责委员会或工作组的秘书处应确保 ISO 和 IEC 的利益相关方了解这一信息。

A.3 准备阶段

项目负责委员会或工作组应确保 ISO 和 IEC 的利益相关方有机会参与起草工作草案。

A.4 委员会阶段

一旦工作草案可以作为委员会草案分发，项目负责委员会或工作组的秘书处应将其报送给上级委员会或 ISO 和/或 IEC 技术管理局进行投票、评议、批准其进入征询意见阶段。

A.5 征询意见阶段

A.5.1 CEO 办公室应将修改后指南草案英语和法语本分发给所有的国家成员体进行为期 16 周的投票。

A.5.2 如果反对票不超过投票的 1/4，该指南草案则被批准作为指南出版。计票时不包括弃权票。

如果是 ISO/IEC 指南，其草案应提交给 ISO 和 IEC 的国家成员体批准。需要两个组织的国家成员体都批准，该文件方可作为 ISO/IEC 指南出版。

如果 ISO 或 IEC 两个组织中只有一个组织满足这一条件，该指南只可以用批准组织的名义出版，除非项目负责委员会或工作组决定采用 A.5.3 规定的程序。

A.5.3 如果指南草案未获批准，或被批准但附有需进一步完善的评议意见，项目负责委员会或工作组的负责人可决定提交修正草案进行为期 8 周投票。修正草案的批准条件与 A.5.2 相同。

A.6 出版阶段

出版阶段应由项目负责委员会或工作组归属组织的 CEO 办公室负责。

如果是 ISO/IEC 联合工作组，出版的责任应由两个首席执行官之间的协议决定。

A.7 指南的撤销

指南项目的负责委员会或工作组决定该指南是否应撤销。指南的正式撤销应由技术管理局（TMB）按其正常的程序进行批准。

附录 B
（规范性）
ISO/IEC 联络和分工程序

B.1 引言

依据 1976 年 ISO/IEC 协议[1]，ISO 和 IEC 共同形成了一个国际标准化体系。为了使这个体系有效运作，两个组织一致同意下列两个组织的技术委员会和分委员会之间的协调和分工程序。

B.2 总体考虑

ISO 和 IEC 之间的分工是以一致同意的原则为基础，这一原则是指与电气和电子工程领域国际标准化有关的所有问题由 IEC 负责，其他领域由 ISO 负责；对于电气和非电气技术界定目前尚不明确的国际标准化相关事宜的职责分配问题则应由两个组织相互协商解决。

在设立新的 ISO 或 IEC 技术委员会时，或是由于现有技术委员会的活动，有可能产生协调和工作分配的问题。

现有下列几个层面的协调和工作分配协议。对于提出的事宜，只有在较低层面解决的努力失败后，方可提交到较高层面。

a) 正式联络：为 ISO 和 IEC 委员会之间的正常合作而进行的正式沟通；

b) 组织磋商：包括技术专家、首席执行官代表之间的磋商，针对其技术协调可能对两个组织的未来活动产生比当前更大影响的事宜；

c) 工作分配的决定：

—— 由技术管理局做出，或，必要时；

—— 由 ISO/IEC 联合技术咨询局（JTAB）做出。

对 ISO 和 IEC 两个组织都有影响且已证明 ISO 技术管理局和 IEC 标准化管理局不可能做出一致决定的问题，应提交 ISO/IEC 联合技术咨询委员会（JTAB）做出决定（见 1.3.1）。

B.3 设立新的技术委员会

当分别向 ISO 国家成员体或 IEC 国家成员体提出设立新技术委员会的提案时，该提案也应提交给另一个组织征求意见和 / 或达成协议，协商的结果可能出现以下两种

---

[1] ISO 理事会决议 49/1976、50/1976 和 IEC 行政管理通函 13/1977 号。

情况：

a) 一致同意该项工作应在其中一个组织中进行；

b) 意见有分歧。

出现 a) 情况时，可根据一致的意见采取正式行动，设立新的技术委员会。

出现 b) 情况时，应安排一次相关领域的专家会议，由两个组织的首席执行官代表参加，旨在达成令人满意的分工（即在组织层面）协议。如果可在这一层面达成协议，可由适宜的组织采取正式行动实施该协议。

如果经过协商后仍不能达成协议，可由任一组织将该问题提交 ISO/IEC 联合技术咨询委员会（JTAB）。

B.4 ISO 和 IEC 技术委员会之间工作的协调与分配

B.4.1 TC 层面的正式联络

ISO 和 IEC 各技术委员会之间的大部分协调需求都通过正式技术联络的安排得到成功解决。对于这些安排，当一个组织提出请求时，另一个组织应予以履行。正式联络安排的请求由 CEO 办公室控制，提出请求的组织应明确其所需要的联络类型，诸如：

a) 全部或有选择地交换委员会文件；

b) 联络代表定期或有选择地参加会议；

c) 参加所选的 ISO 和 IEC 技术委员会常设协调（或指导）委员会活动；

d) 组建联合工作组（JWG）。

B.4.2 协议的具体内容

B.4.2.1 通过把某工作领域委托给两个组织中的一个，不断努力将 IEC 和 ISO 之间的重叠性工作降至最低。

在受委托的工作领域中，IEC 和 ISO 应通过 JTAB 就如何将另一组织的观点和利益予以充分考虑的问题达成一致意见。

B.4.2.2 已建立如下 5 种合作工作方式：

方式 1——信息式

一个组织全权负责某一工作领域，并让另一方完全了解工作进展情况。

方式 2——主辅式

一个组织应在工作中起主导作用，另一个组织则在工作进展中以书面形式做出适宜的贡献。这种关系也包括充分交换信息。

方式 3——分包式

一个组织被全权委托负责完成某个确定的项目，但由于另一个组织具有专业特长，一部分工作被分包给第二个组织负责制定。为了保证分包的这部分工作能够正确地融入工作计划的主要部分，双方需做出必要的安排。为此，作为该标准化任务主要承包方的组织负责征询意见和批准阶段。

方式 4——协作式

一个组织在活动中起主导作用，但其工作会议接受另一组织的联络代表。这些联络代表有权参与讨论，但没有表决权。通过这种联络达到信息充分畅通。

方式 5——联合式

联合工作组和联合技术委员会确保召开联席会议，在完全平等参与的原则下共同处理标准，解决问题。

两个组织的技术委员会之间的联合工作组应按照 1.12.6 规定开展工作。

B.4.2.3 对于可能重叠的领域，ISO 和 IEC 的分工将按相关方商定的工作进程或计划要求列出，并作为双方协议的附录。

该协议的结果是两个组织同意引用对方主管领域的相应标准。

如果被引用的标准已更新，引用该标准的机构有责任关注被引用标准的更新情况。

B.4.2.4 对于一个组织已经承担职责并拟将其分包给其他组织的工作，在确定该工作目的时应充分考虑参加分包工作的利益。

B.4.2.5 征询意见和批准阶段必要的程序应由特定标准化任务的受委托组织来完成，除非两个组织的技术管理局一起作出其他决定。

B.4.2.6 对于按方式 5（联合式）制定的标准，应根据牵头管理的组织的规则，在 ISO 和 IEC 中同步进行委员会、征询意见和批准阶段。承担该项目管理职责的委员会/组织应在草案分发日期前两周将委员会、征询意见和批准阶段用的草案提交给另一个组织。

B.4.2.7 若在一个组织内征询意见草案没有满足批准准则（见 2.6.3），那么：

—— 参加联合工作组的委员会官员可以从 2.6.4 c）给出的选项中选择一项；

—— 或者，在例外情况下，如果参加联合工作组的 ISO 和 IEC 委员会官员与 CEO 办公室达成协议，该项目可以作为批准征询意见草案的组织的单一标志标准继续进行。至此，联合工作组自动解散。

B.4.2.8 如果根据 2.7.3，最终国际标准草案未被批准，那么：

—— 参加联合工作组的委员会可以从 2.7.7 给出的选项中选择一项，注意，在 IEC 不允许传播第二份最终国际标准草案时，应请求 TMB 降低文件等级；

—— 或者，在例外情况下，如果参加联合工作组的 ISO 和 IEC 委员会官员与 CEO 办公室达成协议，该标准就可以作为批准最终国际标准草案的组织的单一标志标准出版。至此，联合工作组自动解散。

B.4.2.9 按方式 5（联合式）由 ISO 和 IEC 联合工作组制定的标准由承担管理职责的委员会所在组织出版。该组织分配标准的参考编号并拥有该标准的版权。该标准配有另一组织的标志并且两个组织都可以销售。这种国际标准的前言应明确所有负责制定的委员会。对于承担管理职责的委员会在 IEC 的那些标准，其前言还要给出 ISO 的投票结果。ISO 主导的文件编号是从 1 到 59999，IEC 主导的文件编号是从 60000 到 79999。如果是多部分标准，一些部分由 ISO 负责和一些部分由 IEC 负责，文件编号分配为 80000 系列（例如，ISO 80000-1，IEC 80000-6）。

B.4.2.10 按方式 5（联合式）制定的标准，其维护程序通常是承担管理职责的委员会所在组织中现行使用的程序。

B.4.2.11 在项目制定期间，如有理由改变运作方式，两个相关的委员会应提出建议并提交给两个技术管理局参考。

B.4.3 秘书处的合作

两个组织相关的技术委员会/分委员会秘书处应就本协议的实施进行合作，根据正常程序，应对进行的工作建立完全的信息流，并应要求相互提供工作文件。

附录 C
（规范性）
对制定标准的提案的论证

C.1 概述

C.1.1 由于涉及大量的财力和人力资源，而且应根据需要分配这些资源，因此，在任何标准化活动开始之前，明确需求、确定拟制定标准的目的及可能受影响的利益是非常重要的，而且这将有助于确保制定的标准恰当地涵盖所要求的要素，在其影响的行业具有市场相关性。因此，任何新的活动都应在其开始前进行合理的论证。

C.1.2 很显然，无论根据本附录的规定对提案做出何种结论，开展任何新工作的前提条件都应是有足够数量的利益相关方明确表示愿意安排必要的人力和资金，并积极参加这项工作。

C.1.3 本附录给出了提出和论证新工作的规则，以便提案能够提供尽可能清晰、明确的工作目的和范围，旨在确保相关方正确分配标准化资源，并且最有效地利用这些资源。

C.1.4 本附录不包括实施和监督本附录中包含的指南的程序规则，也不涉及为此应建立的管理机制。

C.1.5 本附录主要针对拟启动各种新工作的提案方，但也可作为一种工具帮助对提案进行分析或提出意见的人员以及负责对提案做出决定的机构。

C.2 术语和定义

C.2.1 新工作提案

新技术活动领域或新工作项目的提案。

C.2.2 新技术活动领域提案

向某组织提出的该组织现有委员会（如技术委员会、分委员会或项目委员会）未包括的工作领域的标准制定提案。

C.2.3 新工作项目提案

向某组织提出的在该组织现有委员会（如技术委员会）领域内的标准或一系列相关标准的制定提案。

C.3 通用原则

C.3.1 任何新工作提案都应在其所提交组织的范围之内。

注：例如，ISO 的提案对象在其章程中予以规定，IEC 的提案对象在其章程第 2 条中予以规定。

C.3.2 论证 ISO 和 IEC 新工作的文件应提供该提案市场相关性的重要案例。

C.3.3 论证 ISO 和 IEC 新工作的文件应提供可靠的资料作为 ISO 或 IEC 国家成员体在知情的前提下投票的基础。

C.3.4 在 ISO 和 IEC 体系内，提案方有义务提供合适的文件以支撑上述 C.3.2 和 C.3.3 阐述的原则。

C.4 提议新技术活动领域或新工作项目时需要说明的要素

C.4.1 新技术活动领域和新工作项目提案应包括下列信息（C.4.2 至 C.4.13）。

C.4.2 标题

标题应清楚并简洁地表明提案所要涵盖的新技术活动领域或新工作项目。

示例 1：（新技术活动领域提案）"机床"。

示例 2：（新工作项目提案）"电工产品——基本环境测试程序"。

C.4.3 范围

在 JTC 1 中，还考虑文化和语言适应性以及可获取性等其他因素。

C.4.3.1 对于新技术活动领域

范围应精确地界定活动领域的界限。范围不应重复该组织管理工作的通用目标和原则，而应指明相关的具体领域。

示例 "通过切削材料或加压操作加工金属、木材和塑料的各种机床标准化。"

C.4.3.2 对于新工作项目

范围应清楚表明所提议的新工作项目涵盖的内容，必要时为明确起见，应说明不适用范围。

示例 1：

本标准列举了一系列环境测试程序及其严重性，目的是评估电工产品在预期服务条件下的运行能力。

虽然最初是为这些应用而设计，本标准也可以用于其他需要的领域。

其他个别样品类型专用的环境测试可能被包含在相关的规范里。

示例 2：

渔业和水养殖业领域标准化，包括但不局限于术语、设备及其操作的技术规范、水产养殖场地特性描述和适宜的物理、化学和生物条件的维护、环境监控、数据报告、溯源性跟踪及废物处理。

不包括：

—— 食品的分析方法（ISO/TC 34 的范围）；

—— 人员防护服（ISO/TC 94 的范围）；

—— 环境监控（ISO/TC 207 的范围）。

C.4.4 提议的初始工作计划（仅针对新技术活动领域提案）

在 JTC 1 中，工作计划应列入总体战略业务计划中，并保持更新状态。

C.4.4.1 提案的工作计划应符合并明确反映标准化活动的目标，并应指明提案的主题之间的关系。

C.4.4.2 工作计划中的每个项目都应采用拟被标准化的主题和内容（例如，就产品而言，项目可以是产品类型、特性、其他要求、拟提供的数据、测试方法等）加以明确。

C.4.4.3 工作计划中的具体项目可附补充论证。

C.4.4.4 提案的工作计划还应建议新工作项目的优先顺序和目标日期（当提议一系列标准时，应建议其优先顺序）。

C.4.5 表明拟制定的可交付使用的文件的优选类型

如果是新技术活动领域的提案，可查看 C.4.4。

C.4.6 国际、区域和国家层面的有关现行文件列表

应列出所有已知的相关文件（比标准和法规），无论其来源如何，并附上其重要性的说明。

C.4.7 与现有工作的关系及其影响

C.4.7.1 应提供一份说明，描述提议的工作与现有工作的关系和对现有工作的影响，尤其是现行 ISO 和 IEC 的可交付使用的文件。提案方需解释其工作与明显类似的工作有何差异或解释如何将重复和冲突降至最低。

C.4.7.2 如果在其组织的其他委员会或在另一组织的范围内已经有看上去类似或相关的工作，提案的范围应将提议的工作与其他工作区分开。

C.4.7.3 提案方应表明是否可以通过扩大现有委员会的范围或设立新的委员会来处理其提案。

C.4.8 相关国家的参与

C.4.8.1 对于新技术活动领域的提案，应提供该提案的主题对其国家商业利益非常重要的相关国家的名单。

C.4.8.2 对于现有委员会内的新工作项目提案，应提供那些已经不是该委员会 P 成员，但该提案的主题对其国家的商业利益非常重要的相关国家名单。

C.4.9 合作与联络

C.4.9.1 应提供在可交付使用的文件制定中约定为联络组织的相关外部国际组织或内部机构（除 ISO 和 / 或 IEC 委员会外）的名单。

C.4.9.2 为避免与其他机构冲突或重复工作，表明所有可能冲突或重复之处很重要。

C.4.9.3 与其他利益相关方的所有沟通结果也应包含在内。

C.4.10 受影响的利益相关方

应提供一个简洁的说明，识别受影响的利益相关方的类别（包括中小企业）并描述他们如何从所提议的可交付使用的文件中获益或受到影响。

C.4.11 基础文件（仅用于新工作项目提案）

C.4.11.1 若提案方认为一个现行完善的文件可以作为标准接受（经修正或不经修正），应附适当理由说明，在提案上附上该文件。

C.4.11.2 所有新工作项目议案都应附上现有文件，作为 ISO 或 IEC 可交付使用的文件、或提议的大纲、目录的初始基础。

C.4.11.3 如果所附的现行文件受版权保护或包含版权保护的内容，提案方应确保已经获得适宜的书面许可，允许 ISO/IEC 使用那些受版权保护的内容。

C.4.12 领导层承诺

C.4.12.1 如果是新技术活动领域提案，提案方应表明其组织是否已做好准备来承担所需要的秘书处工作。

C.4.12.2 如果是新工作项目提案，提案方还应提名一位项目负责人。

C.4.13 目的及论证

C.4.13.1 拟制定的标准的目的和论证应表达清楚，对拟包括在标准中的每个要素（诸如特性）标准化的需求应予以论证。

C.4.13.2 如果提议一系列的新工作项目，其目的和论证是共同的，即可起草一个共用提案，包括拟说明的所有要素并列举每个项目的标题和范围。

C.4.13.3 请注意下列列出的各项是可用于支撑提案的目的和论证的文件的一系列建议或想法。提案方应考虑这些建议，但不限于此，也不需要严格地符合这些建议。最重要的是，提案方编制并提供的立项目的和论证，既要与他们的提案最相关，又要提供重要的商业案例来说明其提案的市场相关性和需求。编制完善并举证充分的立项的和论证文件有助于提案得到更充分的关注，并有利于其最终在 ISO 和 IEC 系统中成功立项。

—— 一份简洁的说明，描述提案需要解决的业务、技术、社会或环境问题，最好与 ISO 或 IEC 相关委员会的战略工作计划相关联。

—— 全球相关性文件，说明经济、技术、社会或环境问题的范围或幅度，或新的市场。这可包括对所产生的标准进行未来销售趋势的评估，表明未来使用和全球相关性。

—— 技术效益：一份简洁的说明，描写提案在支撑新兴技术与体系的连贯性、合并技术的融合、互操作性、竞争技术的解决、未来的创新等方面的技术影响。

—— 经济效益：一份简洁的说明，描述提案在以下方面的趋势，即消除贸易壁垄，改善国际市场准入，支持公共采购，提高广大企业（包括中小企业）的商业效率和/或最终通过灵活、经济可行的方法以符合国际和区域的规则/惯例等。也可以提供一个简单的成本/效益分析，将产生可交付使用的文件的成本和对全世界企业产生的预期经济效益进行比较分析。

—— 社会效益：一份简洁的说明，描述预期从提议的可交付使用的文件中获得的预期社会效益。

—— 环境效益：一份简洁的说明，描述从提议的可交付使用的文件中获得的预期环境效益或更广泛可持续发展效益。

—— 一份简洁的说明，描述提议的可交付使用的文件的预期用途，例如，该可交付使用的文件是否是支撑合格评定的必要条件，或只是作为指南或者推荐的最佳实践；可交付使用的文件是否是管理体系标准；可交付使用的文件是否拟在技术法规中使用或引用；可交付使用的文件是否拟用来支撑与国际条约和协议有关的法律案件。

—— 一份简洁的指标说明，供委员会跟踪评估已出版标准的效益，能否使C.4.10中提及的利益相关方获益。

—— 一份评估最终的可交付使用的文件的前景说明；对IEC而言，是否符合IEC全球相关性政策：http//www.iec.ch/members_experts/refdocs/ac_cl/AC_200817e_AC.Pdf；对ISO而言，是否符合ISO全球相关性政策：http://www.iso.org/iso/home/standards_development/governance_of_technical_ work.htm；如相关，是否符合ISO/TMB有关可持续发展和可持续性的建议（见下面的注）。

注：对于ISO，ISO/TMB确认下列建议：1）若一个委员会（任一行业）制定一项有关可持续性/可持续发展的标准时，该标准必须保持在该委员会的工作范围内；2）委员会还应尽早地将项目的标题和范围通报给TMB；3）承担该项工作的委员会应在该标准的引言中阐明其宗旨；4）最广泛使用的可持续发展的定义来自联合国的布伦特兰可持续发展委员会：既满足当代人的需求，而又不损害子孙后代满足其需求的能力的发展。

—— 一份评估该提案是否符合与公共政策制定有关或支持公共政策制定的ISO和IEC标准制定的原则的说明，（对于ISO，见《ISO技术工作程序合集》的附录SO；对于IEC和ISO，见《采用和引用ISO/IEC标准支撑公共政策》：http//www.iso.org/sites/policy/）以及最终可交付使用的文件与公共政策的潜在关系，包括一份有关符合相关法律法规，便利市场准入的说明。

附录 D
（规范性）
秘书处资源和秘书资格

D.1 术语和定义

D.1.1 秘书处

经多方同意并指定，负责向某个技术委员会或分委员会提供技术和管理服务的国家成员体。

D.1.2 秘书

秘书处任命的负责管理秘书处所提供的技术和管理服务的人员。

D.2 秘书处资源

指定为秘书处的国家成员体应认识到，无论其在自己国家为提供所需的服务做出什么安排，最终负责秘书处正常运作的是国家成员体本身。承担秘书处职能的国家成员体应视情况成为 ISO 服务协议的缔约方或 IEC 基础协议的缔约方。

因此，秘书处应具备足够的行政管理和财政手段或支持以确保：

a) 用英语和 / 或法语进行文字处理的设备，提供电子文本的设备和必要的文件复印设备；在 JTC 1 中，用英语进行文字处理的设备，提供机读文本的设备和必要的文件复印的设备；

b) 准备足够的技术图表；

c) 识别和使用收到的用官方语言编写的文件，必要时进行翻译；

d) 更新并持续监督委员会及其附属机构（如有）的结构；

e) 接收和快速分发函件和文件；

f) 具备足够的电话、传真及电子邮件等通信设施；

g) 连接互联网；

h) 根据需要，与承办会议的国家成员体协商安排会议期间的笔译、口译和服务设施；在 JTC 1 中，除 4.3 中的规定外，不要求提供笔译、口译和服务安排和设备；

i) 秘书参加任何需要其出席的会议，包括技术委员会和 / 或分委员会会议、编辑委员会会议、工作组会议以及必要时与主席的协商；

j) 秘书可获取基础国际标准（见《ISO/IEC 导则》的第 2 部分有关参考文件和起

草来源），以及相关领域的国际标准、国家标准和/或相关文件；

k） 必要时，秘书可访问能够对委员会领域内技术问题给予指导的专家。

尽管首席执行官争取派其代表参加技术委员会第一次会议、参加有新秘书处的技术委员会会议以及为解决问题而希望他们到会的所有技术委员会或分委员会会议，但是 CEO 办公室不能承担长期或临时为某一秘书处开展工作。

1） 在 JTC 1 中，秘书处的电子文件分发职责应满足常备文件 23 "TC 1 文件的获取控制：开发和限制访问"中的规定。

尽管首席执行官尽量派其代表参加技术委员会第一次会议，设立新秘书处的技术委员会会议，以及任何拟定解决问题的技术委员会或分委员会会议，CEO 办公室不能承诺永久或临时性承担秘书处工作。

D.3 对秘书的要求

被任命为秘书的人员应：

a） 通晓英语和/或法语；在 JTC 1 中，要求秘书通晓英语。

b） 熟悉《章程和议事规则》，并视情况，熟悉 ISO/IEC 导则（见《ISO/IEC 导则》独立补充部分）；在 JTC 1 中，还要求熟悉 JTC 1 补充文件合集、JTC 1 常备文件和 JTC 1 决议。

c） 能够就程序或起草文件向委员会和附属机构提出建议，必要时，先与 CEO 办公室协商。

d） 了解理事局或技术管理局关于技术委员会总体活动的决定，特别是了解关于其归口委员会的活动决定。

e） 是一名优秀的组织者，在技术和管理工作上训练有素，以便组织和开展委员会工作并促进委员会成员和附属机构（如有）积极参与活动。

f） 熟悉 CEO 办公室提供的文件，特别要熟悉使用电子工具和服务。

建议新任命的技术委员会秘书处早日访问位于日内瓦的 CEO 办公室，以便与相关的工作人员讨论程序和工作方法。

附录 E
（规范性）
语言使用通则

E.1 国际环境中意见的表达与交流

国际上通常至少使用两种语言来出版可交付使用的文件。许多理由可以说明使用两种语言的优势，例如：

—— 采用语法和句法不同的两种语言表达一个给定的概念，能使意思更清楚更准确。

—— 如果仅以一种语言起草的文本为依据达成一致意见，则当把这个文本翻译成另一种语言时可能会出现一些困难。有些问题可能不得不重新讨论，如果最初一致同意的文本不得不进行改动，则可能会造成时间的延误。如果一个文本已经以第一种语言通过了，之后再转化为第二种语言，常常会出现一些表述上的困难，如果两种文本同时起草、同时修正，这些困难即可避免。

—— 要确保国际会议富有成效，重要的是在达成的协议中杜绝出现意义不明确的表达法，并且必须消除由于语言本身的误解而对协议产生异议的风险。

—— 使用从两个语系中所选的两种语言，扩大了可能被指定参加会议的代表人数。

—— 如果已有两个完美统一的版本，那么就更容易用其他语言来正确地表达某个概念。

E.2 技术工作中使用的语言

官方语言为英语、法语和俄语。

技术委员会在工作和通信中默认使用英语。

为了上述目的，俄罗斯联邦国家成员体负责提供所有俄语的口译和笔译工作。

在 IEC，每个可交付使用的文件的"前言"中应标明本文件制定时使用的一种语言。适用国际电工词汇（IEV）和/或数据库标准的特例除外。

E.3 国际标准

国际标准由 ISO 和 IEC 采用英语和法语两种语言出版（有时是多语种版，也包括俄语和其他语言，尤其是术语标准）。这些版本的国际标准具有同等效力，每个版本都被视为原版。

标准的技术内容从起草程序开始就用英语和法语两种语言表述会很有优势，这样两种版本可同时研究、修正和批准，始终确保语言的等效性（也可见《ISO/IEC 导则 第 2 部分》"语言版本"条款）。

这可由下列机构完成：

—— 由秘书处完成，或在外部帮助下，由秘书处负责；

—— 或者，由归口技术委员会或分委员会的编辑委员会完成；

—— 或者，由母语为英语或法语的国家成员体根据这些国家成员体与相关秘书处达成的协议完成。

当决定出版一个多语种国际标准（例如，术语标准）时，俄罗斯联邦国家成员体负责该文本的俄语部分；同样，当决定用非官方语言出版含有术语或材料的国际标准时，其母语相关的国家成员体负责选择术语或负责起草文本中拟使用这些语言的部分。

E.4 技术委员会制定的其他出版物

其他出版物可以只用一种官方语言出版。

E.5 技术委员会和分委员会会议文件

E.5.1 与会议议程相关的草案和文件

会前起草和分发的文件如下：

a） 议程草案

议程草案应由归口秘书处以会议语言（默认语言为英语）起草并分发。

b） 列入议程的委员会草案

列入会议议程的委员会草案最好以会议语言（默认语言为英语）提供。

征询意见草案应有英语和法语两种版本。ISO 理事会或 IEC 标准化管理局的指南适用于其中一种语言版本不能适时提供的情况。

与会议议程上的事项有关的其他文件（各种建议、评议意见等）可以只用一种语言（英语或法语）起草。

E.5.2 会议期间起草和分发的文件

会议期间起草和分发的文件如下：

a） 会议期间通过的决议

每次会议开始时，由秘书和一名或多名母语为英语和/或法语的代表（只要可能）组成临时起草委员会，编辑所有提出的决议。

b） 每次会后整理的简要记录（如有）

如果准备这种会议记录，则应用英语或法语起草，最好是在特别起草委员会的帮助下（必要时）用两种语言起草。

E.5.3 会后起草和分发的文件

每次技术委员会和分委员会会议后，相关秘书处应起草会议报告。会议报告可以只使用一种语言（英语或法语），并将通过的正式决议，最好用英语和法语，作为附录附上。

E.6 使用非英语或法语起草的文件

其母语既不是英语也不是法语的国家成员体可将秘书处分发的所有文件翻译成本国语言，以便于本国专家对这些文件进行研究，或帮助他们选派的代表参加技术委员会和分委员会会议。

如果一种语言为两个或多个国家成员体所通用，那么其中一个国家可以在任何时间主动将技术文件翻译成该种语言，并向同一语系的其他国家成员体提供副本。

上述两小节的条款可由秘书处依据自身需求实施。

E.7 技术会议

E.7.1 目的

技术会议的目的是对议事日程中的所有事项尽可能达成一致，并尽力确保所有的代表相互理解。

E.7.2 讨论内容口译成英语和法语

虽然基础文本可以提供英语和法语两种版本，但还必须根据情况决定以一种语言表达的发言是否要由下列人员口译成另一种语言：

—— 自愿的代表；

—— 秘书处或东道国成员体的工作人员；

—— 合格的口译员。在 JTC 1 中，除非 4.3 规定的情况，在讨论时不使用英语和法语的口译员。

还应注意，要使那些母语既不是英语也不是法语的代表能够充分理解会议内容。

对是否有必要口译技术会议上所讨论的内容做出规定是不切实际的，重要的是让所有与会代表能领会讨论的内容，逐字逐句口译每句话并不完全必要。

鉴于上述考虑，除没有必要提供口译的特殊情况外，下列作法是适宜的：

a) 对于要做出程序性决定的会议，可由秘书处的成员或自愿的代表提供简单的口译；

b) 在工作组会议上，只要可能，工作组成员应在工作组召集人的倡议或授权下安排工作组成员之间的必要的口译。

为了使负责会议的秘书处能够对口译做出必要的安排，代表们在告知参加会议的同时应将其能够使用的语言和在口译方面上能够提供的帮助告知秘书处。

在主要使用一种语言的会议上，为了帮助使用其他语言的代表，应尽量采用下列作法：

a) 应当用两种语言宣布对某一主题形成的决定之后，然后再进入下一个主题；

b) 凡是批准对现行文本的一种语言版本进行更改时，应给代表留出时间考虑这一更改对另一种语言版本的影响；

c) 如果代表有要求，应提供另一种语言的讲话摘要。

E.7.3 其他语言的发言口译成英语和法语

在技术委员会或分委员会会议上，如果某一与会者在特殊情况下希望用非英语或法语的其他语言发言，在会议具备口译的条件下，会议主席有权批准此项要求。

为了给所有参加技术委员会和分委员会会议的专家提供平等表达其观点的机会，建议灵活运用本条款。

附录 F
（规范性）
项目制定的选择方案

F.1 选择方案的简化流程图

项目阶段	正常程序	与提案同时提交的草案	"快速程序"[1)	"技术规范"[2)	技术报告[3)	可公开提供的规范[4)
提案阶段（见2.3）	受理提案	受理提案	受理提案[1)	受理提案		受理提案[7)
准备阶段（见2.4）	起草工作草案	*工作组研究[5)*		起草草案		批准PAS草案
委员会阶段（见2.5）	制定和受理委员会草案	*制定和受理委员会草案[5)*		受理草案	受理草案	
征询意见阶段（见2.6）	起草和受理征询意见草案	起草和受理征询意见草案	受理征询意见草案			
批准阶段（见2.7）	*批准FDIS[6)*	*批准FDIS[6)*	*批准FDIS[6)*			
出版阶段（见2.8）	出版国际标准	出版国际标准	出版国际标准	出版技术规范	出版技术报告	出版PAS

用虚线圈的斜体阶段可以省略。
1) 见F.2。
2) 见F3.1。
3) 见3.3。
4) 见3.2。
5) 根据对新工作项目提案的投票结果，准备阶段和委员会阶段可被省略。
6) 如征询意见草案得到批准且没有反对票，则可省略此阶段。
7) 有关PAS提案的细节见ISO和IEC补充部分。

以下表格适用于JTC 1：

阶段名称	阶段描述	标准（见第2章）	快速程序（见F.2）	JTC 1可公开提供的规范（见F.3）	技术规范（见3.1）	技术报告（见3.3）	修改单（见2.10.3）	技术勘误表（见2.10.2）
00 预研阶段（见2.2）	起草提案	起草 NP	—	—	起草 NP		起草 NP	—
10 提案阶段（见2.3）	受理提案	受理 NP	—	—	受理 NP		受理 NP	
20 准备阶段（见2.4）	起草工作草案	起草 WD	—	—	起草 WD		起草 WD	起草缺陷报告（DR）
30 委员会阶段（见2.5）	制定和受理委员会草案	制定和受理 CD	—	—	制定和受理 DTS	制定和受理 DTR	制定和受理 CDAM	制定和受理 DCOR
40 征询意见阶段（见2.6）	制定和受理征询意见草案	制定和受理 DIS	提交和受理 DIS	提交和受理 DIS			制定和受理 DAM	—
50 批准阶段（见2.7）	批准最终草案	批准 FDIS	批准 FDIS	批准 FDIS	—		批准 FDAM	
60 出版阶段	出版文件	出版 IS	出版 IS	出版 IS	出版技术规范	出版技术报告	出版修改单	出版技术勘误表

F.2 "快速程序"

F.2.1 采用快速程序的提案可按下述步骤操作。

F.2.1.1 在JTC 1中，仅JTC 1的P成员和A类联络组织可以提交采用快速程序的提案。

相关技术委员会或分委员会的任何P成员及A类联络组织可提议将任何来源的现行标准作为征询意见草案提交投票。提案方应在提交提案前得到原组织的同意，是否采用快速程序提交现行标准的准则可由提案方自行决定。

在JTC 1中，JTC 1秘书处按照F.4.1的要求，将收到的所有采用快速程序的提案提交给ITTF。鼓励采用快速程序的提案方就将文件分配给某个分委员会提出建议。采用快速程序文件的提案方应提交同意担任该项目编辑的个人姓名。提案方还应提交类似于PAS解释报告的解释报告（见F.3）。

对于初次出版，不要求文件采用ISO/IEC规定的格式，可以采用原始格式发布。

作为出版协议的一部分，出版物的形式（如原始文件的重印或参考 ISO/IEC 封面分发）将由 ITTF 和提案方确定。但是，后续修订应采用《ISO/IEC 导则　第 2 部分》规定的格式。

在 JTC 1 中，对现行国际标准的修正不采用快速程序。

F.2.1.2　ISO 或 IEC 理事局承认的国际标准化机构可建议将该机构制定的标准作为最终国际标准草案提交投票。

F.2.1.3　与 ISO 或 IEC 达成正式技术协议的组织，在征得相应的技术委员会和分委员会的同意后，可提议将该组织制定的标准草案作为征询意见草案在技术委员会或分委员会内提交投票。

F.2.2　提案应由首席执行官受理，并应采取下列行动：

a）与起草提案文件的组织共用解决版权和 / 或商标问题，以便不受限制地自由复印，并分发到各国家成员体，并建议该组织注意提案文件应遵行 ISO/IEC 知识产权政策，具体见 2.13 和 2.14。

b）对于 F.2.1.1 和 F.2.1.3 的情况，与相关秘书处磋商后评估具体哪一个技术委员会或分委员会能够胜任提案文件所涉及的主题；如果现有的技术委员会不能解决提案文件所涉及主题的，首席执行官应将提案提交技术管理局，技术管理局可要求首席执行官将文件提交到征询意见阶段，并设立一个特别工作组来解决后续产生的问题。

c）保证与其他国际标准没有明显的矛盾。

d）依据 2.6.1 作为征询意见草案（F.2.1.1 和 F.2.1.3），或依据 2.7.1 作为最终国际标准草案（F.2.1.2 的情况）分发提案文件，同时向技术委员会或分委员会指明（在 F.2.1.1 和 F.2.1.3 的情况下）提案文件所属的领域。在 JTC 1 中，分委员会的指派建议以及推荐的项目编辑姓名也一并分发。

F.2.3　表决期限和批准条件应分别按照 2.6 征询意见草案或 2.7 最终国际标准草案的规定，如果没有技术委员会参与，则最终国际标准草案的批准条件应是反对票少于投票总数的 1/4。

在 JTC 1 中，除了对标准技术内容的投票外，JTC 1 国家成员体应有机会对项目指派给具体的分委员会进行评论。但是，对分委员会工作指派的意见不应影响对技术内容的表决。如果对分委员会的任务指派有疑义，或者采用快速程序的文件不适合任何现有分委员会，则 JTC 1 秘书处可承担该任务，直到最终确定了承担任务的分委员会。

在 JTC 1 中，采用快速程序制定文件的提案方有权在文件发布前的任何时间从快速程序中撤回文件。

F 2.3.1 在 JTC 1 中，可使用投票决议会议（见 F.5）评议收到的关于快速通道投票查询草案（DIS）的意见。

F.2.4 对于征询意见草案，如果符合批准条件，则标准草案应进入批准阶段（2.7）；如果不符合批准条件，该提案未通过，依据 F.2.2 b)，由归属的技术委员会或分委员会决定采取进一步行动。对于最终国际标准草案，如果符合批准条件，则该文件应进入出版阶段（2.8）；如果不符合批准条件，该提案未通过，依据 F.2.2 b) 由该 FDIS 归属的技术委员会 / 分委员会应决定采取进一步行动，或者如果没有技术委员会参与，则由原组织与 CEO 办公室讨论决定采取进一步行动。

在 JTC 1 中，委员会领导层可决定是否略过 FDIS 投票直接进行出版（见 2.6.4）。

在 JTC 1 中，收到投票结果和任何评议意见后，技术委员会或分委员会主席应与其秘书处和项目编辑合作并与 CEO 办公室协商采取下列措施：

e) 满足 2.6.3 的批准标准，且无技术变更进入出版阶段（见 2.8）。

f) 满足 2.6.3 的批准标准，但是，有技术变更的，将修改后的征询意见草案注册为最终国际标准草案。

g) 不满足 2.6.3 的批准标准；

1) 分发修订后的征询意见草案用于投票（见 2.6.1），或

注 2：修订后的征询意见草案分发后的投票期限为 8 周，应相关委员会一名或多名 P 成员的请求，投票期限可延长至 12 周。在 JTC 1，修订后的征询意见草案分发期可延长至 12 周。

2) 分发修订后的委员会草案用于评议或

3) 下次会议时商讨征询意见草案和评议意见。评议决议会议可采用电话会议或使用电子设备。

如果标准已经出版，其维护工作应依据 F.2.2 b)，由该文件归属的技术委员会 / 分委员会负责。如果没有技术委员会参与，当原组织决定需要修改标准时，应重复上面规定的批准程序。

在 JTC1 中，F.4 描述了采用快速程序提交的文件的转换和采用过程。

如果标准已经出版，其维护工作应依据 F.2.2 b)，由该文件归属的技术委员会 / 分委员会负责。如果没有技术委员会参与，当原组织决定需要修改标准时，应重复上面规定的批准程序。

F.3 国际标准的起草和采用——JTC 1 PAS 转换过程

JTC 1 常备文件 9 "可公开提供的规范转换为国际标准的指南"给出了可公开提供的规范转换为国际标准的规定。

F.3.1 概念

JTC 1 PAS 转换过程基于以下关键概念：

可公开提供的规范（PAS）

如果技术规范满足某些可以转换为国际标准的准则，则称为可公开提供的规范（PAS）。制定这些准则（见 F.3.3）旨在确保高质量、协调一致和妥善处理知识产权（IPR）相关事宜。

PAS 顾问

JTC 1 任命的人员，协助 PAS 发起方和/或承认的 PAS 提交方创建和处理其提交文件，并提供持续的帮助。

PAS 发起方

PAS 发起方制定并拥有可公开提供的规范，且考虑将其转换为国际标准的任何组织。此种组织应具备的形式并无基本限制条件，但是组织的章程应反映出其开放性和 PAS 的制定过程。相关模板请见常备文件 9 "可公开提供的规范转换为国际标准的指南"。

承认的 PAS 提交方

PAS 发起方应向 JTC 1 申请承认其为 PAS 的提交方，以进行 PAS 的转换。一旦获批，承认的 PAS 提交方的状态将在最初两年内保持有效，并有可能进一步延长（见 F.3.4.1）。

说明报告

提交 PAS 时，必须附上由 PAS 发起方起草的说明报告。该报告提供了提交 PAS 所需的所有信息，特别包含对 PAS 准则符合程度的声明。还应清楚地规定 PAS 中使用的技术概念。JTC 1 已经制定了包含在说明报告中的准则清单。

PAS 转换投票

将 PAS 和说明报告一起提交进行投票。

F.3.2 适用性

这些程序适用于将 PAS 转换为国际标准。相比快速程序（见 F.2），期望这些程序用于处理来源更广泛的一类文件。

F.3.3 PAS 准则

JTC 1 制定了判断是否承认特定组织以及是否将其规范转换为国际标准的依据。潜在提交方也可利用这些准则确定其规范对标准化过程的适用性。PAS 准则大致分为两类，并涉及以下主题：

- 与组织相关的准则包括：

- 合作立场；
- 组织的特征；
- 知识产权。
- 与文件相关的准则包括：
  - 质量；
  - 一致性；
  - 符合性；
  - 维护。

详情见 JTC 1 常备文件 9 "可公开提供的规范转换为国际标准指南"。

F.3.4 程序

根据上述 F.3.1 中给出的概念，本部分描述了 PAS 的转换过程。JTC 1 坚持秉承全过程透明原则，并在其网站（www.jtc1.org）上展示了所有提案的当前状态。积极鼓励在 PAS 提交方和 JTC 1 以及 JTC 1 国家成员体之间进行公开对话（通过网站或任何其他可用方式）。

F.3.4.1 PAS 提交方的承认

将现有或即将发布的规范提交到转换过程中的相关 PAS 发起方应向 JTC 1 秘书处申请承认为 PAS 提交方。此类申请应附有计划提交的初始 PAS 的标识，以及 PAS 发起方关于组织相关准则的声明（见下文）。完成的文件应提交 JTC 1 的 P 成员进行为期 12 周的投票。一经批准，作为承认的 PAS 提交方，赋予 PAS 发起方将规范提交到转换过程的权利，期限为两年，并有可能进一步延长至五年（见下文）。如存在以下情况，将终止承认作为 PAS 提交人：

- 未能通过 JTC 1 国家成员体投票确认 PAS 提交方的身份；或
- PAS 发起方未能在预期期限内将规范提交给 JTC 1 进行转换（见 F.3.4.2）。

PAS 发起方具有提交承认申请的主动权。任何 JTC 1 国家成员体、JTC 1 分委员会、JTC 1A 类联络组织或 PAS 顾问均可协助 PAS 发起方与 JTC 1 沟通。

由于 JTC 1 P 成员的投票需要 12 周，因此宜在计划首次提交 PAS 之前及时提交承认申请。虽然对申请的格式没有特殊要求，但宜：

- 规定申请的总体范围；
- 确定计划提交的初始 PAS 及其范围；
- 针对 JTC 1 常备文件 9 "可公开提供的规范转换为国际标准指南"中包含的组织承认准则的所有强制性要素进行阐述。

申请书中还应说明 PAS 提交方对转换后的 PAS 提交文件进行维护的期望。JTC 1

的维护目的是避免当前 JTC 1 转换后的 PAS 修订版与 PAS 提交方发布的原始规范修订版之间出现分歧。因此，该申请包含提交方将如何与 JTC 1 合作维护该标准的说明。虽然 JTC 1 负责标准的维护，但这并不意味着 JTC 1 必须亲自进行维护。JTC 1 可批准提交方的维护方案，只要有适当的 JTC 1 代表参与，即指定提交方的维护组为 JTC 1 维护组。

在组织作为已批准的 JTC 1 PAS 提交方的资格到期前 6 个月，JTC 1 秘书处应邀请提交方审查其作为 PAS 提交方的未来意图，并请 PAS 提交方考虑以下选项申请承认：

- 修订（初始申请存在重大变更，如范围、程序的变更）；
- 撤回（终止）；或
- 重新确认（在不发生重大变化的情况下保持当前状态）。

如果 PAS 提交方选择修订，则必须向 JTC 1 秘书处提交文件，阐述 JTC 1 常备文件 9"可公开提供的规范转换为国际标准指南"中问题答案的变更。如果 PAS 提交方选择重新确认，则应确定拟提交的后续 PAS。为使 JTC 1 及时对修订或确认做出反应，应在其作为 PAS 提交方的资格到期前 12 周内提交必要的文件。JTC 1 秘书处应根据修订或重新确认的请求，发起为期 12 周的信函投票。未能响应秘书处对 PAS 提交方资格的审查邀请，将自动导致本阶段结束时 PAS 提交方资格的终止。

F.3.4.2　PAS 提交文件

PAS 发起方一经承认，即可在申请中确定的范围内向 JTC 1 秘书处提交 PAS。当向 JTC 1 秘书处提交 PAS 时，承认的 PAS 提交方还应提交说明报告和声明，说明承认条件没有改变，或说明已经发生的变更的性质。说明报告应阐述 JTC 1 常备文件 9"可公开提供的规范转换为国际标准指南"中所有强制性要素。

如果承认的 PAS 提交方已获批行使维护职责，则 PAS 提交方应在说明报告中再次确认其履行 JTC 1 维护组职责的承诺。

包括说明报告在内的所有提交文件均应以电子版形式提交。

首次提交应不迟于初始承认后 6 个月。在不迟于 6 个月期限结束前 6 周，应 PAS 发起方的请求，经 JTC 1 主席和秘书处批准，该期限可再延长 6 个月。PAS 发起方未能在预期期限内提交规范，将导致其承认状态终止。

所提交规范的格式不受 JTC 1 的规定限制。如果比较方便，鼓励承认的 PAS 提交方采用接近 ISO/IEC 格式的体例，以便在修订时简化后期校准过程。

JTC 1 秘书处在检查了提交方的承认状态和申请的完整性后，应根据 F.4.1 将规范提交 ITTF。

鉴于说明报告对于成功转换的重要性，承认的 PAS 提交方在起草报告期间以及整个转换期间，向 JTC 1 PAS 顾问、JTC 1 国家成员体、分委员会或 A 类联络组织咨询。咨询过程可以包括对提交材料的审查。

如果承认的 PAS 提交方不能履行最终国际标准的维护职责，鼓励其提出建议，将文件分配给特定分委员会。该建议（如无建议，JTC 1 秘书处可给出建议）应与投票一起分发给 JTC 1 P 成员，但该建议不影响投票。如果分委员会的任务分配存在问题，或者文件不适合所有现有的分委员会，则 JTC 1 秘书处应履行常规任务分配职责，直到最终确定分委员会的任务。

如果承认的 PAS 提交方已获批履行最终国际标准的维护职责，则由 JTC 1 秘书处发起投票，JTC 1 秘书处应履行 F.4 中规定的所有职责。

F.4 根据 JTC 1 快速程序或 JTC 1 PAS 转换过程提交文件

F.4.1 JTC 1 秘书处根据快速程序或 PAS 提交要求，连同说明报告和相关文件一并提交 ITTF。

F.4.2 ITTF 应采取下列措施：

- 与 PAS 提交方或采用快速程序的提交方一起解决版权保护和/或商标问题，以便提交的文本可以在 ISO/IEC 内自由复制和发行。
- 咨询 JTC 1 秘书处，评估 JTC 1 为提案标准所涵盖主题的归口管理委员会，并确定与其他 ISO/IEC 标准没有明显矛盾。
- 以国际标准草案（DIS）的形式分发提案标准的文本，连同说明报告和相关文件，并表明该标准属于 JTC 1 的范围。

F.4.3 投票的期限应为 12 周，其中 8 周为翻译期（见 JA.7）。为了能够获得许可，DIS 必须满足 JA.5.1 中规定的批准条件。对于 JTC 1 快速程序和 JTC 1 PAS 2.6.3 a)"技术委员会或分委员会"应理解为 JTC 1。

F.4.4 JTC 1 秘书处在收到 ITTF 关于 DIS 已登记的通知后，应将 DIS 编号、标题和投票期限通知承担该项目的分委员会秘书处，并向分委员会秘书处发送 DIS 副本及其所附的说明报告。JTC1 秘书处还应将处理 DIS 投票结果的分委员会通知 ITTF，以便 ITTF 可直接将答复表和随同投票的意见和建议发送给分委员会秘书处和 JTC 1 秘书处。

F.4.5 考虑到快速程序和 JTC 1 PAS 转换过程的重要性，JTC 1 秘书处还应将启动快速程序或 PAS 投票、投票结果和负责后续工作的 JTC 1 分委员会告知 JTC 1 国家成员体和联络组织，以及授权为承认的 PAS 提交方的组织。

F.4.6 快速程序的提案方或 PAS 提交方应收到投票文件的副本。

F.4.7 在收到 JTC 1 秘书处已委派负责处理 DIS 的分委员会的通知后，该分委员会秘书处应通知作为分委员会成员的 JTC 1 国家成员体，并计划召开可能的投票决议会议

（见 F.5）。

F.4.8 收到 DIS 投票结果和评议意见后，分委员会秘书处应将该材料分发给分委员会、JTC 1 国家成员体以及 PAS 或快速程序提交方。

委员会领导层可决定是否略过 FDIS 投票直接进行出版（见 2.6.4）。

收到投票结果和任何评议意见后，技术委员会或分委员会主席应与其秘书处和项目编辑合作并与 CEO 办公室协商采取下列行动步骤：

h) 满足 2.6.3 的批准标准，且无技术变更进入出版阶段（见 2.8）。

i) 满足 2.6.3 的批准标准，但是，为了将修改后的征询意见草案注册为最终国际标准草案，须将技术变更包含在内。

j) 不满足 2.6.3 的批准标准；

1) 分发修订后的供投票征询意见草案（见 2.6.1）或

注：修订后的征询意见草案分发后的投票期限为 8 周，应相关委员会一名或多名 P 成员的请求，投票期限可延长至 12 周。在 JTC 1，修订后的征询意见草案分发期可延长至 12 周。

2) 分发修订后的供评议委员会草案或

3) 下次会议时商讨征询意见草案和评议意见。评议决议会议可采用电话会议或使用电子设备。

如果收到了除编辑性更正以外的评议意见，并且无法通过修改投票文件接纳评议意见，则召开投票决议会议（BRM），见 F.5。

F.4.9 投票决议会议（如果在投票结束后顺利举行）结束后，可能出现以下情况：

a) 在出版之前，除编辑性更正外，未对原 DIS 进行任何修改；

b) 除编辑性更正外的其他修改已在投票决议会议上达成一致：在这种情况下，项目编辑应起草修订后的 DIS，并将其发送给分委员会秘书处，由分委员会秘书处转交 ITTF 进行 FDIS 投票。FDIS 的投票期为 8 周。

F.4.10 JTC 1 秘书处在收到 ITTF 已登记为 FDIS 的通知后，应将 FDIS 的编号、标题和投票日期通知负责该项目的分委员会秘书处，并应向分委员会秘书处发送一份 FDIS 和针对 DIS 的评议意见（如有）处理表副本。ITTF 将直接向分委员会秘书处和 JTC 1 秘书处发送回复表和随同投票的评议意见。

F.4.11 如果满足 2.7.3 的要求，ITTF 将文本作为国际标准发布。对于初次发布，不要求文件采用 ISO/IEC 格式，可以用原格式发布。作为出版物协议的一部分，出版物的形式（例如重新打印原始文件或分发附加 ISO/IEC 封面的文件）由 ITTF 和 PAS 或快速程序提案方确定。然而，后续修订应采用《ISO/IEC 导则 第 2 部分》规定的格式。

F.4.12 如果无法符合 FDIS 批准要求的文本（见 2.7.3），则该提案失败。在这种情况下，JTC 1 应向提交方说明导致否决的原因。根据这些信息，提交方可以选择重新提交修改后的规范作为新的快速程序提案或 PAS 提交。

F.4.13 投票结束后，各责任方的活动时间如下：

- 在 DIS 和 FDIS 投票后，ITTF 应立即将投票结果发送给 JTC 1 秘书处和分委员会秘书处，分委员会秘书处应立即将投票结果分发给作为分委员会成员的 JTC 1 国家成员体，以及对提案投票的非分委员会成员的国家成员体；
- 将投票结果分发给作为分委员会成员的 JTC 1 国家成员体的 2 个半月之后，分委员会秘书处应尽快召开投票决议会议（BRM）（必要时）；
- 分委员会秘书处应在投票决议会议后 4 周内分发会议的最终报告和修改的 DIS 文本。

F.4.14 如果提交的标准获得许可，将按照 ISO 和 IEC 的版权和其他知识产权政策发布。该标准的维护将由 JTC 1 或由 PAS 提交方维护组（经 JTC 1 委托）根据 JTC 1 规则进行处理。

PAS 或快速程序提交方可自行决定在发布前的任何时候从转换过程中撤回文件。PAS 或快速程序提交方也有权要求文件在整个转换过程中保持不变。这种要求应在说明报告中明确阐述。

F.5 JTC 1 PAS 和快速程序投票决议会议

F.5.1 投票决议会议的目的和范围

在 JTC 1 中，投票决议会议（BRM）旨在评审就 JTC 1 PAS 或快速程序投票（见 F.4）的征询意见草案（DIS）收到的评议意见；此外，还应对这些意见进行处理，以获得最大共识。在某些情况下，分委员会秘书处可决定不必举行投票决议会议，并将意见的处理结果直接交给项目编辑。

F.5.2 归口分委员会在投票决议会议上的职责

JTC 1 通常将征询意见草案（DIS）指派给某个分委员会。如果 DIS 未指派给特定的分委员会，JTC 1 秘书处将负责所指派的任务。

归口秘书处应：

- 安排投票决议会议，在评议意见分发两个半月之后举行，以评议有关 DIS 的评议意见；
- 为投票决议会议任命召集人；
- 将投票决议会议的日期、地点和召集人告知有资格参会的人员。

在投票决议会议召开 8 周之前，归口秘书处应将会务信息和议程以及召集人的通知一并发送 JTC 1 秘书处，以便分发给下文 F.5.4 所列的各方。

F.5.3 评议意见处理建议

指派给 DIS 的项目编辑应在 ISO 意见汇总表（最后一列）上给出意见处理建议（DoC）。

在投票决议会议开始 4 周之前，责任秘书处应向下文 F.5.4 中列出的接收者分发意见处理建议文件。

F.5.4 接收者和有资格参会者

责任秘书处应通过 ITTF 向下列有资格出席投票决议会议的人员或提名代表提供意见处理建议（DoC）：

- 国家成员体代表；
- ISO 中央秘书处和 IEC 中央办公室的代表；
- 分委员会主席；
- 分委员会委员会主管；
- 指派的项目编辑；
- 投票决议会议召集人；
- 国际标准草案提交方；
- JTC 1 A 类联络组织代表。

F.5.5 会议程序

即使与 JTC 1 或相关分委员会的其他会议一起召开，投票决议会议也应单独举行。投票决议会议可通过电话会议或电子方式以及面对面的方式举行（见"会议"常备文件 19）。

任命的召集人应进行点名。

投票决议会议记录应列出 JTC 1 国家成员体代表团团长（HoD）清单，如需投票，还应列出其他与会者及其角色。

投票决议会议应尽可能解决 DIS 投票中提出的所有意见，以提高最终文件的协调一致性。

在最终意见处理报告上，对于每一条意见，项目编辑应记录投票决议会议协调一致的处理结果，如果未能达成协调一致，则应记录与会者（有资格对 DIS 投票的 JTC 1 国家成员体代表）多数支持的建议。

当所有的投票意见都已处理得当，并且得到会议的批准时，则实现了投票决议会议的目标。

分委员会秘书处应在会议结束后 4 周内或经 ITTF 允许发行：

- 包括投票决议会议上商定的所有变更的 DIS 修订版；
- 在投票决议会议上批准的意见处理报告；
- 投票决议会议报告，包括与会者及其角色的列表，最终意见处理报告的引用以及进一步处理国际标准草案的建议。

这些文件也应呈送 ITTF，以便进一步分发给上述接收者。

F.6 JTC 1 中快速程序和 PAS 提交流程

在 JTC 1 中，图 F.1 展示了 F.2、F.3、F.4 和 F.5 中所述的快速程序和 PAS 提交流程。

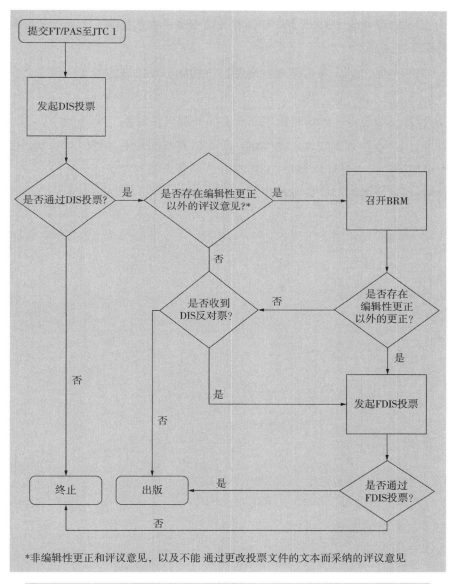

*非编辑性更正和评议意见，以及不能通过更改投票文件的文本而采纳的评议意见

图 F.1 采用快速程序制定国际标准以及通过提交 PAS 转换国际标准的流程图

附录 G
（规范性）
维护机构

G.1 技术委员会或分委员会在制定需要有维护机构的国际标准时，应在早期阶段通知首席执行官，以便 ISO/TMB 或 IEC 理事局在出版国际标准之前做出决定。

G.2 ISO 技术管理局或 IEC 理事局根据相关技术委员会的提案，指定与国际标准有关的维护机构及其成员。

G.3 维护机构的秘书处应尽可能地归属制定该国际标准的技术委员会或分委员会的秘书处。

G.4 首席执行官应负责与维护机构工作有关的外部组织联系。

G.5 维护机构的程序规则应经 ISO/TMB 或 IEC 理事局批准，任何与更新国际标准或发布修改单相关的授权都应由 ISO/TMB 或 IEC 理事局特别授予。

G.6 维护机构提供服务的任何收费应由理事局授权。

附录 H
（规范性）
注册机构

H.1 技术委员会或分委员会在制定需要有注册机构的国际标准时，应在早期阶段通知首席执行官，以便进行必要的协商，并允许技术管理局在出版国际标准前做出决定。

H.2 技术管理局根据相关技术委员会的提案，指定与国际标准有关的注册机构。

H.3 注册机构应是具备资格并在国际上得到认可的机构，如果没有这样的组织，可由技术管理局决定将注册任务交给 CEO 办公室完成。

H.4 注册机构应按要求在工作中清楚地表明其是由 ISO 或 IEC 指定（例如，在被指定机构的信笺抬头添加适当的用语）。

H.5 依据相关国际标准的条款，承担注册职能的注册机构不应要求 ISO 或 IEC 或其成员提供财务贡献。但是，如果得到理事局的授权，并不排除注册机构收取服务费。

附录 I
（规范性）
ITU-T/ITU-R/ISO/IEC 共用专利政策实施指南

最新版 ITU-T/ITU-R/ISO/IEC 共用专利政策实施指南（包括 Word 文本或 Excel 表格）参见 ISO 网站：http//www.iso.org/iso/home/standards_developement/governance_of_technical_work/patents.htm：

也可参见 IEC 网站：http：//www.iec.ch/members_experts/tools/patents/patent_policy.htm

ITU-T/ITU-R/ISO/IEC 共用专利政策实施指南

概要

ITU-T/ITU-R/ISO/IEC 共用专利政策实施指南主要用于阐明和促进专利政策的实施,副本见附录 1 和各组织网站。

专利政策鼓励尽快披露和认定起草中的建议书 / 可交付使用的文件的版权,从而提高起草标准的效率,并避免潜在的版权问题。

历史版本

1.0 版	最初版本	2007 年 3 月 1 日
2.0 版	第 1 版	2012 年 4 月 23 日
3.0 版	第 2 版	2015 年 6 月 26 日
4.0 版	第 3 版	2018 年 11 月 2 日

目 录

第Ⅰ部分　共用指南

Ⅰ.1　目的

Ⅰ.2　术语解释

Ⅰ.3　专利披露

Ⅰ.4　专利陈述和许可声明表

Ⅰ.4.1　声明表的目的

Ⅰ.4.2　联系人信息

Ⅰ.5　召集会议

Ⅰ.6　专利信息数据库

Ⅰ.7　专利权的转让或转移

第Ⅱ部分　各个组织的专用条款

Ⅱ.1　ITU 专用条款

Ⅱ.2　ISO 和 IEC 专用条款

附件 1：ITU-T/ITU-R/ISO/IEC 共用专利政策

附件 2：ITU-T 或 ITU-R 建议书 |ISO 或 IEC 可交付使用的文件的专利陈述和许可声明表

附件 3：ITU-T 或 ITU-R 建议书的通用专利陈述和许可声明表

ITU-T/ITU-R/ISO/IEC 共用专利政策实施指南

第三次修订 2018 年 11 月 2 日生效

第 I 部分　共用指南

1　目的

ITU［在其电信标准化局（ITU-T）和无线电通信局（ITU-R）］、ISO 和 IEC 许多年前就制定了专利政策，其目的是为了给他们的技术机构参与者在遇到专利权问题时提供简单明了的实用指导。

考虑到技术专家通常不熟悉专利法的复杂问题，因此 ITU-T/ITU-R/ISO/IEC 共用专利政策（以下简称"专利政策"）的操作部分起草成核查表的形式，涵盖某一建议书 / 可交付使用的文件要求完全或部分使用或实施专利许可时，可能出现的三种不同的情况。

ITU-T/ITU-R/ISO/IEC 共用专利政策实施指南（以下简称"指南"）的目的在于说明该专利政策并为其实施提供便利。专利政策的副本见附件 1，亦可参见各组织网站。

该专利政策鼓励尽早披露和确认那些可能与正在制定的建议书 / 可交付使用的文件有关的专利。这样做将可能提高标准制定的效率，并且可能避免潜在的专利权问题。

各组织不宜介入有关建议书 / 可交付使用的文件的专利相关性或必要性评价，不宜干涉专利许可谈判，不宜参与解决关于专利的争端；所有这些都宜如同过去的做法，留给相关处理。

各组织的专用条款包含在本文件的第 II 部分中。然而，这些专用条款不应与共用专利政策或指南相矛盾。

2　术语解释

贡献　contribution

技术机构提交审议的任何文件。

免费　free of charge

并不意味着专利权人放弃专利有关的全部权利。更确切地说，是金钱补偿的问题，即专利权人不寻求将任何金钱补偿作为专利许可的一部分（无论这类补偿被称为专利使用费，还是一次性许可费等）。不过，虽然专利权人承诺不收取任何数量的金钱，但专利权人仍然有权要求相关建议书 / 可交付使用的文件的实施者签署一项许可协议，包含其他合理的条款和条件，例如有关适用法律、使用领域、担保等。

组织　organizations

ITU、ISO 和 IEC。

专利　patent

基于发明的专利、实用新型和其他类似的法定权利（包括上述任何权利的申请）中包含的和确认的那些权利要求，并且仅指对建议书 / 可交付使用的文件的实施很必要的那些要求。必要专利是指为实施特定的建议书 / 可交付使用的文件所必需的专利。

专利权人　patent holder

拥有、控制和 / 或有能力许可专利的个人或实体。

互惠　reciprocity

只有当未来的专利许可申请人承诺免费实施同一份相关建议书 / 可交付使用的文件，或在合理的条款或条件下许可其专利，该申请人才能要求专利权人向其授予专利许可。

建议书 / 可交付使用的文件　recommendations/deliverables

ITU-T 和 ITU-R 建议书统称为"建议书"，ISO 可交付使用的文件和 IEC 可交付使用的文件统称为"可交付使用的文件"。在附件 2 的《专利陈述和许可声明表》（以下简称"声明表"）中，建议书 / 可交付使用的文件统称为"文件类型"。

技术机构　technical bodies

ITU-T 和 ITU-R 的研究组、分组和其他小组，以及 ISO 和 IEC 的技术委员会、分委员会和工作组。

3　专利披露

专利政策的第 1 段就强调，各组织工作的参与方[1]任何当事人从一开始就应提请注意任何已知的专利或任何已知的正在处理的专利申请，无论是自己组织的或是其他组织的。

在这种情况下，"一开始"意味着应在建议书 / 可交付使用的文件制定期间尽可能早地披露上述信息。在制定第一个文本草案时也许不能做到这一点，因为此时的文本也许还不够清晰或其后还需要进行重大修改。此外，应该真诚地尽最大的努力来提供

---

[1] 对于 ISO 和 IEC，这包含了在标准制定过程中各阶段的标准草案的接收者。

这方面的信息，但并不要求进行专利检索。

除上述要求外，没有参与技术机构的任何方也可以提请各组织注意任何已知的专利，无论是他们自己的和/或任何第三方的。

在披露他们自己的专利时，专利权人必须使用本指南第4部分规定的《专利陈述和许可声明表》（简称"声明表"）。

任何提请注意第三方专利的信息都宜以书面形式寄给有关组织。然后，如果适用，相关组织的局长/CEO将要求未来的专利权人提交一份声明表。

专利政策和本指南也适用于在建议书/可交付使用的文件批准后披露的或提请各组织注意的任何专利。

无论专利的确认发生在是在建议书/可交付使用的文件批准之前还是批准之后，如果专利权人不愿意按照专利政策的2.1或2.2授予许可，各组织将迅速通告负责受影响的建议书/可交付使用的文件的技术机构，以便采取适宜的措施。这类措施包括（但不限于）复审该建议书/可交付使用的文件或其草案，以便消除潜在的冲突，或者进一步检查并澄清引起冲突的技术考虑。

4　专利陈述和许可声明表

4.1　声明表的目的

为了给各组织的专利信息数据库中提供清楚的信息，专利权人必须使用声明表。这个表格可以通过各组织的网站找到（附件2给出了声明表，供参考）。对于ITU，声明表必须提交给电信标准化局（TSB）或无线电通信局（BR）的局长，对于ISO或IEC，必须提交给CEO。声明表的目的是确保专利权人以标准化的形式向各自的组织提交其发表的声明。

声明表为专利权人提供一种为实施特定建议书/可交付使用的文件所需要的专利权有关的许可声明的途径。具体地讲，通过提交这种声明表，提交方声明其愿意按照专利政策（通过选择表格中的选项1或2）/或者不愿意按照专利政策（通过选择表格中的选项3）授予其所有的并且为部分或完全使用或实施某个特定建议书/可交付使用的文件需要的专利许可。

如果专利权人选择了声明表的选项3，那么，对于引用的相关ITU建议书和/或ISO或IEC可交付使用的文件，ITU、ISO和IEC要求专利权人提供允许专利标识的附加信息。

如果专利权人希望对于同一个建议书/可交付使用的文件标识多个专利并将其分类在声明表的不同选项中，或者如果专利权人将某个复杂专利的不同申请归类在同一个声明表中的不同选项中，则使用多重声明表是合适的。

声明表中包含的信息如有明显错误，例如标准或专利引用编号的印刷错误，可予更正。声明表中包含的许可声明持续有效，直到它被另一个包含对申请许可人更有利的许可条款和条件的声明表取代为止，表现在：

（a）承诺从选项 3 改成选项 1 或 2；

（b）承诺从选项 2 改成选项 1；

（c）或者，取消勾选的一个或多个包含在选项 1 或 2 中的子选项。

4.2 联系信息

在填写声明表时，应注意提供长期有效的联系信息。只要可能，通常应提供"姓名和部门"以及电子邮件地址。只要可能，当事人，特别是多国组织，最好在提交的所有声明表上给出同一个联系人。

为了在各组织的专利信息数据库中保持最新信息，要求对过去提交的声明表的任何变更或修改，特别是与联系人有关的信息变更，通知到各组织。

5 主持会议

早期披露专利有利于提高建议书/可交付使用的文件在制定过程的效率。因此，每个技术机构在制定建议书/可交付使用的文件的过程中，都要求披露任何已知的对提议的建议书/可交付使用的文件必要的专利。

如果合适，技术机构的主席将在每次会议的适当时间，询问是否有人知道为使用或实施考虑中的建议书/可交付使用的文件可能需要使用的专利。应该在会议报告中记录提问情况以及任何肯定的答复。

只要相关组织没有收到专利权人选择专利政策 2.3 的说明，即可运用本组织适宜的相应规则批准该建议书/可交付使用的文件。希望在技术机构讨论时考虑建议书/可交付使用的文件中包括的专利内容，然而，就主张的任何专利的必要性、范围、有效性或具体的许可期限，技术机构可以不承担负责。

6 专利信息数据库

为了促进标准制定过程和建议书/可交付使用的文件的应用，各组织都应向公众提供一个专利信息数据库，包含了以声明表的形式通报给各组织的信息。专利信息数据库可能包含特定专利信息，或可能没有包含此种信息而是包含针对某个具体建议书/可交付使用的文件符合专利政策的陈述。

专利信息数据库的准确性和完备性未经认证，仅仅反映已经通报给各组织的信息。同样，专利信息数据库可以看成是树起的用于提醒用户的一面旗帜，这些用户可能希望与那些已经向各组织提交声明表的实体联系，以便确定为使用或实施某个具体建议书/可交付使用的文件是否必须获得专利使用许可。

7 专利权的转让或转移

指导专利权转让或转移的规则包含在专利陈述和许可声明表中（见附件 2 和附件 3）。通过遵守这些规则，专利权人在转让或转移后完全解除了有关许可承诺的所有义务和责任。本规则不要求专利权人在转让发生后去迫使受让人执行许可承诺。

第 II 部 分各组织的专用条款

II.1 ITU 的专用规定

ITU-1 一般性专利陈述和许可声明表

任何人都可以提交《一般性专利陈述和许可声明表》，这种表格可以从 ITU-T 和 ITU-R 的网站上找到（附件 3 中的表格供参考）。这个表格的目的是为专利权人提供自愿选择，可以用它就其任何贡献中包含的受专利保护的内容做出一般性许可声明。具体地讲，当专利权人向组织提交的贡献中包含的任何提案的部分或全部被纳入建议书，并且被纳入的部分包含已经提交申请的专利，使用或实施这些建议书需要获得专利权人许可。专利权人通过提交表格，声明其拥有的专利的许可意愿。

这个《一般性专利陈述和许可声明表》并不是代替每个建议书需提交的那份"专用"声明表（第 I.4）的替代品，而是希望它能提高专利权人对符合专利政策的响应能力和早期披露。因此，除了现有的关于其贡献的一般性专利陈述和许可声明表外，专利权人在适当的时间（例如，当知道其拥有一项用于特定建议书的专利时），还应提交一份"专用"专利陈述和许可声明表：

—— 对于其向组织提交的任何贡献中含有的并被包含在某一建议书中的专利，任何这样的"专用"专利陈述和许可声明表可以包含与《一般性专利陈述和许可声明表》相同的许可条款和条件，也可以含有"专用"声明表（见 I.4.1）中规定的对申请许可人更有利的许可条款和条件；

—— 对于包含在某一建议书中但专利权人并未贡献给组织的那些专利，任何这样的"专用"专利陈述和许可声明表可以包含声明表的三个选项中的任何一种（见 I.4.1），与现行《一般性专利陈述和许可声明表》中的承诺无关。

《一般性专利陈述和许可声明表》持续有效，直到它被另一份包含对申请许可人更有利的许可条款和条件的《一般性专利陈述和许可声明表》取代为止，表现在：

（a）承诺从选项 2 改成选项 1；

（b）或者，取消勾选一个或多个包含在选项 1 或 2 中的子选项。

ITU 专利信息数据库还包含《一般性专利陈述和许可声明表》的记录。

ITU-2 通告

所有新的和修订的 ITU-T 和 ITU-R 建议书的封面应增加文字说明，在适宜的地方，强烈要求用户查阅 ITU 专利信息数据库。其措辞如下：

"ITU 提请注意，本建议书的使用或实施可能涉及使用已主张的知识产权。ITU 不负责所主张的知识产权的证据、有效性或适用性，无论它们是 ITU 成员宣称的还是本建议书制定过程以外的其他当事人宣称的。

截至本建议书批准之日，ITU（已经／尚未）收到在实施本建议书时可能要求的受专利保护的知识产权的通知。不过，请实施者注意，这可能不代表最新信息，因此强烈要求去查阅 ITU 专利信息数据库。"

II.2 ISO 和 IEC 的专用条款

ISO/IEC-1 对可交付使用的文件草案征求意见

所有提交征求意见的草案都应在其封面页上包含如下文字："请本草案的收件人将其评议意见，连同其了解的任何有关专利权的通知一起提交，并提供支撑文件。"

ISO/IEC-2 通告

在起草过程中没有标识专利权的已发布文件应在前言中包含以下通告：

"本文件的某些内容有可能涉及专利权问题，对此应引起注意，ISO（和／或）IEC 不负责识别任何这样的专利权问题。"

在起草过程中已标识专利权的已发布文件应在引言中包含以下通知：

"国际标准化组织（ISO）［和／或］国际电工委员会（IEC）提请注意以下事实：据称，遵守本文件可能涉及使用（……条……）中规定的关于（……主题内容……）的专利。

ISO［和／或］IEC 对本专利权的证据、有效性和范围不承担负责。

该专利权人已经向 ISO［和／或］IEC 保证，他／她愿意在合理、无歧视的条款和条件下与全世界申请者谈判许可授予事宜。为此，该专利权人向 ISO［和／或］IEC 登记其声明。可以从以下联系方式获得信息：

专利权人姓名……

地址……

提请注意，除了上面标识的专利权外，本文件的某些要素可能是专利权的主题。ISO［和/或］IEC不应承担标识任一或全部这类专利权的责任。"

ISO/IEC-3 国家采用

ISO、IEC 和 ISO/IEC 可交付使用的文件的专利声明仅适用于声明表中注明的 ISO 和/或 IEC 文件。声明不适用于经过修改的文件（例如通过国家或区域采用）。但是，实施符合相同的国家和区域采用和各自的 ISO 和/或 IEC 可交付使用的文件，可依据向 ISO 和/或 IEC 提交的此类可交付使用的文件声明。

附件 1
ITU-T/ITU-R/ISO/IEC 共用专利政策

下面是有关专利的"行为准则",这些专利不同程度涵盖 ITU-T 建议书、ITU-R 建议书、ISO 可交付使用的文件和 IEC 可交付使用的文件的主题内容(本文件中,ITU-T 建议书和 ITU-R 建议书统称为"建议书",把 ISO 可交付使用的文件和 IEC 可交付使用的文件统称为"可交付使用的文件")。"行为准则"中的规则是简单明了的。建议书 / 可交付使用的文件是由技术专家而不是专利专家起草的,他们可能不一定很熟悉诸如专利之类知识产权的复杂的国际法状况。

建议书 / 可交付使用的文件不是约束性的,其目的是确保在世界范围内技术和系统的兼容性。为了达到满足所有参与者共用利益的目的,必须确保每个人都易于获得建议书 / 可交付使用的文件及其应用和使用等。

因此,随之而来的是,全部或部分被纳入建议书 / 可交付使用的文件的专利必须是每个人能容易获取的,没有不适当的限制。总之,满足这个要求是行为准则的唯一目的。由于各案协议可能有差异,因此与专利有关的详细安排(专利许可、专利使用费等)留给相关方去处理。

行为准则可以归纳如下:

1 ITU 电信标准化局(ITU-TSB)、ITU 无线电通信局(ITU-BR)以及 ISO 和 IEC 的 CET 办公室不负责就专利或类似权利的证据、有效性或范围给出权威的或全面的信息,但是非常需要披露最充分的可获取信息。因此,任何参与 ITU、ISO 或 IEC 工作的参与方都应该从一开始就分别提请 ITU- TSB 局长、ITU-BR 局长或 ISO 和 IEC 的 CEO 办公室注意其所知晓的自己组织的或是其他组织的任何专利或任何正在处理的专利申请,尽管 ITU、ISO 或 IEC 不能够验证任何这类信息的有效性。

2 如果某建议书 / 可交付使用的文件已制定,并且上述 1 中言及的信息已经披露,那么可能出现三种不同状况:

2.1 专利权人愿意与其他方在无歧视基础上以合理的条款和条件谈判免费专利许可事宜。这类谈判留待相关方在 ITU-T/ITU-R/ISO/IEC 外进行。

2.2 专利权人愿意与其他方在无歧视基础上以合理的条款和条件谈判专利许可事宜。这类谈判留待相关方在 ITU-T/ITU-R/ISO/IEC 外进行。

2.3 专利权人不愿意遵循上述 2.1 或 2.2 的条款;在这种情况下,建议书 / 可交付使用的文件不应包含与该专利有关的条款。

3 无论适用哪种情况（2.1、2.2 或 2.3），专利权人都必须使用相应的"专利陈述和许可声明表"提供书面陈述，分别在 ITU-TSB、ITU-BR 或 ISO 或 IEC 的 CEO 办公室备案。这种陈述除了针对表格中所选方框对应的情况提供信息之外，不得再包含附加的条款、条件、或任何其他免责条款。

附件 2
ITU-T 或 ITU-R 建议书 /ISO 或 IEC 可交付使用的
文件的专利陈述和许可声明表

ITU-T 或 ITU-R 建议书 /ISO 或 IEC 可交付使用的文件的专利陈述和许可声明
此声明不代表实际授予许可。

请把表格按下列每种文件类型的指示返回相关组织：

国际电信联盟（ITU）电信标准化局局长	国际电信联盟（ITU）无线电通信局局长	国际标准化组织（ISO）秘书长	国际电工委员会（IEC）秘书长
地址：Place des Nations CH1211 Geneva 20，Switzerland	地址：Place des Nations CH1211 Geneva 20，Switzerland	地址：8 chemin de Blandonnet CP 401 1214 Vernier，Geneva Switzerland	地址：3 rue de Varembé CH1211 Geneva 20 Switzerland
传真：+41227305853	传真：+41227305785	传真：+41227333430	传真：+41229190300
电邮：tsbdir@itu.int	电邮：brmail@itu.int	电邮：patent.statements@iso.org	电邮：inmail@iec.ch

专利权人：	
依法登记的名称	
申请许可的联系方式：	
名称和部门	
地址	
电话	
传真	
电邮	
网址（可选填）	
文件类型： □ ITU-T 建议书（*）□ ITU-R 建议书（*）□ ISO 可交付使用的文件（*）□ IEC 可交付使用的文件（*）（请把所填写的表格返回相关组织）□ 共用文本或双文本（ITU-T 建议书 /ISO/IEC 可交付使用的文件（*））（如果是共用文本或双文本，请把所填写的表格返回给三个组织：ITU-T, ISO, IEC）□ ISO/IEC 可交付使用的文件（*）（如果是 ISO/IEC 可交付使用的文件，请把所填写的表格返回 ISO 和 IEC）	
(*) 文件号	

(*) 文件标题	

许可声明：
专利权人确认其持有已批准和／或正在申请的专利，要使用这些专利则需要执行上述文件，并在此根据 ITU-T/ITU-R/ISO/IEC 共用专利政策声明如下（仅能选一个框打勾）：

☐ 1. 专利权人愿意在全球范围内无歧视地免费向不限制数量的申请人授予专利许可，并在其他合理的条款和条件下许可其生产、使用和销售上述文件的实施。
谈判事宜留待相关方在 ITU-T、ITU-R、ISO 或 IEC 以外进行。
如果专利权人愿意在互惠条件下为上述文件授予许可，亦请标记此处_____。
如果专利权人保留如下权利：即只有申请人愿意在合理的条款和条件下（但不免费）许可其自己的专利（为实施上述文件要求使用的专利），专利权人才在合理的条款和条件下（但不免费）向申请人授予其专利许可，亦请标记此处_____。

☐ 2. 专利权人愿意在全球范围内无歧视地向不限制数量的申请人授予专利许可，并在其他合理的条款和条件下许可其生产、使用和销售上述文件的实施。
谈判事宜留待相关方在 ITU-T、ITU-R、ISO 或 IEC 以外进行。
如果专利权人愿意在互惠条件下为上述文件授予许可，亦请标记此处_____。

☐ 3. 专利权人不愿意依照上述选项 1 或选项 2 的规定授予许可。
在这种情况下，必须向 ITU、ISO 和／或 IEC 提供下列资料，作为本声明的一部分：
——获批的专利号或专利申请号（如正在申请中）；
——表明上述文件中受影响的部分；
——涵盖上述文件的专利说明。

免费：
"免费"并不意味着专利权人放弃专利有关的全部权利。更确切地说"免费"是指金钱补偿的问题；即专利权人不寻求将任何金钱补偿作为许可协议的一部分（无论这类补偿被称为专利使用费，还是一次性许可费等）。不过，虽然专利权人承诺不收取任何数量的金钱，但专利权人仍然有权要求上述同一份文件的实施者签署一项许可协议，其中包含其他合理的条款和条件，例如，有关适用法律、使用领域、担保等。
互惠：
"互惠"是指只有当未来的许可申请人承诺免费实施同一份相关建议书／可交付使用的文件，或在合理的条款和条件下许可其专利，该申请人才能要求专利权人向其授予专利许可。
专利：
"专利"是指基于发明的专利、实用新型和其他类似的法定权利（包括上述任何权利的申请）中包含和确认的权利要求，并且仅指对上述同一种文件的实施很必要的那些要求。必要专利是指为实施特定的建议书／可交付使用的文件所必需的专利。
专利权的转让／转移：
根据 ITU-T/ITU-R/ISO/IEC 共用专利政策的 2.1 或 2.2 做出的许可声明应被解释为约束所有与转移的专利有利益关系的继任者的留置权。认识到这种解释可能不适用所有司法管辖区，那些根据共用专利政策已经提交许可声明的专利权人（在专利声明表中选择选项 1 或 2）在通过这样的许可声明转移专利所有权时，应在相关转移文件中包含适当的条款，以确保此类转移的专利的许可声明对受让人具有约束力，并确保受让人在未来转移时也将同样包含适当的条款，目的是约束所有相关继任者。

专利信息［建议填写选项 1 和 2，但不做要求；ITU、ISO 和 IEC 要求填写选项 3（见"注"）］			
序号	状态［已批准 / 申请中］	国家	批准的专利号或申请号（如正在申请中）	文件标题
1				
2				
3				
4				
5				
6				
7				
8				
9				
10				

☐ 如果在附加的页面上提供了附加的专利信息，在此方框中打勾。

注：在选择选项 3 的时，还应提供上述选项 3 框内所列出的附加信息。

签名（仅需签在最后一页）:	
专利权人	
授权人姓名	
授权人头衔	
签名	
地点、日期	

表格：2018 年 11 月 2 日

附件 3
ITU-T 或 ITU-R 建议书的通用专利陈述和许可声明表

ITU-T 或 ITU-R 建议书的通用专利陈述和许可声明表

此声明不代表实际授予许可

请把表格返回相关处局：

国际电信联盟（ITU）
电信标准化局
局长
地址：Place des Nations
CH-1211 Geneva 20 Switzerland
传真：+41 22 730 5853
电邮：tsbdir@itu.int

国际电信联盟（ITU）
无线电通信局
局长
地址：Place des Nations
CH-1211 Geneva 20
Switzerland
传真：+41 22 730 5785
电邮：brmail@itu.int

专利权人：
依法登记的名称
申请许可的联系方式：
名称和部门
地址
电话
传真
电邮：
网址（可选填）
许可声明： 如果 ITU-T / ITU-R 建议书中包含上述专利权人提交的贡献文件的部分或全部提案，且所包含部分包括已提交专利的项目且其专利用途要求实施 ITU-T / ITU-R 建议书，则根据 ITU-T / ITU-R / ISO / IEC 公用专利政策，上述专利权人特此声明如下（仅能选一个框打勾）：

☐ 1. 专利权人愿意在全球范围内无歧视地免费向不限制数量的申请人授予专利许可，并在其他合理的条款和条件下许可其生产、使用和销售相关 ITU-T/ITU-R 建议书的实施。
谈判事宜留待相关方在 ITU-T/ITU-R 以外进行。
如果专利权人愿意在互惠条件下为上述 ITU-T/ITU-R 建议书授予许可，亦请标记此处。
如果专利权人保留如下权利：即只有申请人愿意在合理的条款和条件下（但不免费）许可其自己的专利（为实施上述 ITU-T/ITU-R 建议书要求使用的专利），专利权人才在合理的条款和条件下（但不免费）向申请人授予其专利许可，亦请标记此处。

☐ 2. 专利权人愿意在全球范围内无歧视地向不限制数量的申请人授予专利许可，并在其他合理的条款和条件下许可其生产、使用和销售相关 ITU-T/ITU-R 建议书的实施。
谈判事宜留待相关方在 ITU-T/ITU-R 以外进行。
如果专利权人愿意在互惠条件下为上述 ITU-T/ITU-R 建议书授予许可，则也标记此处。

<u>免费：</u>
"免费"并不意味着专利权人放弃专利有关的全部权利。更确切地说，"免费"是指金钱补偿的问题，即专利权人不寻求将任何金钱补偿作为许可协议的一部分（无论这类补偿被称为专利使用费，还是一次性许可费等）。不过，虽然专利权人承诺不收取任何数量的金钱，但专利权人仍然有权要求上述同一份文件的实施者签署一项许可协议，其中包含其他合理的条款和条件，例如，有关适用法律、使用领域、担保等。

<u>互惠：</u>
"互惠"是指只有当未来的许可申请人承诺免费实施同一份相关建议书/可交付使用的文件，或在合理的条款和条件下许可其专利，该申请人才能要求专利权人向其授予专利许可。

<u>专利：</u>
"专利"是指基于发明的专利、实用新型和其他类似的法定权利（包括上述任何权利的申请）中包含和确认的权利要求，并且仅指对上述同一种文件的实施很必要的那些要求。必要专利是指为实施特定的建议书/可交付使用的文件所必需的专利。

<u>专利权的转让/转移：</u>
根据 ITU-T/ITU-R/ISO/IEC 共用专利政策的 2.1 或 2.2 做出的许可声明应被解释为约束所有与转移的专利有利益关系的继任者的留置权。认识到这种解释可能不适用所有司法管辖区，那些根据共用专利政策已经提交许可声明的专利权人（在专利声明表中选择选项 1 或 2）在通过这样的许可声明转移专利所有权时，应在相关转移文件中包含适当的条款，以确保此类转移的专利的许可声明对受让人具有约束力，并确保受让人在未来转移时也将同样包含适当的条款，目的是约束所有相关继任者。

签名：	
专利权人	
授权人姓名	
授权人头衔	
签名	
签署地点、日期	

表格：2015 年 6 月 26 日

附录 J
（规范性）
技术委员会和分委员会的制定范围

J.1 引言

技术委员会或分委员会的范围是一份明确界定该委员会工作界限的声明。具有以下（但不限）功能：

—— 帮助对工作领域的疑问和建议人找到合适的委员会；

—— 防止两个或多个 ISO 和 / 或 IEC 委员会的工作项目重复；

—— 能够帮助防止超越上级委员会授权的活动领域。

J.2 范围的制定

1.5.10 中规定了技术委员会和分委员会制定范围的基本规则。

范围要素的顺序应是：

—— 基本范围；

—— 在 ISO 中，横向功能（适用时）；

—— 在 IEC 中，横向和 / 或团体安全功能（适用时）；

—— 在 JTC 1 中，横向功能（适用时）；

—— 不适用范围（如果有）；

—— 备注（如果有）。

J.3 基本范围

技术委员会的范围不应涉及国际标准化的总体目标或重复那些管理所有技术委员会的原则。

特殊情况下，如果认为对委员会范围的理解很重要，可以包括说明材料。这类材料应以"备注"的形式给出。

J.4 不适用范围

如果有必要说明某些主题超越该技术委员会的范围，则应列出这些主题，由"不包括……"等词引出。

应清楚地规定不适用范围。

如果不适用范围含在一个或多个其他现有 ISO 或 IEC 技术委员会范围内，还应明确这些委员会。

示例 1 "不包括：那些含在 ISO/TC……范围中的……"

示例 2 "不包括：……（ISO/TC……）……（IEC/TC……）等领域的特定项目的标准化"。
没有必要提及明显的不适用范围。
示例 3 "不包括：其他 ISO 或 IEC 技术委员会涵盖的产品"。
示例 4 "不包括：……电气设备和仪器的规格，这些领域属 IEC 委员会的范围"。

J.5 产品相关委员会的范围

产品相关委员会的范围应清楚表明该委员会拟包括的领域、应用范围或市场部门便易于确定一个特定的产品是否在其领域、应用范围或市场部门之内。

示例 1 "用于……的……和……的标准化"。
示例 2 "建筑用材料、组件和设备的标准化和……和……以及……的服务和维护用设备的操作"。

可以通过表明产品的目的或说明产品的特点来明确范围的界限。

范围不宜列举该委员会涵盖的产品类型，因为这样做可能暗示其他类型产品的标准由或正由其他委员会负责。然而，如果有意这样做，最好列出那些不适用范围的项目。

列举诸如术语、技术要求、抽样方法、测试方法、标识、标志、包装、尺寸等说明了范围对那些特定要素的限制，而且其他要素可以由其他委员会负责进行标。因此，拟进行标准化的产品要素不应包含在范围中，除非有意将该范围限定在特定要素。

如果范围中未提及任何要素，就意味着该主题全部由该委员会涵盖。

注：涵盖范围并不一定意味着有制定标准的需求。只表示在需要的情况下，该委员会（而不他委员会）将制定任一要素的标准。

关于不必列举的要素，举例如下：

示例 3 "分类、术语、抽样、物理、化学或其他测试方法、规格等的标准化"。

不管是产品的类型或要素，都不应在范围内提及优先顺序，因为这些优先顺序工作计划中标明。

J.6 非产品相关委员会的范围

如果委员会的范围拟限定在某些与产品不相关或仅间接与产品相关的要素，其应只表明拟包括的要素（例如，安全颜色和符号、非破坏性试验、水质）。

术语这个词作为一个可能进行标准化的要素，只有当该要素是委员会负责的唯务时才提及。如果情况不是如此，提及术语则是多余的，因为这一要素是任一标活动的逻辑组成部分。

附录 K
（规范性）
项目委员会

K.1 提案阶段

不属于现有技术委员会范围的新工作项目提案，应由一个被授权提出新的工作提案的机构（见 2.3.采用适当的形式提交并进行充分论证（见 2.3.2）。

CEO 办公室可以决定将提案在分发进行投票前退回给提案方做进一步研制。在情况下，提案方应按建议修改或提供不做修改的论证。如果提案方不做修改并且按原提交的提案文本分发进行投票，技术管理局将决定采取适当的行动。这可能拦截该提案直到修改完成或接受按收到的文件进行投票。

在所有情况下，CEO 办公室都可以将评论和建议纳入提案表格中。

有关提案论证的详细情况，见附录 C。

如果提议设立制定管理系统标准的项目委员会，见附录 L。

提案应提交给技术管理局秘书处，由其安排把提案提交给所有国家成员体进行投票。

鼓励提案方标明该项目委员会第一次会议的日期（见 K.3）。

如果该提案不是由一个国家成员体提交的，那么在向国家成员体提交时应包括承担项目委员会秘书处的信息。

投票应在 12 周内反馈。

提案的通过需满足：

—— 参加投票的国家成员体的 2/3 多数赞成；
—— 至少有 5 个批准新工作项目的提案并提名技术专家的国家成员体承诺积极参与。

K.2 项目委员会的设立

技术管理局应对新工作项目投票的结果进行复审，如果符合批准准则，就应设目委员会（其参考编号应为技术委员会/项目委员会次序中的下一个可用号码）。

项目委员会的秘书处应指派给提交该提案的国家成员体，如果该提案不是源自成员体，技术管理局应在收到的申请中决定秘书处的指派。

批准新工作项目提案并提名技术专家的国家成员体，应被注册为该项目委员会的 P

成员。批准新工作项目提案但没有承诺积极参与的国家成员体，应被注册为 O 成员反对票但表示如果新工作项目提案通过就愿意积极参与的国家成员体，应被注册为 P 成员。投反对票且没有表示愿意积极参与的国家成员体，应被注册为 O 成员。

CEO 办公室应向国家成员体宣布项目委员会的设立及其成员。

国家成员体将被邀请通过通知 CEO 办公室来确认 / 改变其成员身份。

秘书处将联系在新工作项目提案或国家成员体评论中确定的所有可能的联络组织邀请他们表明是否有兴趣参与这项工作，以及对哪个联络类型有兴趣。联络申请现有程序处理。

K.3　项目委员会的第一次会议

召开项目委员会会议的程序应按照第 4 章进行，例外情况是，如果提交提案通报了第一次会议的日期，可以使用 6 周的通知期限。

项目委员会的主席应是新工作项目提案中提名的项目负责人，如果在新工作项案中没有提名项目负责人，则应由秘书处提名。

第一次会议应确认新工作项目的范围。如果有必要修订（为了说明而不是扩展的目的），修订的范围应提交技术管理局批准。会议还应确认项目计划，在 ISO 的程序，并决定开展工作所需要的任何分支机构。

如果确定该项目需要细分来制定两个或多个出版物，这是可行的，只要该工作分项目完全处于原新工作项目提案的范围内。如果不在其范围内，则需要制定一工作项目供技术管理局考虑。

注：不要求项目委员会制定战略业务计划。

K.4　准备阶段

应按 2.4 开展准备阶段工作。

K.5　委员会、征询意见、批准和出版阶段

应按 2.5 ~ 2.8 开展委员会、征询意见、批准和出版阶段工作。

K.6　项目委员会的解散

相关标准一旦出版，项目委员会即应解散。

K.7　项目委员会制定标准的维护

承担秘书处的国家成员体应按照 2.9 的程序承担标准维护的责任，除非项目委员会换为技术委员会（见 1.10），在这种情况下，该技术委员会应承担该标准的维护责任。

附录 L
（规范性）
管理体系标准提案

L.1 概述

当提议制定新的管理体系标准（MSS）——包括行业专用 MSS，应当按照本附录附件 1 的内容进行合理性研究（JS）。

注：如现有的 MSS 的制定已经得到批准，且范围已经确定（除非第一次制定时未确定范围），对该 MSS 进行修正时不需要进行 JS。

在条件允许的情况下，提案方应尽力确定将用于构成新的或修订的 MSS 系列的全部出版物，并为每个可交付使用文件准备一个 JS。

L.2 术语和定义

本附录适用以下术语和定义。

L.2.1

管理体系 management system

见本附录的附件 2（3.4）中的定义。

L.2.2

管理体系标准 management system standard

MSS

针对管理体系（L.5.1）的标准。（原文有误，核对是否应为"L.2.1"，目前按照原文"L.5.1"处理）

注 1：就本文件而言，本定义亦适用于其他 ISO 可交付使用文件（如 TS、PAS）。

L.2.3

通用 MSS generic MSS

能够广泛适用于不同经济部门、不同类型和规模的组织以及不同地理、文化和社会条件的 MSS。

L.2.4

行业专用 MSS sector-specific MSS

将通用 MSS 用于特定的经济或商业部门时，提供附加要求或补充性指导的 MSS。

L.2.5

A 型 MSS type A MSS

提出要求的 MSS。

示例：
管理体系要求标准（规范）
管理体系行业专用要求标准

L.2.6

B 型 MSS type B MSS

提供指导的 MSS。

示例：
关于使用管理体系要求标准的指南
关于建立管理体系的指南
关于改善／提高管理体系的指南

L.2.7

高层次结构 high level structure

HLS

ISO/TMB/JTCG "MSS 联合技术协调小组"的工作成果，其中涉及高层次结构（HLS）、相同的子条款标题、相同的内容以及共同的术语和核心定义。见本附录附件 2。

L.3 提交合理性研究的义务

所有 MSS（包括行业专用 MSS（L.2.4），见附录 M）提案及其合理性研究（JS）应由相关 TC/SC/PC（或 IEC SyC）领导层确认，并应在 NP 投票前将 JS 提交 TMB（或其 MSS 特别工作组）进行评估和批准。相关 TC/SC/PC 秘书处负责确认所有 MSS 提案，以确保所有 MSS 提案都能进行 JS（无论是否知晓）或提交 ISO/TMB 进行评估。

若已提交并批准了特定的 A 型 MSS 的 JS，对该 MSS 提供指导的 B 型 MSS 不需要进行 JS。

例如，ISO/IEC 27003：2010《信息技术 安全技术 信息安全管理体系实施指南》不需要提交 JS，因为 ISO/IEC 27001：2013《信息技术 安全技术 信息安全管理体系 要求》已经提交了 JS 并获得了批准。

L.4 未提交 JS 的情况

在 NP 投票前未提交 TMB 评估的 MSS 提案将首先提交 TMB 进行评估，在 TMB 公布评估结果之前不得进行新的投票（暂停项目）。建议在将 JS 提交 TMB 之前，先获得 TC/SC/PC（和／或 IEC SyC）成员对该 JS 的认可。

注：未提交过 JS 但已发布的 MSS 在修订时将作为新的 MSS 处理，即在任何工作开始前提交并批准一个 JS。

L.5 本附录的适用性

上述程序适用于所有 ISO 和 IEC 可交付使用的文件，包括 IWA。

L.6 一般原则

所有新 MSS 项目（或已发布但尚未完成 JS 的 MSS 项目）均应进行 JS（见 L.1 和 L.3 注释）。下列一般原则用于指导评估拟议的 MSS 的市场相关性和 JS 的准备。本附录附件 1 中的评价准则问题基于下列原则。这些问题的答案将作为构成 JS 的一部分。只有在遵守下列所有原则的情况下，才宜发起、发展和维持 MSS。

1） 市场相关性——任何 MSS 都宜满足主要用户和受影响方的需求，并为其增加价值。

2） 兼容性——宜该保持不同的 MSS 之间和 MSS 系列之内的兼容性。

3） 主题覆盖范围——通用 MSS（L.5.3）宜具有足够的应用覆盖范围以消除对行业专用差异的需求或使需求最小化。（原文有误，请核对是否应为"L.2.3"，目前按照原文"L.5.3"处理）

4） 灵活性——MSS 宜适用于所有相关部门和文化以及任何规模的组织。MSS 不宜限制组织进行竞争性的添加或区别于其他组织，或者超出标准地加强其管理体系。

5） 自由贸易——MSS 宜根据世贸组织《技术性贸易壁垒协议》的原则，允许货物和服务的自由贸易。

6） 合格评定的适用性——宜评估第一方、第二方、第三方或其任意组合的合格评定的市场需求。所产生的 MSS 宜明确说明其范围内合格评定的适用性。MSS 宜有利于联合审计。

7） 排除范围——MSS 不宜包括直接相关的产品（含服务）规范、测试方法、性能水平（即极限设置）或对执行组织生产产品的其他形式的标准化。

8） 易用性——确保用户可以轻松实现一个或者多个 MSS。MSS 宜易于理解、表达明确、不受文化偏见的影响、易于翻译并适用于一般业务。

L.7 合理性研究过程和标准

L.7.1 概述

本条款描述了合理性研究（JS）过程，用于论证和评估 MSS 提案的市场相关性。本附录的附件 1 提出了需要在合理性研究中处理的一系列问题。

L.7.2 合理性研究过程

JS 过程适用于所有 MSS 项目。JS 过程包括：

a） 由（或代表）MSS 项目的提出者进行 JS；

b） 由 TMB（或 ISO 的 ISO/TMB MSS 特别工作组）对 JS 进行批准。

在 JS 过程之后，将按照 ISO 或 IEC 正常投票程序适当批准新工作项。

L.7.3 合理性研究标准

依据附录 C 和上述一般原则，应将一组问题（见本附录的附件 1）作为论证和评估一个拟议 MSS 项目的标准，应并由提议人回答。此问题清单并不详尽，宜提供与此项目相关的其他资料。JS 宜阐明已考虑到所有的问题。如果认定它们与某一特定情况无关或者不恰当，宜清楚阐述做出这一决定的理由。为客观评估某一 MSS 独特层面的市场相关性，可能需要考虑其他问题。

L.8 关于 MSS 制定过程和结构的指南

L.8.1 概述

制定 MSS 的影响将与以下方面相关：

—— 这些标准对商业实践的深远影响；

—— 全球范围内支持这些标准的重要性；

—— 许多（如果不是全部）国家成员机构参与的实际可能性；和

—— 对兼容一致的 MSS 的市场需求。

考虑到以上影响，本条款提供除 ISO/IEC 导则中其他条款规定程序外的指导。

所有的 MSS（不论是 A 型或 B 型，通用或特定部门的 MSS）原则上应结构一致、内容和术语相同，以易于使用和相互兼容。原则上也应遵循本附录附件 2 中给出的指导和结构。

B 型 MSS 对同一 MSS 系列的另一种 MSS 提供指导，宜遵循相同的结构（即条款编号）。如涉及提供指导的 MSS（即 B 型 MSS），必须清楚界定它们的职能，以及它们与提供需求的 MSS（A 型 MSS）的关系，例如：

—— 关于使用要求标准的指南；

—— 关于建立／实施管理体系的指南；

—— 关于改善／提高管理体系的指南。

当提议行业专用 MSS：

—— 宜与通用 MSS 兼容且一致；

—— 应遵循附录 M 中规定的规则和原则；

—— 负责通用 MSS 的相关委员会可能有需要满足的额外要求或需要遵循的程序（见附录 M）；

—— 可能需要与其他委员会以及 ISO 合格评定委员会（CASCO）和 IEC CAB 就合格评定问题进行磋商。

就特定部门的文件而言，宜明确界定它们的职能及与通用 MSS 的关系（如额外的特定部门的要求和说明，或恰当情况下的两者）。

特定部门的文件宜始终清晰展示（如使用不同的排版样式）所提供的特定部门信息的种类。

注1：由于特殊情况导致相同内容或任何要求不能应用于特定 MSS 时，应通过 TMB 秘书（tmb@iso.org）或 IEC/SMB 秘书处向 TMB 报告（见 L.9.3）。

L.8.2　MSS 制定过程

L.8.2.1　概述

除 JS 外，MSS 的制定宜遵循与其他 ISO 和 IEC 可交付使用的文件相同的要求（见第 2 条）。

L.8.2.2　设计规范

为确保维持本标准的意图（如 JS 所示），可在拟订工作草案之前制定一份设计规范书。

归口委员会将决定是否需要设计规格。如认为有必要，归口委员会将决定适合 MSS 的格式和内容，并宜设立必要的组织来执行这项任务。

设计规范通常应解决以下问题：

- 用户需求——确定标准的用户及其相关需求，以及这些用户的成本和收益。
- 范围——标准的范围和用途、标题和适用范围。
- 兼容性——如何实现这个系统内部以及与其他 MSS 系列之间兼容，包括识别具有类似标准的通用元素，以及如何将这些元素纳入推荐的结构（见本附录的附件 2）。
- 一致性——与 MSS 系列中制定（即将制定）的其他文档的一致性。

注：大多数（如果不是全部）有关用户需求和范围的信息都可以从 JS 中获得。

设计规范宜确保：

a) JS 的结果正确转化为 MSS 的要求；

b) 识别并解决与其他 MSS 的兼容性和一致性问题；

c) 在制定过程中的适当阶段存在证实最终 MSS 的基础；

d) 设计规范书的批准为 TC/SC 和 / 或 IEC SyC 成员在整个项目中的所有权提供依据；

e) 考虑通过 NP 投票阶段得到的意见；

f) 考虑到任何限制因素。

制定 MSS 的委员会应根据设计规范监控 MSS 的制定，以确保在项目过程中不会发生偏差。

L.8.2.3　生成可交付使用的文件

L.8.2.3.1　监控输出

在起草过程中，宜监控输出以保证与其他 MSS 兼容及易于使用，要覆盖以下问题：
—— 高层次结构（HLS）、相同的子条款标题、相同的内容、共同术语和必须明确的核心定义（语言和表达方面）；和
—— 避免重叠和矛盾。

L.8.2.4 MSS 制定过程的透明度

与大多数其他类型的标准相比，MSS 范围更广。它覆盖了人类努力的大片领域，同时影响着广大用户利益。

因此，MSS 编写委员会宜对标准的制定采用高度透明的方法，确保：
—— 明确识别了参与标准制订进程的可能性；和
—— 所有各方都理解所使用的制定过程。

委员会宜提供项目全过程进展情况的相关资料，包括：
—— 迄今为止项目的状况（包括讨论中的条目）；
—— 为获取未来信息的联系点；
—— 全体会议的公报和新闻稿；
—— 常见问题和答案的常规清单。

提供资料时，需考虑到参与国现有的分发设施。

如果预计 A 型 MSS 的用户可能会证明其符合性，则 MSS 宜以书面形式，以确保制造商或供应商（第一方或自我声明）、用户或买方（第二方）或独立机构（第三方，或称认证方或注册方）能将评估该 MSS 的符合性。

应最大限度地利用 ISO 中央秘书处或 IEC 中央办公室的资源，以提高项目透明度，委员会还应考虑设立一个专用的开放访问网站。

委员会应让国家成员机构参与 MSS 制定，以提高国家对 MSS 项目的认识，为不同的相关和受影响的各方（包括审核机构、认证机构、企业和用户团体），适当提供草案，并根据需要提供额外的专用信息。

委员会宜确保各参与成员（特别是来自发展中国家的成员）随时可以获取关于有关正在制定中的 MSS 内容的技术资料。

L.8.2.5 标准的解释过程

委员会可建立来自用户的有关标准解释问题的处理过程，并用合适的方法将解释结果提供给他人。这种机制可有效解决在早期阶段可能产生的误解，识别出在以后的修订过程中可能需要改进的与标准相关的表述。在 ISO 中，此类过程被称为"委员会专用程序"（见前言 f）。在 IEC 中，委员会应称其为"解释单"过程（见 IEC 补充部分）。

L.9 用于管理体系标准的高层次结构、相同核心内容和通用术语及核心定义

L.9.1 概述

本文件的目的是通过提供一个统一和一致同意的高层次结构、相同的核心内容和通用术语及核心定义来提高 MSS 的一致性。这一目的是使所有 A 型 MSS（和需要时 B 型 MSS）保持一致，加强这些标准的兼容性。每个 MSS 都将根据需要增加额外"领域专用"要求。

注：L.9.1 和 L.9.4 中使用的"领域专用"用于表明一个管理体系标准涉及的特定主题，如能量、质量、记录、环境等。

本文件的预期读者是技术委员会（TC），分委员会（SC）和项目委员会（PC）（和 IEC SyC）及参与 MSS 制定的其他人员。

用以制定新的 MSS 和未来修订现行标准的通用方法将会提高这些标准对用户的价值。对于那些选择实施能够同时满足两个或者多个 MSS 要求的单一（有时亦称"聚合"）管理体系的组织，这种方法非常实用。

本附录的附件 2 列出构成未来和可能修订的 A 型 MSS 和 B 型 MSS 核心的高层次结构、相同核心内容和通用术语及核心定义。

本附录的附件 3 陈述了本附录附件 2 的使用指南。

L.9.2 使用

MSS 包括本附录附件 2 所示的高层次结构和相同核心内容。通用术语及核心定义或直接纳入 MSS 中，或规范性引用包含这些内容的国际标准。

注：高层次结构包括顺序固定的主要条款（第 1 章~第 10 章）及其标题。相同核心内容包括编号的分条款（及其标题）以及分条款中的内容。

L.9.3 不适用性

如果由于例外情况导致高层次结构或任何相同核心内容、通用术语及核心定义不能适用于管理体系标准，TC/PC/SC 需通过以下方式说明原因，以供评审：

a） 在提交 DIS 同时向 ISO 中央秘书处或 IEC 中央办公室提供初始偏离报告；

b） 在提交最终标准文本进行出版时通过 ISO/TMB 秘书处（tmb@iso.org）向 TMB 或 IEC/SMB 秘书处提供最终偏离报告。

TC/PC/SC（或 IEC SyC）应该使用 ISO 或 IEC 评论模板提供他们的偏离报告。

注 1：最终偏离报告可能为初始偏离报告的更新版。

注 2：TC\PC\SC（或 IEC SyC）尽力避免高层次结构或任何核心内容、通用术语及核心定义的不适用性。

L.9.4 使用本附录的附件 2

对本附录附件 2 中增加的领域专用内容的管理如下：

1. 由各 TC、PC、SC（或 IEC SyC）或制定特定管理体系标准的其他团体补充领域专用内容。

2. 领域专用内容不影响协调、不反对和不削弱高层次结构、相同核心内容、通用术语及核心定义的意图。

3. 在相同内容子条款（或次子条款等）前或后插入补充的子条款或次子条款（等），并重新编号。

注1：不允许有悬置段。（见《ISO/IEC 导则 第 2 部分》）。
注2：注意需要使用交互引用的情况。

4. 在本附录的附件 2 内增加或插入领域专用内容。增加内容示例如下：

a） 新的箭头；

b） 领域专用说明性内容（如注或实例），以阐明要求；

c） 在相同内容内的子条款（等）中加入领域专用的新段落；

d） 补充内容，以加强本附录附件 2 中的现有要求。

5. 考虑到上述第 2 点，通过在相同核心内容中补充内容来避免相同核心内容和领域专用内容之间的重复要求。

6. 起草过程初期就要区分领域专用内容和相同核心内容。这有助于在编制和投票期间识别不同类型的内容。

注1：通过不同颜色、字型、字号、使用斜体或括号等方式加以区分。
注2：对区分内容的识别无需放进已发布的版本中。

7. 对于"风险"概念的理解可能比本附录附件 2 中 3.9 的定义更具体。在这种情况下，可能需要一个领域专用定义。领域专用术语和定义有别于核心定义，例如（×××）风险。

注：以上所述也能适用于其他一些定义。

8. 通用术语及核心定义将按标准的概念体系纳入领域专用管理体系标准的术语和定义清单。

L.9.5 实施

制定任何新的管理体系标准和修订任何现行管理体系标准都应遵循如下顺序：高层次结构、相同核心内容、通用术语及核心定义。

L.9.6 指南

见本附录的附件 3 中的支撑指南。

附件 1
（规范性）
论证准则问题

1. 概述

在合理性研究（JS）中拟阐述的问题清单符合 L.6 列出的原则。这一清单并不详尽。需提供与情况相关但未包含在这些问题中的补充性信息。

在编制 JS 时应适当考虑每一条一般原则，提案提出者在回答与原则有关的问题之前，需针对每条原则提供一般理由。

MSS 提案提出者在编制 JS 时需适当注意的原则是：

1. 市场相关性；
2. 兼容性；
3. 主题覆盖范围；
4. 灵活性；
5. 自由贸易；
6. 符合性评定的适用性；
7. 排除范围。

注：没有任何问题直接涉及原则 8（易用性），但是"易用性"宜指导可交付使用文件的制定。

有关 MSS 提案的基本信息

1	提出 MSS 的目的是什么？范围如何？提出的文件是指南文件还是规定要求文件？
2	提出建议的目的和范围包括产品（含服务）规范、产品试验方法、产品性能水平或其他形式的指南或与实施组织生产或提供产品直接相关的要求吗？
3	是否有一个或者多个委员会或非 ISO 和非 IEC 组织能够负责拟议的 MSS？如果有，给出其名称。
4	确定了相关参考资料吗？如，现行指南或已确定的惯例。
5	是否有技术专家支持其标准化工作？这些专家是来自不同地理区域的相关方面的直接代表吗？
6	从所需专家和会议的次数/时间方面来看，预计制定本文件需要做出哪些努力？
7	预期 MSS 是一份指南性文件、合同规范还是某一组织的管理性规范？

原则1：市场相关性

8	已经识别了所有受影响的相关方吗？例如： a) 组织（各种类型和规模）：组织内批准实施 MSS 和实现 MSS 符合性的决策者； b) 客户/终端用户，即支付或使用来自一个组织的产品（含服务）的个人或团体； c) 供方组织，如产品的制造商、销售商、零售商或卖主，或服务、信息的提供商； d) MSS 服务提供者，如 MSS 认证机构、审核机构或咨询者； e) 管理机构； f) 非政府组织。
9	对本 MSS 有何需求？这些需求存在于地方、国家、区域还是全球的层面上吗？这些需求适用于发展中国家吗？这些需求适用于发达国家吗？拥有 ISO 或 IEC 文件会有什么附加值（例如，促进不同国家的组织之间的交流）？
10	某些行业有这样的需求吗？是一般性需求吗？如果是，有哪些？小、中或大型组织有这样的需求么？
11	这种需求重要吗？会继续有这种需求吗？如果有，完成拟议的 MSS 的目标日期能满足这种需求吗？确定了可替代的可行方法吗？
12	描述如何确定这些需求及其重要性。列出已知的受影响方及其所在的重要地理或经济区域。
13	对拟议的 MSS 有已知或预期的支持吗？列出已表明支持该 MSS 的团体。对拟议的 MSS 有已知或预期的反对吗？列出已表明反对该 MSS 的团体。 有反对吗？列出已表明反对的那些团体。
14	预测为组织带来什么样的利益和成本？为适用的小、中和大型组织带来的利益和成本有哪些差别？ 描述如何确定利益和成本。提供有关组织的地理或经济重点、工业领域和规模方面的有效信息。提供已知的来源及其基础（如，已证明的范例）、场所、设想和条件（如，推测或理论上的条件）以及其他相关信息。
15	预测对其他受影响方（包括发展中国家）带来哪些利益和成本？ 描述如何确定这些利益和成本。提供已表明的受影响方的相关信息。
16	对社会带来哪些预期的价值？
17	识别了任何其他风险吗（例如，时限或对某一特定企业带来的意外后果）？

原则2：兼容性

18	是否存在与其他现行或拟制定的 ISO、IEC、非 ISO 或非 IEC 国际标准、或国家或区域标准重叠或冲突（或者对这些标准有何补充价值）？是否有其他寻求解决确定需求的公共或专用措施、指南、要求和法规（例如技术文件、已证实的惯例、学术或专业研究或已知的任何其他团体。）？
19	MSS 或相关的合格评定活动（如审核、认证）有可能补充、代替全部或部分、协调和简化、重叠或重复、冲突或损害上述已确认的现有活动吗？考虑采用哪些措施确保兼容性，解决冲突或避免重复？
20	拟议的 MSS 有可能促进还是阻碍 MSS 在国家或区域层面或者通过行业部门的迅速扩大？

原则 3：主题覆盖范围

21	MSS 是针对某个特定部门吗？
22	MSS 是否会参照或纳入现有的非行业特定的 MSS（例如：ISO 中 9000 条质量管理标准）？如果是，MSS 的制定是否符合 ISO/IEC 行业政策（见《ISO/IEC 导则 第 2 部分》），以及任何其他相关的政策及指导程序（如相关 ISO 委员会可提供的政策及指引程序）？
23	已采取哪些步骤来消除或减少对通用 MSS 的特定部门偏差的需要？

原则 4：灵活性

24	MSS 是否允许组织进行竞争性的增加、区分或鼓励其管理体制标准之外的创新？

原则 5：自由贸易

25	MSS 将如何促进或影响全球贸易？MSS 可以建立或阻止技术性贸易壁垒吗？
26	MSS 能否建立或阻止针对中小企业或大型组织的技术壁垒？
27	MSS 能否为发展中国家或发达中国家建立或阻止技术性贸易壁垒？
28	如果拟议的 MSS 拟用于政府法规，是否可能增加、复制、取代、加强或支持现行的政府法规？

原则 6：合格的适用性

29	如果预期用途是出于合同或监管目的，有哪些潜在的方法可以证明一致性（如，第一方、第二方或第三方）？MSS 是否使组织能够灵活地选择证明一致性的方法，并适应其操作、管理、物理位置和设备中的更改？
30	如果第三方注册/认证是一个潜在的选择，那么组织预期的利益和成本是什么？MSS 是否会促进与其他 MSS 的联合审计或平行评估？

原则 7：排除范围

31	拟议的目的或范围是否包括产品（含服务）规格、产品测试方法、产品性能级别，或与实施组织生产或提供的产品直接相关的其他形式指导或要求？

附件 2
（规范性）
高层次结构、相同核心文本、通用术语及核心定义

注：在相同的文本提案中，XXX= 需要插入的 MSS 规程特定限定词（如能源、道路交通安全、IT 安全、食品安全、社会安全、环境、质量）。本附件中斜体文字 * 为标准起草人员的咨询意见注释。

概述

起草说明　领域专用

本文本的编制采用《ISO/IEC 导则》第 1 部分中附录 L 的附件 2 中的"高层次结构"（即条款顺序、相同核心文本和通用术语与核心定义），旨在提高 ISO 和 IEC 管理体系标准的一致性，促进其在需要满足两个及以上此类标准要求的组织中的实施。

文中（第 1 章～第 10 章）使用阴影部分字体突出显示高层次结构（HLS）。黑色字体表示 ISO 或 IEC 专用领域文本。删除线用来表示 HLS 文本内同意删除的内容。使用阴影部分文本和删除线仅为了便于分析，不会在国际标准草案编制阶段之后使用。

1　范围

　　起草说明：领域专用。

2　规范性引用文件

　　起草说明：应使用条款标题。领域专用。

3　术语与定义

　　起草说明 1：应使用条款标题。术语与定义可位于标准或一个单独的文件中。

　　通用术语和核心定义以及其他领域专用的术语与定义应予以说明。

　　术语与定义最好按照概念层次（即系统顺序）排列，最好不要按照字母顺序排列。

　　本文件适用下列术语与定义。

　　起草说明 2：下列术语与定义作为管理体系标准"通用文本"不可或缺的一部分。可根据需要增加其他术语与定义。为实现每个标准的目的，可增加或修改注。

　　起草说明 3：定义中的斜体字表示对本条款所定义的另一个术语的交叉引用，该引用术语的编号在括号中给出。

　　起草说明 4：在本条款中出现"×××"文本的位置，应根据使用这些术语与定义的上下文插入适当的信息。例如："一个 ××× 目的"可替换为"一个信息安全目标"。

---

* 译者注：本附件中中文斜体字部分对应英文中的灰底斜体字（原 ISO/IEC 文件为蓝色斜体字。）

3.1

组织　organization

为实现目标（3.8），由职责、权限和相互关系构成其自身功能的一个人或一组人。

注1：组织的概念包括但不限于：个体经营者、公司、股份有限公司、商号、企业、当局、合伙关系、慈善组织或机构，或上述组织的部分或组合，无论是否法人、公有或私有。

3.2

相关方（优先术语）interested party（preferred term）

利益相关方（许用术语）stakeholder（admitted term）

能够影响决定或活动，受决定或活动影响或认为自己受到决定或活动影响的个人或组织（3.1）。

3.3

要求　requirement

表明的需求或期望，包括一般默示的或强制的。

注1："一般默示"即对于组织和相关方，考虑这些隐含的需求和期望是出于惯例或普遍做法。
注2：明确需求是指明确指出的需求，例如文件化信息中指出的需求。

3.4

管理体系　management system

组织的相互关联或相互作用的一套要素（3.1），用于建立政策（3.7）和目标（3.8），以及实现这些目标的过程（3.12）。

注1：一个管理体系可以涉及一个或多个领域。
注2：系统要素包括组织的结构、角色和职责、计划和操作。
注3：一个管理体系的范围可以包括整个组织、组织专用和已确定的职能、组织专用和已确定的部分，或者跨组织群一项或多个一项功能。

3.5

高层管理者　top management

在最高层级指导或控制一个组织的一个或一组人员。

注1：高层管理者有权在组织内委托权力或提供资源。
注2：如果管理体系（3.4）的范围仅涵盖组织的一部分，则高层管理者是指指导或控制该部分组织的人员。

3.6

有效性　effectiveness

计划活动的实现程度和计划结果的实现程度。

3.7

政策　policy

组织（3.1）的目的和方向，与其高层管理者（3.5）正式表达的一致。

3.8

目标　objective

达到的结果。

注 1：可以是一个战略性、战术性或行动的目标。

注 2：目标可以涉及不同的领域（如金融、健康与安全，以及环境目标），且适用于不同的层次（如战略、全组织范围、项目、产品和过程（3.12））。

注 3：目标可以用其他方式表达，例如作为预期成果、目的、操作准则、×××目标，或使用其他具有类似含义（目的、目标或指标）的文字。

注 4：为取得具体成效，在×××管理体系下，×××目标由组织设定，符合×××政策。

3.9

风险　risk

不确定的影响。

注 1：偏离预期的效果—积极或消极的。

注 2：不确定性是缺乏事件、其后果或类似情况的相关信息、理解或知识的状态，即使是部分不确定。

注 3：风险的特征通常是指潜在的"事件"（见 ISO 指南 73 的定义）和"后果"（见 ISO 指南 73 的定义），或其组合。

注 4：风险通常指事件（包括环境变化）的结果和发生的"可能性"（见 ISO 指南 73 的定义）的组合。

3.10

能力　competence

为达到预期结果，应用知识和技能的能力。

3.11

文件化信息　documented information

需要一个组织控制和维护的信息及其所载媒介。

注 1：文件化信息可以是任何格式和媒介，并可以来自任何来源。

注 2：文件化信息可指：

——管理体系（3.4），包括相关过程（3.12）；

——为组织的运转而建立的信息（文件资料）；

——取得成果的证据（记录）。

3.12

过程　process

将输入转化为输出的一系列相互关联或相互作用的活动。

3.13

绩效　performance

可测量的结果。

注 1： 绩效可能涉及定量的或定性的结果。

注 2： 绩效可能涉及活动、过程（3.12）、产品（含服务）、体系或组织（3.1）的管理。

3.14

外包（动词） outsource（verb）

安排外部组织（3.1）执行组织的部分职能或过程（3.12）。

注 1： 虽然外包的职能或过程是在管理体系（3.04）覆盖范围内，但是外部组织处在管理体系覆盖范围之外。

3.15

监控 monitoring

确定体系、过程（3.12）或活动的状态。

注 1： 确定状态可能需要检查、监督或密切观察。

3.16

测量 measurement

确定数值的过程（3.12）。

3.17

审计 audit

为获得审计证据并对其进行客观的评价，以确定满足审计准则的程度所进行的系统的、独立的并文件化的过程（3.12）。

注 1： 审计可以是内部审计（第一方）或外部审计（第二方或第三方），也可以是组合审计（结合两个或多个领域）。

注 2： 内部审计由组织本身或由代表其利益的外部方进行。

注 3： "审计证据"和"审计准则"在 ISO 19011 中定义。

3.18

合格 conformity

满足要求（3.3）。

3.19

不合格 nonconformity

未满足要求（3.3）。

3.20

纠正措施 corrective action

消除不合格（3.19）原因，并防止再次发生的行动。

3.21

持续改进 continual improvement

不断提升绩效（3.13）的活动。

4 组织所处的环境

4.1 理解组织及其所处的环境

组织应确定与其宗旨相关并影响其实现×××管理体系预期结果的能力的外部和内部问题。

4.2 理解相关方的需求和期望

组织应确定：

—— 与×××管理体系有关的相关方；

—— 这些相关方的相关要求。

4.3 确定×××管理体系的范围

组织应确定×××管理体系的边界和适用性，以界定其范围。

确定范围时，组织应考虑：

—— 4.1所提及的内、外部问题；

—— 4.2所提及的要求。

范围应可以文件化信息形式获取。

4.4 ×××管理体系

组织应按照本文件的要求建立、实施、保持并持续改进×××管理体系，包括所需的过程及其相互作用。

5 领导能力

5.1 领导能力与承诺

最高管理者应通过以下方式展示其在×××管理体系方面的领导能力和承诺：

—— 确保建立×××方针和×××目标，并确保其与组织的战略方向相一致；

—— 确保将×××管理体系的要求融入组织的业务过程；

—— 确保可获得×××管理体系所需的资源；

—— 就有效×××管理和符合×××管理体系要求的重要性进行沟通；

—— 确保×××管理体系实现其预期结果；

—— 指导并支持员工对×××管理体系的有效性做出贡献；

—— 促进持续改进；

—— 支持其他相关管理人员在其职责范围内展示其领导能力。

注：在本文件中提到的"业务"可广义地解释为那些对组织存在的目的具有核心意义的活动。

5.2 方针

最高管理者应制定×××方针，该方针须：

a) 与组织宗旨相适应的；

b) 为制定 ××× 目标提供框架；

c) 包括满足适用要求的承诺；

d) 包括持续改进 ××× 管理体系的承诺。

××× 方针应：

—— 可以文件化信息形式获取；

—— 使组织内的员工可沟通；

—— 适当时，可向相关方提供。

5.3 岗位、职责和职权

最高管理者应确保在组织内部分配并沟通相关岗位的责任和权限。

最高管理者应对下列事项分配职责和权限：

a) 确保 ××× 管理体系符合本文件的要求；

b) 向最高管理层报告 ××× 管理体系的绩效。

6 策划

6.1 应对风险和机遇的措施

为 ××× 管理体系策划时，组织应考虑到 4.1 所描述的情况和 4.2 所提及的要求，确定需要应对的风险和机遇，以便：

—— 确保 ××× 管理体系能够实现其预期结果；

—— 预防或减少不良影响；

—— 实现持续改进。

组织应计划：

a) 采取行动应对这些风险和机遇，

b) 如何：

—— 将行动整合并实施到其 ××× 管理体系流程中；

—— 评估这些行动的有效性。

6.2 ××× 目标及实现目标的策划

组织应针对其相关职能和层次建立 ××× 目标。

××× 目标应：

a) 符合 ××× 政策；

b) 可度量（如可行）；

c) 考虑适用性要求；

d) 可监控；

e）可沟通；

f）适当地更新。

组织应保存关于×××目标的文件化信息。

策划如何实现其×××目标时，组织应确定：

—— 要做什么；

—— 需要什么资源；

—— 由谁负责；

—— 何时完成；

—— 如何评价结果。

7 支持

7.1 资源

组织应确定并提供建立、实施、保持和持续改进×××管理体系所需的资源。

7.2 能力

组织应：

—— 确定在其控制下从事影响其×××绩效的工作中人员的必要能力；

—— 基于适当的教育、培训或经历，确保员工能够胜任工作；

—— 适当时，采取措施以培养员工所必需的能力，并评价所采取措施的有效性；

—— 保留适当的文件化信息作为能力的证据。

注：适当的措施可能包括向现有员工提供培训和指导，或重新委派其职务；或聘用、雇佣胜任的人员等。

7.3 意识

在本组织控制下工作的人员应知道：

—— ×××方针；

—— 他们对×××管理体系的有效性做出的贡献，包括提高×××绩效的好处；

—— 不符合×××管理体系要求的可能后果。

7.4 沟通

组织应确定与×××管理体系有关的内外部沟通，包括：

—— 沟通的内容；

—— 何时进行沟通；

—— 与谁进行沟通；

—— 怎样进行沟通。

7.5 文件化信息

7.5.1 概述

组织的 ××× 管理体系应包括：

a） 本文件要求的文件化信息；

b） 组织确定的 ××× 管理体系有效性所必需的文件化信息；

注：不同组织的 ××× 管理体系文件化信息的范围可能不同，取决于：

—— 组织的规模及其活动、过程、产品和服务的类型；

—— 过程的复杂性及其相互作用；

—— 人员的能力。

7.5.2 创建和更新

创建和更新文件化信息时，组织应确保适当的：

—— 识别和描述（如：标题、日期、作者或文献编号）；

—— 形式（如：语言、软件版本、图表）与载体（如：纸质、电子）；

—— 评审和批准，以确保适宜性和充分性。

7.5.3 文件化信息的控制

应控制 ××× 管理体系和本文件要求的文件化信息，以确保：

a） 在需要的时间和场所均可获得并适用；

b） 受到充分的保护（如：防止失密、不当使用或完整性受损）。

为了控制文件化信息适用，组织应采取以下措施：

—— 分发、访问、检索和使用；

—— 存储和保护，包括保持易读性；

—— 变更的控制（如：版本控制）；

—— 保留和处置。

组织确定的为 ××× 管理体系的策划和运行所必需的外部来源文件化信息应被识别并在适当情况下加以控制。

注：访问可能表明关于仅查看文档化信息的权限，或查看和更改文档化信息的许可和权限的决策。

8 运行

8.1 运行策划和控制

起草说明：若无需添加分条款至第 8 章，删除本分条款题目。

组织应通过以下方式策划、实施和控制满足要求所需的过程，并实施 6.1 中确定的行动：

—— 建立过程准则；

—— 按照准则实施过程控制；

—— 保持必要的文件化信息，以确信已按策划实施过程。

组织应控制计划中的意外变更，并审查意外变更的后果，必要时采取行动减轻任何不利影响。

组织应确保外包过程可控。

9 绩效评价

9.1 监控、测量、分析和评价

组织应确定：

—— 需要监控和测量的内容；

—— 适用时，监控、测量、分析与评价的方法，以确保结果有效；

—— 何时应实施监控和测量；

—— 何时应分析、评价监视和测量结果。

组织应保留适当的文件化信息，作为监控、测量、分析和评价结果的证据。

组织应评价其 ××× 绩效并确定 ××× 管理体系的有效性。

9.2 内部审计

9.2.1 内部审计目标

组织应按计划的时间间隔实施内部审计，以提供 ××× 管理体系相关的以下信息：

a) 是否符合：

—— 组织自身对 ××× 管理体系的要求；

—— 本文件的要求；

b) 是否得到有效实施和维护。

9.2.2 组织应：

a) 策划、建立、实施并保持一个或多个内部审核方案，包括实施审核的频次、方法、职责、策划要求和报告。策划、建立、实施和保持内部审计方案时，应考虑相关过程的重要性和以往审计的结果；

b) 规定每次审计的准则和范围；

c) 选择审计员并实施审计，以确保审计过程的客观性与公正性；

d) 确保向相关管理者报告审计结果；

e) 保留文件化信息，作为审计方案实施和审计结果的证据。

9.3 管理评审

最高管理者应按计划的时间间隔对组织的 ××× 管理体系进行评审，以确保其持续的适宜性、充分性和有效性。

管理评审应包括对下列事项的考虑：

a) 以往管理评审所采取措施的状况；

b) 与×××管理体系相关的内外部问题的变更；

c) ×××绩效方面的信息，包括以下方面的趋势：

—— 不符合和纠正措施；

—— 监控和测量的结果；

—— 审计结果；

d) 持续改进的机遇；

管理评审的输出应包括与持续改进机会和×××管理体系变更的任何需要相关的决策。

组织应保存文件化信息，作为管理评审结果的证据。

10 改进

10.1 不符合和纠正措施

当不符合发生时，组织应：

a) 对不符合做出反应，并且适用时：

—— 采取措施控制并纠正不符合；

—— 处理后果。

b) 通过以下评估采取行动消除不符合原因的必要性，以防止不符合再次发生或在其他地方发生：

—— 评审不符合项；

—— 确定不符合的原因；

—— 确定是否存在或可能发生类似的不符合。

c) 实施任何必要的行动；

d) 评审所采取纠正措施的有效性；

e) 如有必要，对×××管理体系进行变更。

纠正措施应与所遇到的不符合项的影响相适应。

组织应保留文件化信息作为下列事项的证据：

—— 不符合的性质和所采取的任何后续措施；

—— 任何纠正措施的结果。

10.2 持续改进

组织应持续改进×××管理体系的适宜性、充分性与有效性。

附件 3
（资料性）
高层次结构、相同核心文本、通用术语及核心定义指南

高层次结构、相同核心文本、通用术语及核心定义指南在以下 URL 中提供：

附录 L 指南文档

（http：//isotc.iso.org/livelink/livelink?func=ll&objId=16347818&objAction=browse&viewType=1）

附录 M
（规范性）
制定行业专用管理标准和管理体系标准（MSS）的政策

M.1 概述

任何技术委员会或分委员会、项目委员会（IEC 体系委员会）或国际研讨会提议制定行业专用管理标准（M.2.2）或行业专用管理体系标准（MSS）（M.2.4）的，均应遵守本附录中规定的导则。其中包括委员会专用政策（M.5），这可不限于行业专用管理标准或管理体系标准。

M.2 术语和定义

M.2.1

通用管理标准 generic management standard

旨在广泛适用于不同经济部门、不同类型和规模的组织以及不同地理、文化和社会条件的管理标准

M.2.2

行业专用管理标准 sector-specific management standard

就特定经济或业务部门如何适用通用管理标准（M.2.1）提供附加要求或补充性指导的管理标准

M.2.3

通用管理体系标准 generic management system standard

通用 MSS generic MSS

旨在广泛适用于不同经济部门、不同类型和规模的组织以及不同地理、文化和社会条件的 MSS

M.2.4

行业专用管理体系标准 MSS sector-sprcific management system standard（MSS）

行业专用 MSS sector-specific MSS

就特定经济或业务部门如何适用通用 MSS（M.2.3）提供附加要求或补充性指导的 MSS

M.3 行业专用管理标准和管理体系标准

任何要求制定行业专用管理标准（M.2.2）或管理体系标准（M.2.4）的新提议

均应：

——透过 ISO 表格 4：新工作项目提案、或 IEC 的表格 NP 完成适当的 ISO 或 IEC 项目批准程序，清楚表明该标准对市场有意义，符合通用标准的要求；

——【就行业专用的 MSS（M.2.4）的制定而言】，清楚表明附录 L 中的所有规则和原则均得到遵守，包括关于合理性研究的批准（见附录 L）；以及

——清楚表明与负责相关通用管理标准或通用 MSS 的委员会联络的有效性；

——必要时，遵守以下适用于特定委员会的政策。

M.4 起草规则

行业专用管理标准（M.2.2）和行业专用的 MSS（M.2.4）应分别遵守以下规则：

a) 将通用管理标准（M.2.1）或通用 MSS（M.2.3）作为规范性引用文件。也可以逐字复制其条款和分条款。

b) 如果在行业专用标准中复制通用管理标准（M.2.1）或通用 MSS（M.2.3）的文本，则应将其与行业专用标准的其他要素区分开来。

c) 对通用管理标准（M.2.1）或通用 MSS（M.2.3）中规定的术语和定义应作规范性引用或逐字复制。

M.5 委员会专用政策

M.5.1 概述

行业专用管理标准（M.2.2）和行业专用 MSS（M.2.4）不应解释、更改或降低通用管理标准或通用 MSS 的要求。

M.5.2 环境

M.5.2.1 术语和定义

下列术语和定义适用于环境政策：

M.5.2.1.1

行业专用环境管理标准　sector-specific environmental management standard

就特定经济或业务部门如何适用通用环境管理标准提供附加要求或补充性指导的标准

示例：将环境管理体系（ISO 14001）或生命周期评估（ISO 14044）适用于农业食品或能源行业

M.5.2.1.2

特定方面的环境管理标准　aspect-specific enviromental management standard

就特定环境方面或其范围内的多个方面如何适用通用环境管理标准提供附加要求或补充性指导的标准

示例：就温室气体（方面）管理适用环境管理体系（ISO 14001）或产品水足迹（方面）适用生命周期评估（ISO 14044）。

M.5.2.1.3

特定要素的环境管理标准　element-specific environmental management standard

就特定要素或其范围内的多个要素如何适用通用环境管理标准提供附加要求或补充性指导的标准

示例：环境管理体系（ISO 14001）中的通信或应急管理（要素）或生命周期评估（ISO 14044）中的数据收集或关键评审（要素）。

M.5.2.2　概述

任何技术委员会、分委员会、项目委员会（IEC 体系委员会）或国际研讨会提议制订特定行业、方面或要素专用的环境管理标准的，均应透过完成适当的项目批准程序，清楚表明该标准对市场有意义，符合通用标准的要求，包括：

—— 通用环境管理体系标准、环境标识、生命周期评估和温室气体管理标准在特定行业、方面或要素专用特别适用的相关 ISO 表格 4：新工作项目提案，以及

—— 在特定行业、方面或要素专用特别适用通用环境 MSS 的附录 L：管理体系标准提案。

批准文件宜具体说明相关通用 ISO 14000 系列标准不足以满足特定行业、方面或要素专用需求的理由，以及提议的新标准将如何有效解决所识别出的问题。提案方宜批判性地评估是否确实需要附加适用特定行业、方面或要素专用的要求，而不是就通用环境管理标准提供补充性指导。

M.5.2.3 任何技术委员会、分委员会、项目委员会（IEC 体系委员会）或国际研讨会提议制订特定行业、方面或要素专用的环境管理标准的，宜考虑和体现发展中国家、转型经济体、各行各业中小企业和组织的需求。

M.5.2.4 ISO/TC 207 将与制订特定行业、方面或要素专用的环境管理标准的技术委员会、分委员会、项目委员会（IEC 体系委员会）或国际研讨会合作开展联合项目，或在必要情况下根据技术管理局的决定领导开展联合项目，以避免重复工作以促进标准的一致和统一。但这并不意味着要在 ISO/TC 207 以外的委员会限制市场相关标准的制定。

M.5.2.5 技术委员会、分委员会、项目委员会（IEC 体系委员会）或国际研讨会制定特定行业、方面或要素专用的环境管理标准的，应：

—— 将适当的通用 ISO 14000 系列环境管理体系、环境审计、环境标识、生命周期评估和温室气体管理标准作为规范性引用文件；

—— 规范性引用通用 ISO 14050 术语和定义；

—— 将复制的 ISO 14000 系列文本区分开来；以及

—— 不应解释、更改或降低通用 ISO 14000 系列环境管理体系、环境审计、环境标识、生命周期评估和温室气体管理标准的要求。

M.5.2.6 对于任何就该特定行业、方面或要素专用的政策提供指导、或解释通用 ISO 14000 系列标准或 ISO 14050 术语和定义、或就特定行业、方面或要素专用文件提供指导的请求，均应提交给 ISO 中央秘书处以及相关的 TC 207 分委员会。

M.5.3 质量

技术委员会、分委员会、项目委员会（IEC 体系委员会）或国际研讨会希望为特别产品或行业 / 经济部门制定质量管理体系要求的，应遵守以下规则：

a) 应完整规范性引用 ISO 9001。也可以逐字复制其条款和分条款。

b) 如果在行业文件中复制 ISO 9001 的文本，则应将其与行业文件的其他要素区分开 [见 d)]。

c) 对 ISO 9000 中规定的术语和定义应作规范性引用或逐字复制。

d) 不论是在确定是否需要编制行业专用要求或指南文件时，还是在文件编制过程中，均应参照经 ISO/TC 176 批准的《质量管理体系 编制特定产品和行业 / 经济部门专用文件的指南和标准》中规定的指南和标准。

对于本行业政策提供指导、或解释 ISO 9000 术语和定义、ISO 9001 或 ISO 9004 的请求，均应提交给 ISO/TC 176 的秘书处。

M.5.4 资产管理

技术委员会、分委员会、项目委员会（IEC 体系委员会）或国际研讨会希望为特别产品或行业 / 经济部门制定资产管理体系要求的，应遵守以下规则：

a) 应完整规范性引用 ISO 55001。也可以逐字复制其条款和分条款。

b) 如果在行业文件中复制 ISO 55001 的文本，则应将其与行业文件的其他要素区分开。

c) 对 ISO 55000 中规定的术语和定义应作规范性引用或逐字复制。

对于行业专用文件提供指导、或解释 ISO 55000 术语和定义或 ISO 55001 的请求，均应提交给 ISO/TC 251 的秘书处。

M.5.5 风险

技术委员会、分委员会、项目委员会（IEC 体系委员会）或国际研讨会希望为特别产品或行业、经济部门制定风险管理要求或指导的，应遵守以下规则：

a) 应完整引用 ISO 31000。也可以逐字复制其条款和分条款。

b) 如果在行业文件中复制 ISO 31000 的文本，则应将其与行业文件的其他要素区分开。

c) 对 ISO 31000 中规定的术语和定义应作规范性引用或逐字复制。

对于行业专用文件提供指导、或解释 ISO 31000 术语和定义的请求,均应提交给 ISO/TC 262 的秘书处。

M.5.6 社会责任

技术委员会、分委员会、项目委员会(IEC 体系委员会)或国际研讨会希望为特别产品或行业 / 经济部门制定社会责任要求或指导的,应遵守以下规则:

a) 应完整引用 ISO 26000。也可以逐字复制其条款和分条款。

b) 如果在行业文件中复制 ISO 26000 的文本,则应将其与行业文件的其他要素区分开。

c) 对 ISO 26000 中规定的术语和定义应作规范性引用或逐字复制。

附录 JA
（规范性）
投票

JA.1 概述

JA.1.1 会议

参加会议的 P 成员只可通过其代表团团长或团长（HoD）指定的人员进行投票。主席没有投票权，对于票数相等的问题应予进一步讨论。

在会议上，除非《JTC 1 补充部分合集》或 JTC 1 常备文件中另有规定，否则所有问题都通过出席会议的 P 成员在会议上投票（赞成、不赞成或弃权）的多数来决定。

如果会议通过电话会议或使用电子方式进行，有关其他要求，请见常备文件 19 "会议"第 3 条和第 4 条。

JA.1.2 信函投票

对于 JTC1 及其分委员会的信函投票，除非融合的 JTC1 补充部分或 JTC1 常备文件中另有说明外，否则所有问题由 P 成员投票（赞成或不赞成）的多数决定。信函投票通过网络投票、电子邮件、传真或者邮寄（如果实在需要）等方式进行。

JTC1 责成其秘书处在已经宣布的截止日期关闭所有信函投票，不接受迟到的投票和评议意见。JTC1 允许在 JTC1 全体会议之间通过在 JTC 1 内进行 8 周信函投票采取行动；可以由 JTC1 主席、JTC1 分委员会或 JTC1 咨询组提出需要批准的这些行动，否则，任何信函投票期不应在发出通知之日起不到 12 周内结束。

JA.1.3 默认投票

在某些情况下，可以确认对于预期不包含任何争议的问题达成共识，并且提前预见委员会内部达成一致。这些问题的分发期限为 4 周。如果在此期间未收到任何异议，则该问题被认为已获批准。如果在此期间任何 JTC1 P 成员反对该问题，JTC 1 委员会主管应立即撤回投票并通过会议投票方式或信函投票方式解决问题。欲对默认投票提出异议的 JTC1 国家成员体请尽快通知归口秘书处，以防止出现不适当的延误。

这些问题可涉及：

- 指定 / 更换注册机构；
- 建立或撤回 C 类联络关系；
- 稳定 / 撤销某项标准的提议；

- 要求免费提供符合既定标准的 ISO / IEC 出版物；
- 修改分委员会的工作计划，包括按 SD-3（详见 1.17.7.2）规定与 ITU-T 建立合作交流或协作小组；
- JTC1 批准的其他事宜。

JA.2 提案阶段——对新工作项目提案的投票

JTC1 P 成员、委员会秘书处、另一技术委员会或分委员会、联络组织、技术管理局或其咨询组及 CEO 都可向分委员会或 JTC1 提交新工作项目提案。对每个新工作项目提案都应进行信函投票表决，即使该项目已经列入会议议程。如果该提案包括与 ITU-T 建立合作交流或合作团队，见 1.17.7.2 获取拟考虑的清单，评估其中的各项以确定理由及哪些需要纳入 NP 文件中。2.3.5 中规定了验收标准。新工作项目提案的投票期限通常是自通知之日起 12 周（见 2.3.4）。

JA.2.1 对 SC 层面的 NP 进行投票

新工作项目提案只应在分委员会内投票一次。

应当注意到，如果在没有事先与分委员会协商的情况下提交新工作项目进行投票，由于缺乏必要的共识和支持，存在投票无法通过的风险。只要不导致无理由的延迟，分委员会主席或秘书处可安排重新提交的新工作项目提案，以便在全体会议或工作组会议上进行讨论，然后再进行投票。

对于在分委员会层面表决的新工作项目提案，分委员会秘书处应将分委员会投票的副本转交 JTC1 秘书处，以便在分发新工作项目提案投票的同时提供信息（见图 1）。JTC1 秘书处应将此分委员会级别投票的副本分发给 JTC1 P 成员和 JTC1 分委员会，以便进行同时审查。

发出分委员会 NP 投票进行 JTC1 同步评审 4 周内，JTC1 P 成员可要求 JTC1 秘书处启动并行的 JTC1 层面的投票。JTC1 P 成员必须给出上述请求的理由，而且这种理由的重点必须放在对新工作项目提案的适当处理上。如果理由的重点只放在新工作项目提案的技术方面，则不能获得许可。如果两个或多个 JTC1 P 成员请求这种 JTC1 层面并行投票，则 JTC1 秘书处应进行投票。JTC1 层面的投票应与分委员会级别的投票相同且截止日期相同。提交的请求理由应伴随 JTC1 层面投票并应发送给有关的分委员会。JTC1 层面投票的批准准则应等同分委员会层面投票的批准准则，但是其参与承诺的要求不适用（2.3.5a）中）。

在完成 JTC1 层面的投票后，只有当分委员会层面投票和 JTC 1 投票均通过时，NP 才会获批。

经批准后，JTC1 秘书处应将其结果（和分配的项目编号）通知其所有国家成员体和分委员会委员会主管。

如果未通过分委员会层面投票，且没有启动 JTC1 投票，则不采取进一步行动。

如果通过分委员会层面投票，但没有启动 JTC1 投票，则新工作项目提案获批。

注：如果在 JTC1 同步评审期间，任何未参加 SC 的 JTC1 P 成员对该 NP 提出评议意见，但没有请求 JTC1 投票，JTC1 P 成员可以向 SC 秘书处提交此类意见，进行分委员会层面的投票。

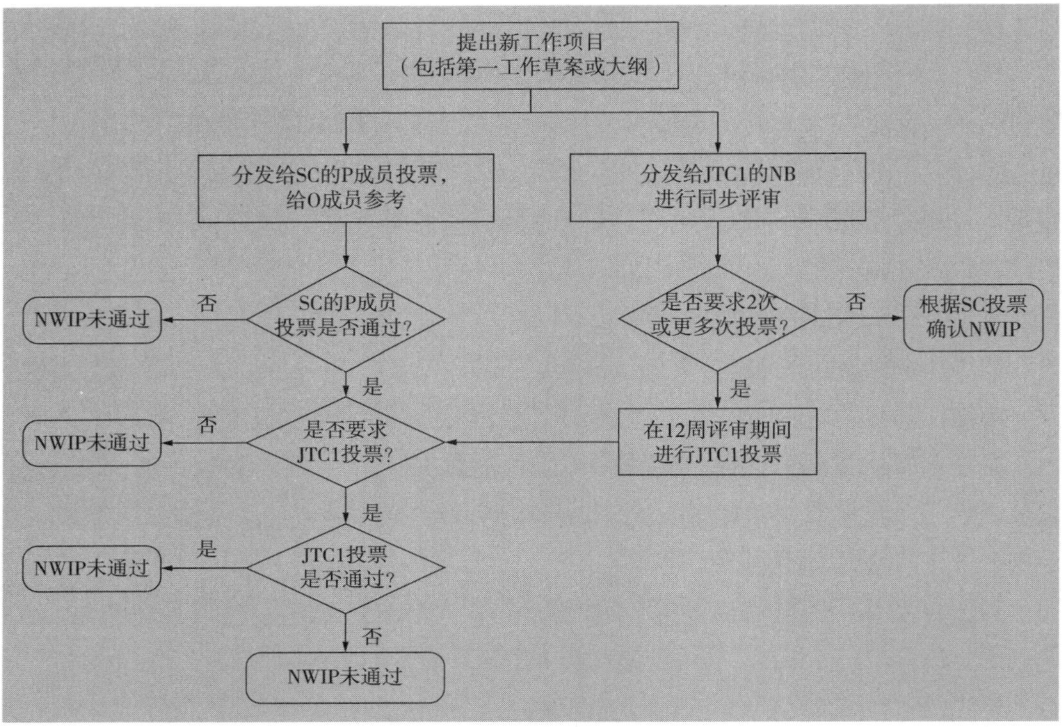

图 A.1　NWIP 在 SC 层面投票流程图

JA.2.2　JTC1 层面的新工作项目提案投票

JTC 1 宜考虑新工作项目提案包括：

- 直接向 JTC1 报告的工作组提出的工作项目；
- 或者在例外情况下，不属于现有分委员会范围的新工作项目建议。

除以上情况外，相应的分委员会宜对新工作项目提案进行投票。对每个新工作项目提案都应通过信函投票进行表决，即便其出现在会议议程上。验收标准如 2.3.5 所述。

JA.3　准备阶段

本阶段不投票。

JA.4　委员会阶段 – 对 CD/CDAM/DTS/DTR 的投票

如果以通信方式处理委员会草案 / 委员会修正草案 / 技术规范草案 / 技术报告草案（CD/CDAM/DTS/DTR），P 成员、联络的技术委员会及组织在指定日期之前提交其评

议意见（P 成员同时须提交其投票）

就委员会草案/委员会修正草案/技术规范草案/技术报告草案而言，该日期宜为通知发布之日起的第 8、12 或 16 周。

CD/CDAM/DTS/DTR 的默认流通时间为 8 周。

若 P 成员放弃对委员会草案/委员会修正草案/技术规范草案/技术报告草案投票，其仍有权在同期或后期阶段对该文件的后续版本进行投票。

应持续对连续的委员会草案/委员会修正草案/技术规范草案/技术报告草案进行审议，直至获得委员会 P 成员的实质性支持或已达成放弃或推迟该项目的决定。

由联合工作组编写的委员会草案/委员会修正草案/技术规范草案/技术报告草案宜由正式参与联合工作的所有分委员会的所有 P 成员进行投票。每位成员应仅有一票。

JA.5 对技术勘误草案的投票

对技术勘误草案（DCOR）的审议须通过通信方式进行。SC P 成员和联络组织须在指定日期之前提交其评议意见（SC P 成员同时须提交其投票）。该日期应为自通知发布之日起至少 12 周。

JA.6 ISO/IEC JTC 1 投票周期概览

下表为用于 ISO/IEC JTC 1 的投票周期概览。

投票类型	投票期限	对应条款
新工作项目提案 -JTC 1 或 SC 投票	通常为 12 周	JA.2
分委员会的新工作项目提案：JTC 1 确认	8 周	JA.2.1
委员会草案	8、12 或 16 周（默认 8 周）	2.5.4；JA.4
技术规范草案/技术报告草案	8、12 或 16 周（默认 8 周）	JA.4；2.5.4
委员会修正草案	8、12 或 16 周（默认 8 周）	JA.4；2.5.4；2.10.3
国际标准草案	12 周（翻译期为 8 周）	2.6.1
快速程序国际标准草案	12 周（翻译期为 8 周）	2.6.1；F.4.3
JTC 1 可公开提供技术规范国际标准草案	12 周（翻译期为 8 周）	2.6.1；F.4.3
修正草案	12 周（翻译期为 8 周）	2.6.1
最终国际标准草案	8 周	2.7.1
快速程序最终国际标准草案	8 周	2.7.1；F.4.8
JTC 1 可公开提供技术规范最终国际标准草案	8 周	F.4.9（b）
最终修正草案	8 周	2.10.3
技术勘误草案	至少 12 周	JA.6
稳定标准撤销建议	4 周	JA 1.4
撤销标准的恢复	12 周（翻译期为 8 周），或 8 周	2.9.6

表续

投票类型	投票期限	对应条款
JTC 1 可公开提供的技术规范提交者认可	12 周	F.3.4.1
JTC 1 可公开提供的技术规范提交者再确认	12 周	F.3.4.1
JTC 1——其他信函投票期	至少 12 周	JA.1.3
JTC 1——默认信函投票	4 周	JA.1.4
JTC 1——两次全会之间的行动	8 周	JA.1.3
分委员会工作变更计划	4 周	JA.1.4
第 2 版和更多版国际标准草案	8 周，最长 12 周	2.6.4
第 2 版和更多版修正草案	8 周，最长 12 周	2.6.4
标准的撤销	8 周	2.9.3
标准的稳定	20 周（正常系统评价投票）	2.9.5
系统复审	20 周	2.9.2

附录 JB
（规范性）
ITU-T 与 ISO/IEC JTC 1 的合作

1.ITU-T 和 ISO/IEC JTC 1 合作指南经由 ISO/IEC JTC 1 和 ITU-T 起草，并且已通过 ISO/TMB、IEC/SMB 和 ITU-T 的批准。

常备文件 3（ITU-T 和 ISO/IEC JTC 1 合作指南）中的文本内容与 ITU-T 建议书 A.23 附录 A 的内容相同。

2. 它延续了同一组织之间关于合作方法的长期协议，其中 ITU-T 建议书和 ISO/IEC JTC 1 制定的 ISO/IEC 国际标准具有共同的文本或相同的技术内容。

3. 除三个组织之间已经使用的正常联络安排外，当在特定工作领域需获得共同的文本或者相同的技术内容时，ITU-T 和 ISO/IEC JTC 1 应采用以下两种更为紧密的合作模式之一：进行协作交流或者建立合作团队。

4. 进行协作交流需要在两个相关组织的连续会议上推进某一文本的技术工作，同时同步各自的评论和批准程序。如果工作相对直接、没有争议，并且共同参加两个组织的会议足以使交流非常有效，则应采用协作交流的方式。同时双方应商定需完成工作的职权范围。

5. 应设立一个单独的合作团队，以推进任何需要通过扩大对话来制定解决方法并达成共识的工作。同时应商定该团队的职权范围，包括其工作范围和需要对每个组织中哪些上级机构进行报告。一旦达成共识，就会同步使用 ITU-T、IEC 和 ISO 中的批准程序来实现发布。合作团队遵循的程序可见 JTC 1 常备文件 3（ITU-T 和 ISO/IEC JTC 1 合作指南）的第 8 条。

6. 在任一合作模式下，经批准的可交付使用文件可发布为通用文本（一份遵循指南附录 Ⅱ 中指定演示风格的 ITU-T 建议书和国际标准），或发布为双文本（ITU-T 建议书和国际标准，其内容在技术上一致但不完全相同）。发布为双文本时，审批流程在时间上不需要精确同步。

7.ITU-T 研究组和 ISO/IEC JTC 1 分委员会应就以下问题达成一致：是否需要联系，是否需在每个领域建立联络关系，进行协作交流或者建立合作团队。若达成一致，可在项目过程更换合作模式。

8. 若任一组织认为应终止双方在某一特定工作领域的合作，其应立即与另一组织讨论该情况。如果无法商定出令人满意的解决办法，ITU-T 或 ISO/IEC JTC 1 可单方面终止某一项目的合作，或决定不发布任何通用文本。如果发生终止情况，两个组织都可以利用之前的合作工作成果，且每个组织都可以使用合作终止前完成的任何成果。

附录 JD
（规范性）
阶段编码的矩阵表示

JD.1 协调阶段编码概述

标准化过程中包含若干明确的步骤或阶段，可以用来描述其过程，表明某个项目在该过程中的进展。总体而言，无论由哪个机构监管标准化过程，通过国际、区域和国家标准化机构操作的正式标准化过程来制定和发布标准的方法都非常相似。因此，在较高层面上，有可能对标准化过程达成共识并具有共同的阶段设置。然而，每个机构的标准化过程均存在差异，导致各机构产生了不同的阶段体系。

协调阶段编码（HSC）体系应用于 ISO 数据库，用来跟踪标准制定项目。其目的是提供一个用于核心数据传输的通用框架。该体系允许在国际、区域和国家层面使用的数据库中用相同方法跟踪给定项目的发展，因此设计了这种能易于适应新要求的矩阵结构。

JD.2 阶段编码矩阵的设计

制定了表示不同组织间通用程序序列的"阶段"系列。这些"阶段"代表标准制定的主要阶段。

在每个阶段内设立"子阶段"系列，使用同"阶段"一致的概念逻辑体系。因此，术语"阶段"和"子阶段"用来标明合成矩阵的相关轴。

每个主阶段和子阶段用两位数 00 到 99 编码，以 10 进位。通用矩阵的每个单元格用 4 位数编码，由其阶段和子阶段坐标组成。对于视觉表述（虽然从数据库操作目的来看没有必要），坐标对由一个点分隔（例如，用 10.20 表示阶段 10、子阶段 20）。

所有未使用的阶段编码留待将来使用，以便识别插入增补的阶段，例如阶段编码 10、30、40、50 和 80。

JD.3 使用本体系的基本指南

—— 其他相关信息（如文件来源或文件类型）宜记录在不同的数据库字段且不宜以阶段编码的形式显示。

—— 没有用于表示某项目停滞在某一特定阶段的子编码。建议使用另一个数据库字数来解决这一问题。

—— HSC 体系允许标准过程的循环特性及当前或早期阶段的重复。在一个项目的生命周期中可能重复的事件可通过重复的相同阶段编码予以记录。

—— 利用项目达到的编码可以在任何点冻结该项目。已中止的项目信息宜记录

— 在一个单独的数据库字段中。

— HSC 体系不涉及记录达到阶段的目标日期或实际日期。

项目阶段的矩阵表示

阶段	子阶段						
	00 注册	20 主要行动开始	60 主要行动完成	90 决策			
				92 重复先前阶段	93 重复当前阶段	98 放弃	99 继续
00 预研阶段	00.00 接收新项目提案	00.20 审查新项目提案	00.60 审查结束			00.98 废弃新项目提案	00.99 批准进行新项目提案投票
10 提案阶段	10.00 注册新项目提案	10.20 启动新项目投票	10.60 投票结束	10.92 返回提案至提交者以作进一步界定		10.98 拒绝新项目	10.99 通过批准的新项目
20 准备阶段	20.00 在TC/SC工作计划中注册新项目	20.20 启动工作草案（WD）研究	20.60 评论期结束			20.98 删除项目	20.99 WD批准注册为CD
30 委员会阶段	30.00 注册委员会草案（CD）	30.20 启动CD研究/投票	30.60 投票/评论期结束	30.92 返回CD至工作组		30.98 删除项目	30.99 CD批准注册为DIS
40 征询意见阶段	40.00 注册国际标准草案（DIS）	40.20 启动DIS投票：五个月	40.60 投票结束	40.92 发布完整报告：返回DIS至TC/SC	40.93 发布完整报告：新DIS投票的决定	40.98 删除项目	40.99 发布完整报告：DIS批准注册为FDIS
50 批准阶段	50.00 注册最终国际标准草案（FDIS）为正式批准	50.20 启动FDIS投票：八周。交送校样至秘书处	50.60 投票结束。返还校样至秘书处	50.92 返回FDIS至TC/SC		50.98 删除项目	50.99 FDIS批准发布
60 出版阶段	60.00 国际标准发布中		60.60 国际标准正式发布				
90 复审阶段		90.20 系统复审国际标准	90.60 复审结束	90.92 修订国际标准	90.93 确认国际标准		90.99 由TC/SC提出撤销国际标准
95 撤销阶段		95.20	95.60	95.92			95.99

附录 JE
（规范性）
图形符号标准化程序

JE.1 引言

本附录阐述了在适当情况下 ISO 文件中出现的所有图形符号在提交及随后的批准和注册中采取的流程。

在 ISO 内部，协调图形符号制定的责任被细分为两个主要领域，分配给了两个 ISO 技术委员会：

—— ISO/TC 145 所有图形符号（技术产品文件用图形符号除外）（见 ISO/TC 145 网站）；

—— ISO/TC 10 技术产品文件（TPD）用图形符号（见 ISO/TC 10 网站）。

此外，还涉及与 IEC/TC 3 进行信息结构、文档和图形符号方面的协调以及与 IEC/TC 3/SC 3C 进行设备用图形符号方面的协调。

图形符号标准化的基本目标是：

—— 满足用户需求；

—— 确保考虑到所有相关 ISO 委员会的利益；

—— 确保图形符号清晰明确，符合一致的设计标准；

—— 确保图形符号不存在重复或不必要的扩散。

新图形符号标准化的基本步骤是：

—— 确定需求；

—— 阐述；

—— 评价；

—— 批准（若适用）；

—— 注册；

—— 出版。

所有步骤宜采用电子方式进行。

—— 新的或修订的图形符号提案可由 ISO 委员会、ISO 委员会的联络员或任何 ISO 成员组织（以下统称"提案方"）提交。

—— 每个批准的图形符号将被分配一个唯一的编号，以便通过注册表对其进行管理和识别。该注册表提供能通过电子格式检索的信息。

—— 如与图形符号的相关要求和指南存在冲突，ISO/TC 145 或 ISO/TC 10 与相关产品委员会应尽早进行联络和对话，解决冲突。

JE.2 除技术产品文件用图形符号外的所有图形符号

JE.2.1 概述

ISO/TC 145 在 ISO 内部负责图形符号（TPD 中的除外）领域标准化的总体协调。主要工作包括：

—— 图形符号及颜色形状领域的标准化，只要这些颜色或形状元素构成了符号意图传达的信息的一部分，例如安全标志；

—— 制定图形符号的准备、协调和应用原则：全面负责审查和协调已存在的、研究中的和即将制定的原则。

不包括字母、数字、标点符号、数学符号及数量和单位符号的标准化。然而，上述元素可能作为图形符号的组成部分。

—— ISO/TC 145 的审查和协调工作适用于所有负责在其特定领域内创建和标准化图形符号的委员会。

ISO/TC 145 分配了以下职责：

—— ISO/TC 145/SC 1：公共信息符号；

—— ISO/TC 145/SC 2：安全标识、标志、形状、符号和颜色；

—— ISO/TC 145/SC 3：设备用图形符号。

ISO/TC 145 与 ISO/TC 10 和 IEC 保持联络，特别是与负责设备用图形符号的 IEC/SC 3C。

表 JE.1 为各协调委员会所负责的图形符号的类别。

表 JE.1 图形符号的类别

	基本信息	定位	目标群体	设计原则	概述	归口委员会
公共信息符号	服务或设施的位置	在公共场所	普通大众	ISO 22727	ISO 7001	ISO/TC 145/SC 1
安全标识（符号）	与人身安全和健康相关	在工作场所和公共场所	a) 普通大众或 b) 经过授权和培训的人员	ISO 3864-1 ISO 3864-3	ISO 7010	ISO/TC 145/SC 2
产品安全标签	与人身安全和健康相关	在产品上	c) 普通大众或 d) 经过授权和培训的人员	ISO 3864-2 ISO 3864-3	—	ISO/TC 145/SC 2
设备用图形符号	与设备相关	在设备上	e) 普通大众或 f) 经过授权和培训的人员	IEC 80416-1 ISO 80416-2 IEC 80416-3	ISO 7000 IEC 60417	ISO/TC 145/SC 3 IEC/TC 3/SC 3C
TPD 符号	（产品代表）	技术产品文件（图示，图表等）	经过培训的人员	ISO 81714-1	ISO 14617 IEC 60617	ISO/TC 10/SC 10 IEC/TC 3

表 JE.2　不同类型的图形符号在其使用环境中的示例

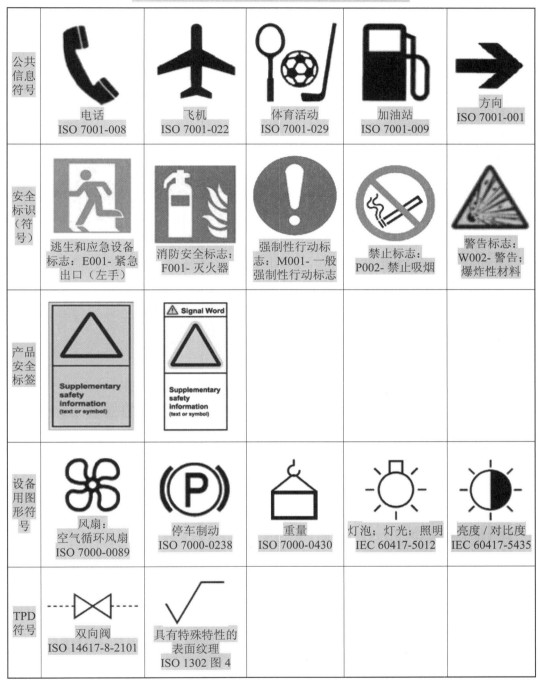

类型	示例				
公共信息符号	电话 ISO 7001-008	飞机 ISO 7001-022	体育活动 ISO 7001-029	加油站 ISO 7001-009	方向 ISO 7001-001
安全标识（符号）	逃生和应急设备标志：E001- 紧急出口（左手）	消防安全标志：F001- 灭火器	强制性行动标志：M001- 一般强制性行动标志	禁止标志：P002- 禁止吸烟	警告标志：W002- 警告；爆炸性材料
产品安全标签					
设备用图形符号	风扇：空气循环风扇 ISO 7000-0089	停车制动 ISO 7000-0238	重量 ISO 7000-0430	灯泡；灯光；照明 IEC 60417-5012	亮度/对比度 IEC 60417-5435
TPD 符号	双向阀 ISO 14617-8-2101	具有特殊特性的表面纹理 ISO 1302 图 4			

JE.2.2　提案的提交

提案方应尽快将相关申请表上的提案提交给相应的 ISO/TC 145 分委员会秘书处，以便分委员会及时审查和评论。强烈建议提案方在 CD 阶段提交。但对于国际标准相关提案，提交时间不得迟于第一个征询意见阶段（即 DIS 或 DAM）。

在提交图形符号提案之前，提案方应：
—— 能够证明对拟议图形符号的需求；
—— 已审查图形符号相关的 ISO 和/或 IEC 标准，以避免混淆和/或与现有标准化图形符号重叠，检查与已标准化的任何相关图形符号或图形符号族的一致性；
—— 根据相关标准和说明（包括设计原则和验收标准）创造拟议的图形符号。

JE.2.3 拟议图形符号的标准化程序

收到提案后，相关的 ISO/TC 145 分委员会应在 8 周内审核申请表，检查申请表是否已正确填写完整，是否已正确提供相关的图形文件。如有必要，将邀请提案方修改申请，并重新提交。

收到正确填写完整的申请表后，应开始正式的审查程序。审查提案与标准化图形符号、相关设计原则和验收标准的一致性。

在正式审查过程完成后，审查结果应与所有建议一起发送给提案方。将酌情邀请提案方修改提案，并重新提交提案以供进一步审查。

应遵循在相关 ISO/TC 145 分委员会网站上列出的程序：
—— ISO/TC 145/SC 1：公共信息符号（www.iso.org/tc145/sc1）；
—— ISO/TC 145/SC 2：安全标识、标志、形状、符号和颜色（www.iso.org/tc145/sc2）；
—— ISO/TC 145/SC 3：设备用图形符号（www.iso.org/tc145/sc3）。

上述网站还提供了提交提案的申请表。

应分配给经 ISO/TC 145 批准的图形符号一个确定的注册号并将其纳入相关的 ISO/TC 145 标准中。

注：在特殊情况下，经 TMB 批准后，未经注册的符号可纳入 ISO 标准。

JE.3 技术产品文件（TPD）中使用的图形符号（ISO/TC 10）

ISO/TC 10 负责技术产品文件（TPD）用图形符号领域的标准化工作。主要工作包括：
—— 与 IEC 合作，维护 ISO 81714-1《产品技术文件用图形符号的设计 第 1 部分：基本规则》；
—— 与 IEC 协调，标准化技术产品文件中使用的图形符号；
—— 建立和维护图形符号数据库，包括注册号管理。

同时包括标准化用于图表和示意图的符号。

ISO/TC 10 已将这些职责分配给 ISO/TC 10/SC 10。ISO/TC 10/SC 10 秘书处由一个维护小组提供支持。

任何确定需要新的或修订 TPD 图形符号的委员会应尽快向 ISO/TC 10/SC 10 秘书处提交其提案，以供审查，并在获得批准后分配一个注册号。

附录 JF
（规范性）
注册机构（RA）政策

JF.1 范围

由 ISO 技术委员会制定的一些国际标准要求分配独特的注册要素，并描述这些要素的分配方法。这些要素本身并不是标准的一部分，而是由指定的注册机构分配的，注册机构还保留了已分配要素的准确记录。注册机构是一个主管机构，拥有必要的基础设施以确保注册要素的有效分配。这些机构由 ISO 指定为特定标准的独特 RA，这创造了事实上的垄断局面。

本注册机构政策是强制性的，应与 ISO/IEC 导则第 1 部分的附录 H 一起阅读。

如果 ISO/CS 发现 RA 标准未遵守本政策，它将停止发布过程，以便在 RA 标准发布前有时间实施本政策。因此，鼓励委员会让 ISO 技术项目主管（TPM）在制定过程中尽早了解要求 RA 的项目，以避免出版延误。

JF.2 定义

JF.2.1 RA 标准：注册机构（RA）提供注册服务的标准。

JF.2.2 注册服务：注册机构在实施 RA 标准时提供的服务并在 RA 标准中说明。

JF.2.3 注册机构（RA）：由 ISO 指定的实体，承担 RA 标准中的注册服务。

JF.2.4 注册代理：注册机构委托进行部分注册服务的第三方（如国家或地区子实体）。即使授权给注册代理，注册服务仍由注册机构负责。

JF.2.5 注册机构协议（RAA）：基于 RA 与代表 ISO 的 ISO 秘书长签署的 RAA 模板的协议，其中详细说明了相关方的职能、角色和法律义务。在发布 RA 标准（包括修订版）之前，应签署 RAA。

JF.2.6 注册要素：唯一要素，虽然在标准中有描述其方法，但注册要素本身不属于标准的一部分。

JF.2.7 技术项目主管（TPM）：ISO/CS 中指定与特定委员会合作的个人。

JF.3 程序

JF.3.1 年表

本注册机构政策按照典型注册机构生命周期的顺序处理注册机构的各方面问题。注意有些阶段可以并行进行。各阶段如下：

—— 确定对 RA 的需求（JF.3.2）

—— 起草 RA 标准（JF.3.3）

—— 选择 RA（JF.3.4）

—— 任命 RA（JF.3.5）

—— 签署 RAA（JF.3.6）

—— 终止 RA（JF.3.7）

—— 实施 RA 标准（JF.3.8）

—— RA 的作用（JF.3.8.1）

—— 委员会的作用（JF.3.8.2）

—— ISO 中央秘书处的作用（JF.3.8.3）

JF.3.2 确定对 RA 的需求

委员会确认其决定，即需要 RA 以决议的方式实施一项标准。

委员会秘书处填写 RA 确认（RAC）表（见 www.iso.org/forms），并于筹备阶段开始之前尽快提交给 TPM。

JF.3.3 起草 RA 标准

所有 RA 标准均应包含以下内容：

—— 生成唯一注册要素的标识方案或机制的描述。

—— 注册服务的描述，及 RA 提供注册服务的责任。

—— 指向 iso.org 页面的链接，用户可以在链接中找到给定标准的 RA 的名称和联系信息。还应提供 RA 网站的链接，其中包含有关注册服务的更多信息。

RA 标准中不应包括以下内容：

—— 根据《ISO/IEC 导则 第 2 部分》第 4 章，RA 标准中不应包含合同或其他法律要素。

—— RA 标准中不应包含有关提供注册服务的程序（如 RA 提供的手册）。这些程序将由 RA 公开提供（如通过网站）。

—— RA 标准中不应提及 RA 的名称。相反，应提供 ISO 网站的链接（见上文）。

—— RA 标准中不应包括对 RA 选择或确认过程的引用。

—— 如果 RA 按照 RAA 的约定将注册服务委托给第三方（如称为"注册代理"的国家或区域子实体），RA 标准可能会提及一些注册服务已被委托的事实。但是，RA 标准中不应包含任何有关注册代理的进一步信息。

TPM 负责与委员会协调，以确保 RA 标准中包含正确的要素。有关 RA 标准中宜包含哪些内容的任何问题都应提交给 TPM。

JF.3.4 选择 RA

在现有 RA 由于某些原因不再是 RA 的情况下，RA 的选择过程适用于新 RA 标准和现有 RA 标准。

在修订的情况下，委员会应决定是否宜使现有 RA 继续，或是否宜启动选择程序以选择其他 RA 候选机构。启动选择程序的决定宜得到合理支持。委员会应通过决议确认其决定。

选择程序应在项目达到 DIS 阶段之前完成。DIS 中包含有关所需注册服务性质的详细信息，这很重要。

委员会确定了选择 RA 的标准，并通过决议确认这些标准。选择程序的最小要素应为：

—— 选择标准——应清楚地解释这些标准并提供足够的详细信息，以便可能入选的 RA 候选机构评估其是否符合标准并在此基础上运用其能力。选择标准中应包括要求 RA 书面提供以下信息：

—— 证据（如章程）可证明它是一个法律实体，这意味着该组织是根据管辖法律成立的组织，因此受到相关规则的约束。

—— 表示愿意承担注册服务的责任。

—— 确认 RA 在技术和财务上能够在国际层面上执行 RA 标准和 RAA 中描述的 RA 服务。

—— 表示愿意签署 RAA。ISO/CS RAA 模板应与 RA 候选机构共享。

—— 确认是否打算将部分注册服务委托给注册代理。

—— 是否会收取服务费用。如果收取费用，则确认所有此类费用将以成本回收为基础。

—— RA 候选机构的公开征集——委员会应采取适当的步骤，将主管 RA 申请方的需求发布到尽可能广的市场，邀请可能参与的组织申请作为 RA 候选机构。

—— 评估——RA 候选机构应以书面形式给予其答复。委员会（或其子机构）应根据选择标准确定客观评估标准，包括给予每个选择标准的相对权重。

—— 记录保存——委员会秘书处应记录保存选择过程中的所有文件，包括候选机构需求，申请，评估，决定等。

之后，委员会应通过决议确认其关于任命某一组织为 RA 的建议。

JF.3.5 任命 RA

委员会在 RAC 中提供的信息（见上文 JF.3.2）用于启动任命 RA 的 TMB 投票及

RA 打算收取费用时需进行的 ISO 理事会投票。在修订的情况下，如果委员会决定同一个 RA 宜继续，则不需要 TMB 或 ISO 理事会的批准（见 JF.3.4）。

ISO/IEC 导则第 1 部分声明在经过 ISO 理事会授权且收费的基础严格按成本回收计算的情况下，RA 可收取注册服务费。

因此，相关信息应完整。对于 JTC 1 RA 标准，还应向 IEC 提供 RAC 表格的副本，因为 RA 任命均应由负责收取费用的 IEC/SMB（和理事局）确认。

JF.3.6 签署 RAA

只有在 TMB（如果收取费用，则由理事会收取）任命 RA（如果是 JTC 1 RA 标准，则涉及到 IEC）后，才能签署 RAA。所有 RA 都必须根据 ISO/CS 模板签署 RAA。应在发布新的或修订的 RA 标准之前签署 RAA。如果未签署 RAA，则不应发布新的或修订的 RA 标准。

由于 RA 标准在整个制定过程中可以变更，因此应向 RA 提供准备发布的 RA 标准版本。RAA 会引用 RA 任务列表。

为确保不同 RA 之间处理的一致性和平等性，任何要求偏离 RAA 模板（ISO/CS 认为其具有重要意义）的情况均应提交 TMB 批准。

JF.3.7 终止 RA

当 RA 辞职或委员会和 ISO/CS 由于 RA 标准撤销、客户投诉等原因认为 RA 不宜再继续提供注册服务时，可能会终止 RA。委员会应通知 TPM 关于其发现的可能导致 RA 终止注册服务的任何问题。

在选择替代 RA 时，宜遵循上述 JF.3.4 中详述的程序，除非委员会已确定了某个符合 JF.3.4 中选择标准的替代 RA 候选机构。额外 RA 候选机构的选择程序将导致 RA 服务中断。

RAA 规定了终止 RA 的通知期限。通知期将用于进行新 RA 的选择程序，以便从一个 RA 无缝过渡到另一个 RA。

JF.3.8 实施 RA 标准

JF.3.8.1 RA 的作用

RA 通过以下方式提供注册服务：

—— 提供 RA 标准中描述的注册服务及

—— 遵守 RAA 的规定。

JF.3.8.2 委员会的作用

虽然 RAA 由 RA 和 ISO 秘书长签署，但秘书长签署的 RAA 将约束 ISO 体系中的所有组成部分，包括 ISO 成员和 ISO 委员会。委员会发挥核心作用。除了为新的和修订的 RA 标准确定对 RA 标准的需求（见 JF.3.2）、起草 RA 标准（见 JF.3.3）和选择 RA（见 JF.3.4）外，委员会主要负责监督 RA，具体如下：

—— 回答问题：RA 可以要求委员会回答有关 RA 标准的问题，并澄清关于其在实施 RA 标准中作用的任何期望。

—— 评估 RA 的年度报告：RAA 要求 RA 提供具有特定要素列表的年度报告。委员会应确保 RA 提供这些年度报告同时应通读这些报告。

—— 监测：除年度 RA 报告外，委员会还应收集行业和用户关于 RA 标准的所有反馈。委员会应根据所有这些要素（RA 报告和其他反馈）向 ISO/CS 报告（见下文）。

—— 向 ISO/CS 报告：每年至少一次，根据在上述监测中收集的信息，委员会应使用提交 TMP 的年度委员会报表（ACR）（见 www.iso.org/forms）向归口 TPM 报告。此类报告的目的是确认 RA 按照 RAA 的规定运行，或提出任何相关问题（问题可能包括：RA 标准不符合行业或用户需求、对注册服务质量的投诉等）。此类报告应至少每年向归口 TPM 提供一次，或委员会认为必要的情况下提供多次。TPM 还可要求委员会提供特别报告。如果报告确定一些问题，则应包括解决这些问题所需的计划纠正措施（见下文）。

—— 纠正措施：

　　—— 委员会：委员会负责实施其职责范围内的纠正措施。这些措施可能包括：修订 RA 标准、向 RA 提供建议和指导、进行审计或在严重情况下向 ISO/CS 建议终止 RAA。

　　—— ISO/CS：将由 TPM 协调属于 ISO/CS 职责范围内的纠正措施（如更新或监督 RAA）。TPM 也可提出纠正措施。

　　—— RA：RA 负责其责任范围内的所有纠正措施，可能包括注册服务和 RAA 中描述的规定。

—— 争议解决：若与 RA 服务客户发生争议，RA 应努力解决。RA 应将其无法解决的争议提交给委员会（如 RAA 所述），委员会应正式通知归口 TPM。委员会（和 ISO/CS）的职责仅限于支持 RA 解决争议。委员会不应接管争议的责任，也不应成为针对 RA 争议的上诉机构，因为这可能无意中给人留下 ISO 负责注册服务的印象。

—— 记录的维护：RA 的所有重要通信和文件都应存档。委员会秘书处负责确保将这些文件保存在电子委员会的单独文件夹中。

委员会可设立一个小组（通常为注册管理小组"RMG"）以帮助他们解决上述问

题。委员会（直接或通过 RMG）不应参与或卷入提供注册服务。

JF.3.8.3 ISO/CS 的作用

委员会与 ISO/CS 的接口是通过归口 TPM 实现的。TPM 的作用包括：

—— 在制定过程中确定 RA 标准（如果委员会没有做这项工作）。

—— 为起草 RA 标准提供指导和建议。

—— 对委员会进行 RA 政策培训。

—— 与各委员会进行协调，以确保符合 RA 政策、注册服务质量、适当处理投诉、满足行业和用户需求，包括解决各委员会提供的年度报告（采用 ACR 表格）中提出的问题，以及对任何纠正措施的实施提供建议和协助（见 JF.3.8.2）。

—— 维护其参与的有关记录。

参 考 文 献

以下是 JTC 1 社区一些重要主题的参考文献链接。

—— JTC 1 的网页（www.jtc1.org）

—— ISO/IEC 导则，第 1 部分和第 2 部分，联合 JTC 1 补充（www.iso.org/directives 和 www.jtc1.org）

—— JTC 1 常备文件（www.jtc1.org）

—— 如何编写标准（http://www.iso.org/iso/how-to-write-standards.pdf）

—— ISO 标准制定活动结对指南（http://www.iso.org/iso/guidance-twining-ld.pdf）

—— 有关 ISO 出版物中规范性引用的政策（https://connect.iso.org/pages/viewpage.action？pageid=27592570）

—— 向 ISO/CS 提交文本和图纸的指南（http://isoc.iso.org/livelink/livelink/open/18862226）

SMB/6799/QP

For IEC use only

2019-08-16

INTERNATIONAL ELECTROTECHNICAL COMMISSION

STANDARDIZATION MANAGEMENT BOARD

SUBJECT

ISO/IEC Directives Part 1 – Consolidated JTC 1 Supplement 2019 – Procedures specific to JTC 1

BACKGROUND

The attached Consolidated JTC 1 Supplement 2019 is based on the ISO/IEC Directives Part 1, Edition 15, 2019. It has been balloted and approved by JTC 1.

The document is in track-changes version, with indications of why changes have been made.

The following clauses have been modified with respect to the previous edition: Foreword, 1.7.3, 1.12.2, 1.13.2, 1.13.6, 1.17.2, 3.2.1, 3.2.2, 3.2.4, Annex B, Annex D, Annex E, Annex F, Annex I, and Annex M. 1. The track changes version of this fifteenth edition should be consulted for the details of the changes made.

The JTC 1 Supplement 2019 is submitted to IEC SMB and ISO/TMB for approval.

ACTION

SMB is invited to approve the item below using the IEC SMB Voting System **before 2019-09-13**.

Item 1: Publication of the ISO/IEC Directives Part 1 – Consolidated JTC 1 Supplement 2019 – Procedures specific to JTC 1

编者注：以下收入的英文为 ISO 提供的授权文件。该文件为修订格式。为便于使用，本书出版时遵照原样式未做调整。

ISO/IEC Directives, Part 1

Directives ISO/CEI, Partie 1

Consolidated JTC 1 Supplement 2019 — **Procedures specific to JTC 1**

Procédures spécifiques à JTC 1

Based on ISO/IEC Directives Part 1 Fifteenth Edition- 2019

International Organization for Standardization (ISO)
Chemin de Blandonnet 8
CP-401 1214 Vernier Geneva
Switzerland
Telephone +41 22 749 01 11
Fax +41 22 733 34 30
Email: central@iso.org

International Electrotechnical Commission (IEC)
3, rue de Varembé
P.O. Box 131
CH - 1211 GENEVA 20
Switzerland
Phone: ++41 22 919 02 11
Fax: +41 22 919 0300
E-info@iec.ch

ISO/IEC Joint Technical Committee 1 (JTC 1) Secretariat
c/o ANSI
25 West 43rd Street, 4th Floor
New York, NY 10036
USA
Phone: +1 212 642 4932
Fax: +1 212 840 2298
Email: lrajchel@ansi.org

0 Introduction (*Consolidated JTC 1 Supplement*)

0.1 What is the *Consolidated JTC 1 Supplement*?

The ISO/IEC Directives Part 1 define the basic procedures to be followed in the development of International Standards and other publications. This *Consolidated JTC 1 Supplement* contains the procedures specific to JTC 1.

Part 1 of the ISO/IEC Directives, together with this *Consolidated JTC 1 Supplement*, provides procedural rules to be followed by ISO/IEC JTC 1. There are, however, other documents which provide further guidance, such as JTC 1 Standing Documents (SD). Forms unique to JTC 1 are found in the JTC 1 Templates folder at

http://isotc.iso.org/livelink/livelink?func=ll&objId=8913214&objAction=browse&sort=name.

0.2 Relationship of the *Consolidated JTC 1 Supplement* to ISO/IEC Directives Part 1

This edition of the *Consolidated JTC 1 Supplement* incorporates the fifteenth edition of the ISO/IEC Directives Part 1, as published in 2019, along with procedures specific to JTC 1.

0.3 The structure of the *Consolidated JTC 1 Supplement*

The clause structure of the *Consolidated JTC 1 Supplement* follows that of Part 1 of the ISO/IEC Directives.

0.4 Obtaining the *Consolidated JTC 1 Supplement*

The ISO/IEC Directives Parts 1 and 2, the Consolidated ISO Supplement, the Consolidated JTC 1 Supplement, and other related documents, are available via https://www.iso.org/directives-and-policies.html. The JTC 1 Standing Documents are available via www.jtc1.org. The IEC Supplement is available via http://www.iec.ch/members_experts/refdocs/iec/isoiecdir-iecsup%7Bed10.0%7Den.pdf

0.5 Contact information for the *Consolidated JTC 1 Supplement*

Comments or questions on the *Consolidated JTC 1 Supplement* should be referred to:

| International Organization for Standardization (ISO)
Chemin de Blandonnet 8
CP 401 - 1214 Vernier, Geneva, Switzerland
Tel.: +41 22 749 01 11
Fax: +41 22 733 34 30
E-mail: central@iso.org | International Electrotechnical Commission (IEC)
3, rue de Varembé
P.O. Box 131
CH - 1211 GENEVA 20
Switzerland
Phone: ++41 22 919 02 11
Fax: +41 22 919 0300
E-info@iec.ch | ISO/IEC Joint Technical Committee 1 (JTC 1) Secretariat c/o ANSI
25 West 43rd Street, 4th Floor
New York, NY 10036
USA
Phone: +1 212 642 4932
Fax: +1 212 840 2298
Email: lrajchel@ansi.org |

Foreword

The ISO/IEC Directives are published in two parts:

— Part 1: Procedures for the technical work

— Part 2: Principles and rules for the structure and drafting of ISO and IEC documents

Furthermore, the International Organization for Standardization (ISO), the International Electrotechnical Committee (IEC) and ISO/IEC Joint Technical Committee (JTC) 1 have published independent supplements to Part 1, which include procedures that are not common.

This part sets out the procedures to be followed within ISO and the IEC in carrying out their technical work: primarily the development and maintenance of International Standards through the activities of technical committees and their subsidiary bodies.

ISO, IEC and ISO/IEC JTC 1 provide additional guidance and tools to all those concerned with the preparation of technical documents on their respective websites (http://www.iso.org/directives; http://www.iec.ch/members_experts/refdocs/ and http://www.jtc1.org).

This ~~fourteenth~~ fifteenth edition of the ISO/IEC Directives Part 1, incorporates all the changes agreed by the respective technical management boards since publication of the ~~thirteenth fourteenth~~ edition in ~~2017~~2018. Procedures which are not common to the ISO/IEC Directives are published separately in the ISO Supplement (also referred to as the Consolidated ISO Supplement), the IEC Supplement or the ISO/IEC JTC 1 Supplement (also referred to as the *Consolidated JTC 1 Supplement*), respectively. The Supplements are to be used in conjunction with this document.

The following clauses have been modified with respect to the previous edition: Foreword, 1.7.3, 1.12.2, 1.13.2, 1.13.6, 1.17.2, 3.2.1, 3.2.2, 3.2.4, Annex B, Annex D, Annex E, Annex F, Annex I, and Annex M. ~~1.5.12, 1.7.5, 1.8.2, 1.12.1, 1.12.4, 1.12.6, 1.14, 1.15.3, 1.15.4, 1.16.2, 1.17, 2.1.2, 2.1.5, 2.2.3, 2.3.6, 2.4.4, 2.5.6, 2.7.1, 2.7.2, 2.7.3, 2.7.5, 2.13, 3.2.2, 4.2.1, 4.3, Annex E.1 and E.2, and Annex F~~ 2.21. The track changes version of this ~~fourteenth~~ fifteenth edition should be consulted for the details of the changes made.

These procedures have been established by ISO and IEC in recognition of the need for International Standards to be cost-effective and timely, as well as widely recognized and generally applied. In order to attain these objectives, the procedures are based on the following concepts.

a) **Current technology and project management**

Within the framework of these procedures, the work may be accelerated and the task of experts and secretariats facilitated both by current technology (e.g. I.T. tools) and project management methods.

b) **Consensus**

Consensus, which requires the resolution of substantial objections, is an essential procedural principle and a necessary condition for the preparation of International Standards that will be accepted and widely used. Although it is necessary for the technical work to progress speedily, sufficient time is required before the approval stage for the discussion, negotiation and resolution of significant technical disagreements.

For further details on the principle of "consensus", see 2.5.6.

c) **Discipline**

National Bodies need to ensure discipline with respect to deadlines and timetables in order to avoid long and uncertain periods of "dead time". Similarly, to avoid re-discussion, National Bodies have the responsibility of ensuring that their technical standpoint is established taking account of all interests concerned at national level, and that this standpoint is made clear at an early stage of the work rather than, for example, at the final (approval) stage. Moreover, National Bodies need to recognize that substantial comments tabled at meetings are counter-productive, since no opportunity is available for other delegations to carry out the necessary consultations at home, without which rapid achievement of consensus will be difficult.

d) **Cost-effectiveness**

These procedures take account of the total cost of the operation. The concept of "total cost" includes direct expenditure by National Bodies, expenditure by the offices in Geneva (funded mainly by the dues of National Bodies), travel costs and the value of the time spent by experts in working groups and committees, at both national and international level.

e) **General principles for voting and decisions**

The committee secretariat shall ensure that all decisions taken by the committee, whether at plenary meeting or by correspondence, are recorded and in ISO, by committee resolutions.

In JTC 1, for votes by correspondence, or during a committee meeting, a simple majority of the P-members voting is required for approval unless otherwise specified in the ISO/IEC Directives Part 1.

In JTC 1, the committee leadership shall ensure that votes submitted in writing, in advance of a committee meeting, are considered at the meeting.

In JTC 1, when a document is out for ballot (NP, CD or any later stage), formal discussion during meetings, or distribution of National Body positions via formal committee distribution channels are prohibited.

In JTC 1, all votes, abstentions are not counted.

In JTC 1, proxy voting is not permitted.

f) **Global relevance of ISO International Standards**

In JTC 1, the ISO procedures for twinning are supported. It is ISO's aim and expectation that each of its International Standards represents a worldwide consensus and responds to global market needs. In order to achieve this aim, it has been recognized that special measures are needed in particular to ensure that the needs of developing countries are taken into account in ISO's technical work. One such measure is the inclusion of specific provisions for "twinning", i.e. partnerships between two National Bodies for the purpose of capacity building, between developed and developing countries, in this ISO JTC 1 Supplement to the ISO/IEC Directives Part 1. (See 1.7, 1.8.1, 1.8.3, 1.9.2, 1.9.3, and 1.9.4 and Annex ST of the Consolidated ISO Supplement..)

Whilst these provisions are necessarily limited to the technical work, "twinning" may occur at multiple levels, in particular to assist the twinned partner in capacity building in developing countries of their standardization, conformity assessment and IT infrastructures, with the aim of the twinned partner ~~their~~ ultimately being self-sufficient in carrying out their activities.

g) Terminology used in this document

NOTE 1   Wherever appropriate in this document, for the sake of brevity the following terminology has been adopted to represent similar or identical concepts within ISO and IEC.

Term	ISO	IEC
National Body	Member Body (MB)	National Committee (NC)
technical management board (TMB)	Technical Management Board (ISO/TMB)	Standardization Management Board (SMB)
Chief Executive Officer (CEO)	Secretary-General	General Secretary
office of the CEO	Central Secretariat (CS)	Central Office (CO)
council board	Council	Council Board (CB)
advisory group	Technical Advisory Group (TAG)	Advisory Committee
Secretary (of a committee or subcommittee)	Committee manager	secretary
For other concepts, ISO/IEC Guide 2 applies.		

In JTC 1, JTC 1 National Bodies are National Bodies that are members of JTC 1, both P-members and O-members.   In JTC 1, the "office of the CEO" is the Information Technology Task Force (ITTF). In this *Consolidated JTC 1 Supplement*, singular terms, such as "technical management board" refer to both the ISO and IEC entities. For example, the use of the term "Chief Executive Officer (CEO)" should be understood to include both the ISO Secretary-General and the IEC General Secretary.

In JTC 1, the term committee manager is used. The name change applies to the TC and SC levels only. Secretary still applies at the Working Group level.

~~In JTC 1, the acronyms commonly used in the *Consolidated JTC 1* Supplement and Standing Documents are listed in JTC 1 Standing Document 18 on Acronyms.~~

For this ~~2018~~ 2019 *Consolidated JTC 1 Supplement*, the following clauses have been substantively modified with respect to the previous edition, in addition to those clauses referenced above:. Reference to a JTC 1 specific procedure has been deleted from 1.13.2, 1.14, 2.1.3.3, 2.1.6, 2.1.9, 2.4, 2.5 and Annex JC. 1.6.1, 1.6.5, 1.8.2, 1.12.1, 1.8.3, 1.12.4, 1.14, 2.4, 2.4.1, 2.4.4, 2.6.4 and 5.1.2. Format changes have been made to 1.17.4 and 2.9.   Twinning is mentioned only once in 1.7. Text has been added for alignment to the Foreword, 2.5.3, 2.7.3, 2.9.2.1, 2.9.5, Annex L and Annex M.

The track changes version of this ~~2018~~ 2019 edition should be consulted for the details of the changes made.

NOTE 2   In addition the following abbreviations are used in this document.

JTAB	Joint Technical Advisory Board
JCG	Joint Coordination Group
JPC	Joint Project Committee
JTC	Joint Technical Committee
JWG	joint working group
TC	technical committee
SC	Subcommittee
PC	project committee
WG	working group
PWI	preliminary work item
NP	new work item proposal
WD	working draft
CD	committee draft
DIS	draft International Standard (ISO)
CDV	committee draft for vote (IEC)
FDIS	final draft International Standard
PAS	Publicly Available Specification
TS	Technical Specification
TR	Technical Report
AMD	amendment
COR	corrigendum
DCOR	draft corrigendum
DR	defect report
SD	standing document
HoD	Head of Delegation
NWI	new work item

1 Organizational structure and responsibilities for the technical work

1.1 Role of the technical management board

The technical management board of the respective organization is responsible for the overall management of the technical work and in particular for:

a) establishment of technical committees;

b) appointment of chairs of technical committees;

c) allocation or re-allocation of secretariats of technical committees and, in some cases, subcommittees; In JTC 1, JTC 1 shall decide on the allocation of the secretariat of a subcommittee in all cases;

d) approval of titles, scopes and programmes of work of technical committees;

e) ratification of the establishment and dissolution of subcommittees by technical committees;

f) allocation of priorities, if necessary, to particular items of technical work;

g) coordination of the technical work, including assignment of responsibility for the development of standards regarding subjects of interest to several technical committees, or needing coordinated development; to assist it in this task, the technical management board may establish advisory groups of experts in the relevant fields to advise it on matters of basic, sectoral and cross-sectoral coordination, coherent planning and the need for new work;

h) monitoring the progress of the technical work with the assistance of the office of the CEO, and taking appropriate action;

i) reviewing the need for, and planning of, work in new fields of technology;

j) maintenance of the ISO/IEC Directives and other rules for the technical work;

k) consideration of matters of principle raised by National Bodies, and of appeals concerning decisions on new work item proposals, on committee drafts, on enquiry drafts or on final draft International Standards.

NOTE 1   Explanations of the terms "new work item proposal", "committee draft", "enquiry draft" and "final draft International Standard" are given in Clause 2.

NOTE 2   For detailed information about the role and responsibilities of the ISO technical management board, see the Terms of reference of the TMB – http://www.iso.org/iso/home/standards_development/list_of_iso_technical_committees/iso_technical_committee.htm?commid=4882545 and for the IEC see http://www.iec.ch/dyn/www/f?p=103:47:0::::FSP_ORG_ID,FSP_LANG_ID:3228,25.

1.2 Advisory groups to the technical management board

1.2.1 A group having advisory functions in the sense of 1.1 g) may be established

a) by one of the technical management boards;

b) jointly by the two technical management boards.

NOTE   In IEC certain such groups are designated as Advisory Committees.

1.2.2 A proposal to establish such a group shall include recommendations regarding its terms of reference and constitution, bearing in mind the requirement for sufficient representation of affected interests while at the same time limiting its size as far as possible in order to ensure its efficient operation. For example, it may be decided that its members be only the chairs and secretaries of the technical committees concerned. In every case, the TMB(s) shall decide the criteria to be applied and shall appoint the members.

Any changes proposed by the group to its terms of reference, composition or, where appropriate, working methods shall be submitted to the technical management boards for approval.

1.2.3 The tasks allocated to such a group may include the making of proposals relating to the drafting or harmonization of publications (in particular International Standards, Technical Specifications, Publicly Available Specifications and Technical Reports), but shall not include the preparation of such documents unless specifically authorized by the TMB(s).

1.2.4 Any document being prepared with a view to publication shall be developed in accordance with the procedural principles given in Annex A.

1.2.5 The results of such a group shall be presented in the form of recommendations to the TMB(s). The recommendations may include proposals for the establishment of a working group (see 1.12) or a joint working group (see 1.12.6) for the preparation of publications. Such working groups shall operate within the relevant technical committee, if any.

1.2.6 The internal documents of a group having advisory functions shall be distributed to its members only, with a copy to the office(s) of the CEO(s).

1.2.7 Such a group shall be disbanded once its specified tasks have been completed, or if it is subsequently decided that its work can be accomplished by normal liaison mechanisms (see 1.16).

1.3 Joint technical work

1.3.1 Joint Technical Advisory Board (JTAB)

The JTAB has the task of avoiding or eliminating possible or actual overlapping in the technical work of ISO and IEC and acts when one of the two organizations feels a need for joint planning. The JTAB deals only with those cases that it has not been possible to resolve at lower levels by existing procedures. (See Annex B.) Such cases may cover questions of planning and procedures as well as technical work.

Decisions of the JTAB are communicated to both organizations for immediate implementation. They shall not be subject to appeal for at least 3 years.

1.3.2 Joint Technical Committees (JTC) and Joint Project Committees (JPC)

1.3.2.1 JTC and JPC may be established by a common decision of the ISO/TMB and IEC/SMB, or by a decision of the JTAB.

1.3.2.2 For JPC, one organization has the administrative responsibility. This shall be decided by mutual agreement between the two organizations.

Participation is based on the one member/country, one vote principle.

Where two National Bodies in the same country elect to participate in a JPC then one shall be identified as having the administrative responsibility. The National Body with the administrative responsibility has the responsibility of coordinating activities in their country, including the circulation of documents, commenting and voting.

Otherwise the normal procedures for project committees are followed (see 1.10).

1.4 Role of the Chief Executive Officer

The Chief Executive Officer of the respective organization is responsible, *inter alia*, for implementing the ISO/IEC Directives and other rules for the technical work. For this purpose, the office of the CEO arranges all contacts between the technical committees, the council board and the technical management board.

Deviations from the procedures set out in the present document shall not be made without the authorization of the Chief Executive Officers of ISO or IEC, or of the ISO/IEC Joint Technical Advisory Board (JTAB), or the technical management boards for deviations in the respective organizations

In JTC 1, the CEOs are represented by the Information Technology Task Force (ITTF).

1.5 Establishment of technical committees

1.5.1 Technical committees are established and dissolved by the technical management board.

1.5.2 The technical management board may transform an existing subcommittee into a new technical committee, following consultation with the technical committee concerned.

1.5.3 A proposal for work in a new field of technical activity which appears to require the establishment of a new technical committee may be made in the respective organization by

— a National Body;

— a technical committee or subcommittee;

— a project committee;

— a policy level committee;

— the technical management board;

— the Chief Executive Officer;

— a body responsible for managing a certification system operating under the auspices of the organization;

— another international organization with National Body membership.

1.5.4 The proposal shall be made using the appropriate form and these are available in electronic format, (typically MS Word), for download from www.iso.org/forms and http://www.iec.ch/standardsdev/resources/docpreparation/forms_templates/, which covers:

a) the proposer;

b) the subject proposed;

c) the scope of the work envisaged and the proposed initial programme of work;

d) a justification for the proposal;

e) if applicable, a survey of similar work undertaken in other bodies;

f) any liaisons deemed necessary with other bodies.

For additional informational details to be included in the proposals for new work, see Annex C.

The form shall be submitted to the office of the CEO.

1.5.5 The office of the CEO shall ensure that the proposal is properly developed in accordance with ISO and IEC requirements (see Annex C), and provides sufficient information to support informed decision making by National Bodies. The office of the CEO shall also assess the relationship of the proposal to existing work, and may consult interested parties, including the technical management board or committees conducting related existing work. If necessary, an ad hoc group may be established to examine the proposal.

Following its review, the office of the CEO may decide to return the proposal to the proposer for further development before circulation for voting. In this case, the proposer shall make the changes suggested or provide justification for not making the changes. If the proposer does not make the changes and requests that its proposal be circulated for voting as originally presented, the technical management board will decide on appropriate action. This could include blocking the proposal until the changes are made or accepting that it be balloted as received.

In all cases, the office of the CEO may also include comments and recommendations to the proposal form.

For details relating to justification of the proposal, see Annex C.

Proposers are strongly encouraged to conduct informal consultations with other National Bodies in the preparation of proposals.

1.5.6 The proposal shall be circulated by the office of the CEO to all National Bodies of the respective organization (ISO or IEC), asking whether or not they

a) support the establishment of a new technical committee providing a statement justifying their decision ("justification statement"), and

b) intend to participate actively (see 1.7.1) in the work of the new technical committee.

The proposal shall also be submitted to the other organization (IEC or ISO) for comment and for agreement (see Annex B).

The replies to the proposal shall be made using the appropriate form within 12 weeks after circulation. Regarding 1.5.6 a) above, if no such statement is provided, the positive or negative vote of a National Body will not be registered and considered.

In JTC 1, the form for replies to the proposals has been replaced by an electronic balloting system. Replies not using the electronic balloting system will not be counted.

1.5.7 The technical management board evaluates the replies and either

— decides the establishment of a new technical committee, provided that

    1) a 2/3 majority of the National Bodies voting are in favour of the proposal, and

    2) at least 5 National Bodies who voted in favour expressed their intention to participate actively,

and allocates the secretariat (see 1.9.1), or

— assigns the work to an existing technical committee, subject to the same criteria of acceptance.

1.5.8 Technical committees shall be numbered in sequence in the order in which they are established. If a technical committee is dissolved, its number shall not be allocated to another technical committee.

1.5.9 As soon as possible after the decision to establish a new technical committee, the necessary liaisons shall be arranged (see 1.15 to 1.17).

1.5.10 A new technical committee shall agree on its title and scope as soon as possible after its establishment, preferably by correspondence.

The scope is a statement precisely defining the limits of the work of a technical committee.

The definition of the scope of a technical committee shall begin with the words "Standardization of ..." or "Standardization in the field of ..." and shall be drafted as concisely as possible.

For recommendations on scopes, see Annex J.

The agreed title and scope shall be submitted by the Chief Executive Officer to the technical management board for approval.

1.5.11 The technical management board or a technical committee may propose a modification of the latter's title and/or scope. The modified wording shall be established by the technical committee for approval by the technical management board.

1.6 Establishment of subcommittees

1.6.1 Subcommittees are established and dissolved by a 2/3 majority decision of the P-members of the parent committee voting, subject to ratification by the technical management board. A subcommittee may be established only on condition that a National Body has expressed its readiness to undertake the secretariat.

1.6.2 At the time of its establishment, a subcommittee shall comprise at least 5 members of the parent technical committee having expressed their intention to participate actively (see 1.7.1) in the work of the subcommittee.

1.6.3 Subcommittees of a technical committee shall be designated in sequence in the order in which they are established. If a subcommittee is dissolved, its designation shall not be allocated to another subcommittee, unless the dissolution is part of a complete restructuring of the technical committee.

1.6.4 The title and scope of a subcommittee shall be defined by the parent technical committee and shall be within the defined scope of the parent technical committee.

1.6.5 The secretariat of the parent technical committee shall inform the office of the CEO of the decision to establish a subcommittee, using the appropriate form. The office of the CEO shall submit the form to the technical management board for ratification of the decision.

1.6.6 As soon as possible after ratification of the decision to establish a new subcommittee, any liaisons deemed necessary with other bodies shall be arranged (see 1.15 to 1.17).

1.7 Participation in the work of technical committees and subcommittees

In JTC 1, the ISO procedures for twinning are supported. It is recognized that National Bodies in developing countries often lack the resources to participate in all committees which may be carrying out work which is important for their national economy. Developing country member bodies are therefore invited to establish P-member twinning arrangements with more experienced P-members from developed countries. Under such arrangements, the lead P-member shall be the developed country member who will undertake to ensure that the views of the twinned P-member are communicated to and taken into consideration by the responsible ISO committee. The twinned P-member shall consequently also have the status of P-member (see note) and be registered as a twinned P-member by the Central Secretariat.

NOTE   It is left to the National Bodies concerned to determine the most effective way of implementing twinning. This may include for example the P-member sponsoring an expert from the twinned National Body to participate in committee meetings or to act as an expert in a working group, or it may involve the P-member seeking the views of the twinned National Body on particular agenda items/documents and ensuring and conveying those comments to the committee, including casting a vote on behalf of the twinned National Body during meetings. In order to ensure the greatest possible transparency, that the twinned member body National Body should provides its positions in writing not only to the twinning partner, but also to the committee secretariat, who should verify that proxy votes cast on behalf of the twinned member are consistent with its written positions.

The details of all twinning arrangements shall be notified to the secretariat and chair of the committee concerned, with the committee members and the office of the CEO being informed accordingly to ensure the greatest possible transparency.

A lead P-member shall twin with only one other P-member in any particular committee.

The twinned P-member shall cast its own vote on all issues referred to the committee for vote by correspondence.

For more information on twinnings, see "Twinning Guidance" a link to which is provided in the Annex "Reference Documents". Annex ST of the Consolidated ISO Supplement for the Twinning Policy.

Consistent with the ISO Statutes and Rules of Procedure, correspondent and subscriber members are not eligible for P-memberships. Correspondent members of ISO may register as observers of committees but do not have the right to submit comments.

1.7.1 All National Bodies have the right to participate in the work of technical committees and subcommittees.

In JTC 1, no more than one National Body per country (either member body of ISO or National Committee of IEC) is permitted to be a member of JTC 1 and similarly only one National Body per country is permitted to be a member of a JTC 1 subcommittee.

In order to achieve maximum efficiency and the necessary discipline in the work, each National Body shall clearly indicate to the office of the CEO, with regard to each technical committee or subcommittee, if it intends

— to participate actively in the work, with an obligation to vote on all questions formally submitted for voting within the technical committee or subcommittee, on new work item proposals, enquiry drafts and final draft International Standards, and to contribute to meetings (**P-members**), or

— to follow the work as an observer, and therefore to receive committee documents and to have the right to submit comments and to attend meetings (**O-members**).

In JTC 1, National Bodies that choose to be P-members of a JTC 1 committee have the additional obligation to vote on all systematic review ballots under the responsibility of that committee.

A National Body may choose to be neither P-member nor O-member of a given committee, in which case it will have neither the rights nor the obligations indicated above with regard to the work of that committee. Nevertheless, all National Bodies irrespective of their status within a technical committee or subcommittee have the right to vote on enquiry drafts (see 2.6) and on final draft International Standards (see 2.7).

In JTC 1, there is only one vote per country.

National Bodies have the responsibility to organize their national input in an efficient and timely manner, taking account of all relevant interests at their national level.

1.7.2 Membership of a subcommittee is open to any National Body, regardless of their membership status in the parent technical committee.

Members of a technical committee shall be given the opportunity to notify their intention to become a P- or O-member of a subcommittee at the time of its establishment.

Membership of a technical committee does not imply automatic membership of a subcommittee; National Bodies shall notify their intended status in each subcommittee

1.7.3 A National Body may, at any time, begin or end membership or change its membership status in any technical committee or subcommittee in IEC by informing the office of the CEO and the secretariat of the committee concerned and in ISO by direct input via the Global Directory, subject to the requirements of clauses 1.7.4 and 1.7.5.

1.7.4 A technical committee or subcommittee secretariat shall notify the Chief Executive Officer if a P-member of that technical committee or subcommittee

— has been persistently inactive and has failed to contribute to 2 successive technical committee/subcommittee meetings, either by direct participation or by correspondence and has failed to appoint any experts to the technical work, or

— In IEC:

has failed to vote on questions formally submitted for voting within the technical committee or subcommittee (see 1.7.1).

— In ISO:

has failed to vote on over 20% (and at least 2) of the questions formally submitted for voting on the committee internal balloting (CIB) within the technical committee or subcommittee over one calendar year (see 1.7.1).

In JTC 1, the ISO policy is followed.

Upon receipt of such a notification, the Chief Executive Officer shall remind the National Body of its obligation to take an active part in the work of the technical committee or subcommittee. In the absence of a satisfactory response to this reminder, and upon persistent continuation of the above articulated shortcomings in required P-member behaviour, the National Body shall without exception automatically have its status changed to that of O-member. A National Body having its status so changed may, after a period of 12 months, indicate to the Chief Executive Officer that it wishes to regain P-membership of the committee, in which case this shall be granted.

NOTE this clause does not apply to the development of Guides.

1.7.5 If a P-member of a technical committee or subcommittee fails to vote on an enquiry draft or final draft International Standard prepared by the respective committee, or in ISO on a systematic review ballot for a deliverable under the responsibility of the committee, the Chief Executive Officer shall remind the National Body of its obligation to vote. In the absence of a satisfactory response to this reminder, the National Body shall automatically have its status changed to that of O-member. A National Body having its status so changed may, after a period of 12 months, indicate to the Chief Executive Officer that it wishes to regain P-membership of the committee, in which case this shall be granted.

NOTE this clause does not apply to the development of Guides.

In JTC 1, the ISO policy for systematic review is followed.

1.8 Chairs of technical committees and subcommittees

1.8.1 Appointment

Chairs of technical committees shall be nominated by the secretariat of the technical committee and approved by the technical management board, for a maximum period of 6 years, or for such shorter period as may be appropriate. Extensions are allowed, up to a cumulative maximum of 9 years.

Chairs of subcommittees shall be nominated by the secretariat of the subcommittee and approved by the technical committee for a maximum period of 6 years, or for such shorter period as may be appropriate. Extensions are allowed, up to a cumulative maximum of 9 years. Approval criterion for both appointment and extension is a 2/3 majority vote of the P-members of the technical committee.

Secretariats of technical committees or subcommittees may submit nominations for new chairs up to one year before the end of the term of existing chairs. Chairs appointed one year before shall be designated as the "chair elect" of the committee in question. This is intended to provide the chair elect an opportunity to learn before taking over as chair of a committee.

1.8.2 Responsibilities

The chair of a technical committee is responsible for the overall management of that technical committee, including any subcommittees and working groups.

The chair of a technical committee or subcommittee shall

a) act in a purely international capacity, divesting him- or herself of a national position; thus s/he cannot serve concurrently as the delegate of a National Body in his or her own committee;

b) guide the secretary of that technical committee or subcommittee in carrying out his or her duty;

c) conduct meetings with a view to reaching agreement on committee drafts (see 2.5);

d) ensure at meetings that all points of view expressed are adequately summed up so that they are understood by all present;

e) ensure at meetings that all decisions are clearly formulated and made available in written form by the secretary for confirmation during the meeting;

f) take appropriate decisions at the enquiry stage (see 2.6);

g) advise the technical management board on important matters relating to that technical committee via the technical committee secretariat. For this purpose, s/he shall receive reports from the chairs of any subcommittees via the subcommittee secretariats;

h) ensure that the policy and strategic decisions of the technical management board are implemented in the committee;

i) ensure the establishment and ongoing maintenance of a strategic business plan covering the activities of the technical committee and all groups reporting to the technical committee, including all subcommittees.

j) ensure the appropriate and consistent implementation and application of the committee's strategic business plan to the activities of the technical committee's or subcommittee's work programme.

k) assist in the case of an appeal against a committee decision.

In case of unforeseen unavailability of the chair at a meeting, a session chair may be elected by the participants.

SC chairs shall attend meetings of the parent committee as required and may participate in the discussion, but do not have the right to vote. In exceptional circumstances, if a chair is prevented from attending, he or she shall delegate to the secretary (or in ISO and IEC another representative) to represent the subcommittee. In the case where no representative from the SC can attend, a written report shall be provided.

1.9 Secretariats of technical committees and subcommittees

1.9.1 Allocation

The secretariat of a technical committee shall be allocated to a National Body by the technical management board.

The secretariat of a subcommittee shall be allocated to a National Body by the parent technical committee. However, if two or more National Bodies offer to undertake the secretariat of the same subcommittee, the technical management board shall decide on the allocation of the subcommittee secretariat.

JTC 1 shall decide on the allocation of the secretariat of a subcommittee in all cases.

For both technical committees and subcommittees, the secretariat shall be allocated to a National Body only if that National Body

a) has indicated its intention to participate actively in the work of that technical committee or subcommittee, and

b) has accepted that it will fulfil its responsibilities as secretariat and is in a position to ensure that adequate resources are available for secretariat work (see D.2).

Once the secretariat of a technical committee or subcommittee has been allocated to a National Body, the latter shall appoint a qualified individual as secretary (see D.1 and D.3).

1.9.2 Responsibilities

The National Body to which the secretariat has been allocated shall ensure the provision of technical and administrative services to its respective technical committee or subcommittee.

The secretariat is responsible for monitoring, reporting, and ensuring active progress of the work, and shall use its utmost endeavour to bring this work to an early and satisfactory conclusion. These tasks shall be carried out as far as possible by correspondence.

The secretariat is responsible for ensuring that the ISO/IEC Directives and the decisions of the technical management board are followed.

A secretariat shall act in a purely international capacity, divesting itself of a national point of view.

The secretariat is responsible for the following to be executed in a timely manner:

a) Working documents:

   1) Preparation of committee drafts, arranging for their distribution and the treatment of the comments received;

   2) Preparation of enquiry drafts and text for the circulation of the final draft International Standards or publication of International Standards;

   3) Ensuring the equivalence of the English and French texts, if necessary with the assistance of other National Bodies that are able and willing to take responsibility for the language

versions concerned. (See also 1.11 and the respective Supplements to the ISO/IEC Directives).

In JTC 1, texts are only required to be prepared in English, except in exceptional instances.

b) Project management

1) Assisting in the establishment of priorities and target dates for each project;

2) Notifying the names, etc. of all working group and maintenance team convenors and project leaders to the office of the CEO;

3) Proposing proactively the publication of alternative deliverables or cancellation of projects that are running significantly overtime, and/or which appear to lack sufficient support;

c) Meetings (see also Clause 4), including:

1) Establishment of the agenda and arranging for its distribution;

2) Arranging for the distribution of all documents on the agenda, including reports of working groups, and indicating all other documents which are necessary for discussion during the meeting (see E.5);

3) Regarding the decisions (also referred to as resolutions) taken in a meeting ensuring that the decisions endorsing working groups recommendations contain the specific elements being endorsed;

— making the decisions available in writing for confirmation during the meeting (see E.5); and

— posting the decisions within 48 hours after the meeting in the committee's electronic folder.

4) Preparation of the minutes of meetings to be circulated within 4 weeks after the meeting;

5) Preparation of reports to the technical management board (TC secretariat), in the IEC within 4 weeks after the meeting, or to the parent committee (SC secretariat);

In JTC 1, see also Standing Document 19 on "Meetings".

d) Advising

Providing advice to the chair, project leaders, and convenors on procedures associated with the progression of projects.

In all circumstances, each secretariat shall work in close liaison with the chair of its technical committee or subcommittee.

The secretariat of a technical committee shall maintain close contact with the office of the CEO and with the members of the technical committee regarding its activities, including those of its subcommittees and working groups.

The secretariat of a subcommittee shall maintain close contact with the secretariat of the parent technical committee and as necessary with the office of the CEO. It shall also maintain contact with the members of the subcommittee regarding its activities, including those of its working groups.

The secretariat of a technical committee or subcommittee shall update in conjunction with the office of the CEO the record of the status of the membership of the committee.

1.9.3 Change of secretariat of a technical committee

If a National Body wishes to relinquish the secretariat of a technical committee, the National Body concerned shall immediately inform the Chief Executive Officer, giving a minimum of 12 months' notice. The technical management board decides on the transfer of the secretariat to another National Body.

If the secretariat of a technical committee persistently fails to fulfil its responsibilities as set out in these procedures, the Chief Executive Officer or a National Body may have the matter placed before the technical management board, which may review the allocation of the secretariat with a view to its possible transfer to another National Body.

1.9.4 Change of secretariat of a subcommittee

If a National Body wishes to relinquish the secretariat of a subcommittee, the National Body concerned shall immediately inform the secretariat of the parent technical committee, giving a minimum of 12 months' notice.

If the secretariat of a subcommittee persistently fails to fulfil its responsibilities as set out in these procedures, the Chief Executive Officer or a National Body may have the matter placed before the parent technical committee, which may decide, by majority vote of the P-members, that the secretariat of the subcommittee should be re-allocated.

In either of the above cases an enquiry shall be made by the secretariat of the technical committee to obtain offers from other P-members of the subcommittee for undertaking the secretariat.

If two or more National Bodies offer to undertake the secretariat of the same subcommittee or if, because of the structure of the technical committee, the re-allocation of the secretariat is linked with the re-allocation of the technical committee secretariat, the technical management board decides on the re-allocation of the subcommittee secretariat. If only one offer is received, the parent technical committee itself proceeds with the appointment.

JTC 1 shall decide on the reallocation of the secretariat of a subcommittee in all cases.

1.10 Project committees

Project committees are established by the technical management board to prepare individual standards not falling within the scope of an existing technical committee.

NOTE   Such standards carry one reference number but may be subdivided into parts.

Procedures for project committees are given in Annex K.

Project committees wishing to be transformed into a technical committee shall follow the process for the establishment of a new technical committee (see 1.5).

1.11 Editing committees

It is recommended that committees establish one or more editing committees for the purpose of updating and editing committee drafts, enquiry drafts and final draft International Standards and for ensuring their conformity to the ISO/IEC Directives, Part 2 (see also 2.6.6).

Such committees should comprise at least

— one technical expert of English mother tongue and having an adequate knowledge of French;

— one technical expert of French mother tongue and having an adequate knowledge of English;

— the project leader (see 2.1.8).

The project leader and/or secretary may take direct responsibility for one of the language versions concerned.

In JTC 1, the working language is English, though a working knowledge of French may be required for certain documents. Technical expertise in French is not required unless a text in French is being developed.

Editing committees shall meet when required by the respective technical committee or subcommittee secretariat for the purpose of updating and editing drafts which have been accepted by correspondence for further processing.

Editing committees shall be equipped with means of processing and providing texts electronically (see also 2.6.6).

In JTC 1, an alternative process is used.

A project editor is assigned responsibility for the editing and updating of working drafts, committee drafts, enquiry drafts, final draft International Standards and TR and TS track documents, and for ensuring their conformity to the ISO/IEC Directives, Part 2 (see also 2.6.6).

A project editor should be identified as early as possible for each standard or other document under development. The project editor is appointed by the subcommittee and shall follow the editing instructions given by the entity responsible for the project.

It is the responsibility of the project editor to maintain the document throughout the stages of technical work, i.e. until publication. The Foreword of the final text of the deliverable shall indicate the JTC 1 subcommittee responsible for the deliverable.

After publication, the project editor should maintain an updated document incorporating all approved corrigenda (COR) and amendments (AMD) so that a revision may be published with minimum delay when appropriate. The Foreword of the revision shall list all amendments and corrigenda incorporated therein.

JTC 1 or its subgroups may establish editing groups to assist the project editor in ensuring the best possible editorial presentation of drafts in conformity with the ISO/IEC Directives, Part 2. An editing group works under the responsibility of the secretariat of JTC 1 or the subgroup that established it.

A project editor shall act in a purely international capacity, divesting him or herself of a national point of view.

Responsibility for any changes of project editors rests with the committee and not with the JTC 1 National Body (or liaison organization).

1.12 Working groups

1.12.1 Technical committees or subcommittees may establish, by decision of the committee, working groups for specific tasks (see 2.4). A working group shall report to its parent technical committee or subcommittee through a convenor appointed by the parent committee.

Working group Convenors shall be appointed by the committee for up to three-year terms ending at the next plenary session of the parent committee following the term. Such appointments shall be confirmed by the National Body (or liaison organization). The convenor may be reappointed for additional terms of up to three-years. There is no limit to the number of terms.

Responsibility for any changes of convenors rests with the committee and not with the National Body (or liaison organization).

The convenor may be supported by a secretariat, as needed. In JTC 1, the nomination of the working group secretary shall be confirmed by his/her National Body.

A working group comprises a restricted number of experts individually appointed by the P-members, A-liaisons of the parent committee and C-liaison organizations, brought together to deal with the specific task allocated to the working group. The experts act in a personal capacity and not as the official representative of the P-member or A-liaison organization (see 1.17) by which they have been appointed with the exception of those appointed by C-liaison organizations (see 1.17). However, it is recommended that they keep close contact with that P-member or organization in order to inform them about the progress of the work and of the various opinions in the working group at the earliest possible stage.

It is recommended that working groups be reasonably limited in size. The technical committee or subcommittee may therefore decide upon the maximum number of experts appointed by each P-member and liaison organization.

Once the decision to set up a working group has been taken, P-members and A- and C-liaison organizations shall be officially informed in order to appoint expert(s). Working groups shall be numbered in sequence in the order in which they are established.

When a committee has decided to set up a working group, the convenor or acting convenor shall immediately be appointed and shall arrange for the first meeting of the working group to be held within 12 weeks. This information shall be communicated immediately after the committee's decision to the P-members of the committee and A- and C-liaison organizations, with an invitation to appoint experts within 6 weeks. Additional projects may be assigned, where appropriate, to existing working groups.

In JTC 1, the parent body shall assign responsibility for the administration of a working group to a convenor, if necessary supported by a secretariat. Any secretariat shall be provided by either a P-member of the parent committee or an organization endorsed by the P-member of the parent committee.. The P-member of the parent committee shall confirm in writing its consent to the arrangement before it can be effected. Convenorships of all WG's shall be for a nominal three-year term ending at the next plenary session of the parent body following the three-year term. The Convenor may be reappointed for additional three-year terms.

1.12.2 The composition of the working group is defined in the ISO Global Directory (GD) or in the IEC Expert Management System (EMS) as appropriate. Experts not registered to a working group in the ISO GD or the IEC EMS respectively, ~~may~~ shall not participate in its work. Convenors may invite a specific guest to participate in a single meeting.

1.12.3 Persistently inactive experts, meaning absence of contributions through attendance to working group meetings or by correspondence shall be removed, by the office of the CEO at the request of the technical committee or sub-committee secretary, from working groups after consultation with the P-member.

1.12.4 On completion of its task(s) — normally at the end of the enquiry stage (see 2.6) of its last project — the working group shall be disbanded by decision of the committee the project leader remaining with consultant status until completion of the publication stage (see 2.8).

1.12.5 Distribution of the internal documents of a working group and of its reports shall be carried out in accordance with procedures described in the respective Supplements of the ISO/IEC Directives.

1.12.6 In special cases a joint working group (JWG) may be established to undertake a specific task in which more than one ISO and/or IEC technical committee or subcommittee is interested. Committees who receive requests to establish JWG shall reply to such requests in a timely manner.

NOTE For specific rules concerning JWGs between ISO committees and IEC committees, see Annex B in addition to the following.

The decision to establish a joint working group shall be accompanied by mutual agreement between the committees on:

— the committee/ organization having the administrative responsibility for the project;

— the convenor of the joint working group, who shall be nominated by a P-member from one of the committees, with the option to appoint a co-convenor from the other committee;

— the membership of the joint working group (membership may be open to all P-members and Category A, and C-liaisons that wish to participate which may be limited to an equal number of representatives from each committee, if agreed.

The committee/organization with the administrative responsibility for the project shall:

— record the project in their programme of work;

— In JTC 1, conduct the call for experts in all committees that are part of the JWG;

— be responsible for addressing comments (usually referred back to the JWG) and ensure that the comments and votes at all stages of the project are compiled and handled appropriately (see 2.5, 2.6 and 2.7) – all comments are made available to the leadership of the committees;

— prepare drafts for the committee, enquiry and approval stages according to procedures given in 2.5, 2.6 and 2.7;

— In JTC 1, send all relevant documents (minutes, working drafts, drafts for the committee, enquiry and approval stages) to the secretariat of the other committee(s) for circulation in their respective committee and/or action;

— be responsible for maintenance of the publication.

Approval criteria are based on the Directives used by the committee with the administrative lead. If the lead committee is a JTC 1 committee, the Consolidated JTC 1 Supplement also applies.

For proposal stage (NP):

- For ISO/ISO JWGs, only one NP ballot is needed – if a NP has already been launched or approved in one committee, it cannot be balloted again in another TC, but two (2) NP are launched for ISO/IEC JWG [one (1) in each organization].

- it is possible to establish a JWG at a later stage, in which case its administrative lead will be confirmed by the TCs concerned.

- once the joint work is agreed, the committee with the administrative lead informs ISO/CS or IEC/CO respectively, of its lead and of the committees participating in the work.

- the other TCs launch a call for experts for participation in the JWG.

For preparatory stage (WD)

- The JWG functions like any other WG: consensus is required to advance to CD.

For committee stage (CD)

- The CD is circulated for review and comment by each committee.

- The final CD requires consensus by all committees, as defined in the ISO/IEC Directives, Part 1

For DIS and FDIS ballots

- National Bodies are requested to consult all national mirror committees involved to define one position. A statement is included on the cover page to draw attention of NSBs.

- For an ISO/IEC JWG, two DIS/FDIS votes are launched, i.e. one in each organization. For an ISO and ISO/IEC JTC 1 JWG, one DIS/FDIS vote is launched

The Foreword identifies all committees involved in the development of the deliverable.

1.13 Groups having advisory functions within a committee

In JTC 1, Standing Document 10 "Advisory and Ad hoc Groups" provides additional information regarding the establishment of advisory groups.

1.13.1 A group having advisory functions may be established by a technical committee or subcommittee to assist the chair and secretariat in tasks concerning coordination, planning and steering of the committee's work or other specific tasks of an advisory nature.

1.13.2 A proposal to establish such a group shall include recommendations regarding its constitution and terms of reference, including criteria for membership, bearing in mind the requirement for sufficient representation of affected interests while at the same time limiting its size as far as possible in order to ensure its efficient operation. Members of advisory groups shall be committee officers, individuals be nominated by National Bodies and/or, as relevant, by A-liaison organizations. The parent committee shall approve the final constitution and the terms of reference prior to the establishment of and nominations to the advisory group.

For chair's advisory groups, consideration shall be given to the provisions of equitable participation.

In JTC 1, advisory groups may decide to invite liaison organizations and external experts to participate.

1.13.3 The tasks allocated to such a group may include the making of proposals relating to the drafting or harmonization of publications (in particular International Standards, Technical Specifications, Publicly Available Specifications and Technical Reports), but shall not include the preparation of such documents.

1.13.4 The results of such a group shall be presented in the form of recommendations to the body that established the group. The recommendations may include proposals for the establishment of a working group (see 1.12) or a joint working group (see 1.12.6) for the preparation of publications.

1.13.5 The internal documents of a group having advisory functions shall be distributed to its members only, with a copy to the secretariat of the committee concerned and to the office of the CEO.

1.13.6 Such a group shall be disbanded once its specified tasks have been completed and agreed by the parent committee.

1.14 Ad hoc groups

In JTC 1, Standing Document 10 on "Advisory and Ad hoc Groups" provides additional information regarding the establishment of ad hoc groups.

Technical committees or subcommittees may establish ad hoc groups, the purpose of which is to study a precisely defined problem on which the group reports to its parent committee at the same meeting, or at the latest at the next meeting.

In JTC 1, working groups may also create ad hoc groups. However, as O-members cannot participate in working groups, they also cannot participate in ad hoc groups of working groups.

The membership of an ad hoc group shall be chosen from the delegates present at the meeting of the parent committee, supplemented, if necessary, by experts appointed by the committee. The parent committee shall also appoint a convenor.

In JTC 1, the membership of ad hoc groups may be extended to experts or delegates not present at the meeting where the ad hoc group was formed (e.g. additional P-members of the parent committee or liaison organization experts or delegates). Ad hoc groups may be comprised of experts or delegates depending on the structure of the group.

An ad hoc group shall be automatically disbanded at the meeting to which it has presented its report.

1.15 Liaison between technical committees

In JTC 1, see Standing Document 15 on "Liaisons" for additional requirements.

1.15.1 Within each organization, technical committees and/or subcommittees working in related fields shall establish and maintain liaison. Liaisons shall also be established, where appropriate, with technical committees responsible for basic aspects of standardization (e.g. terminology, graphical symbols). Liaison shall include the exchange of basic documents, including new work item proposals and working drafts.

==In JTC 1, committees may pass a resolution to decide on the establishment of an internal liaison. Committees receiving requests for internal liaisons cannot refuse such requests and there is no need for the committee receiving the request to pass a resolution confirming its acceptance.==

1.15.2 The maintenance of such liaison is the responsibility of the respective technical committee secretariats, which may delegate the task to the secretariats of the subcommittees.

1.15.3 A technical committee or subcommittee may designate a Liaison Representative or Liaison Representatives to follow the work of another technical committee with which a liaison has been established, or one or several of its subcommittees. The designation of such Liaison Representatives shall be notified to the secretariat of the committee concerned, which shall communicate all relevant documents to the Liaison Representative(s) and to the secretariat of that technical committee or subcommittee. The appointed Liaison Representative shall make progress reports to the secretariat by which s/he has been appointed.

1.15.4 Such Liaison Representatives shall have the right to participate in the meetings of the technical committee or subcommittee whose work they have been designated to follow but shall not have the right to vote. They may contribute to the discussion in meetings, including the submission of written comments, on matters within the competence of their own technical committee and based on feedback that they have collected from their own committee. They may also attend meetings of working groups of the technical committee or subcommittee, but only to contribute the viewpoint of their own technical committee on matters within its competence, and not to otherwise participate in working group activities.

1.16 Liaison between ISO and IEC

1.16.1 Arrangements for adequate liaison between ISO and IEC technical committees and subcommittees are essential. The channel of correspondence for the establishment of liaison between ISO and IEC technical committees and subcommittees is through the offices of the CEOs. As far as the study of new subjects by either organization is concerned, the CEOs seek agreement between the two organizations whenever a new or revised programme of work is contemplated in the one organization which may be of interest to the other, so that the work will go forward without overlap or duplication of effort. (See also Annex B.)

1.16.2 Liaison Representatives designated by ISO or IEC shall have the right to participate in the discussions of the other organization's technical committee or subcommittee whose work they have been designated to follow, and may submit written comments; they shall not have the right to vote.

1.17 Liaison with other organizations

1.17.1 General requirements applicable to all categories of liaisons

In order to be effective, liaison shall operate in both directions, with suitable reciprocal arrangements.

The desirability of liaison shall be taken into account at an early stage of the work.

The liaison organization shall accept the policy based on the ISO/IEC Directives concerning copyright (see 02.13), whether owned by the liaison organization or by other parties. The statement on copyright policy will be provided to the liaison organization with an invitation to make an explicit statement as to its acceptability. The liaison organization is not entitled to charge a fee for documents submitted.

A liaison organization shall be willing to make a contribution to the technical work of ISO or IEC as appropriate. A liaison organization shall have a sufficient degree of representativity within its defined area of competence within a sector or subsector of the relevant technical or industrial field.

A liaison organization shall agree to ISO/IEC procedures, including IPR (see 02.13).

Liaison organizations shall accept the requirements of 2.14 on patent rights.

Technical committees and subcommittees shall review all their liaison arrangements on a regular basis, at least every 2 years, or at every committee meeting.

In JTC 1, and its subgroups, liaison relationships shall be reviewed annually.

1.17.2 Different categories of liaisons

In JTC 1, see also Standing Document 15 on "Liaisons".

1.17.2.1 At the technical committee/subcommittee level (Category A and B liaisons)

The categories of liaisons at the technical committee/subcommittee levels are:

> **Category A:** Organizations that make an effective contribution to the work of the technical committee or subcommittee for questions dealt with by this technical committee or subcommittee. Such organizations are given access to all relevant documentation and are invited to meetings. They may nominate experts to participate in a WG (see 1.12.1).

> **Category B:** Organizations that have indicated a wish to be kept informed of the work of the technical committee or subcommittee. Such organizations are given access to reports on the work of a technical committee or subcommittee.

NOTE　Category B is reserved for inter-governmental organizations.

The procedure for the establishment of Category A and B liaisons is:

- The organization wishing to create a Category A or B liaison shall send an application to the office of the CEO with copies to the technical committee or subcommittee officers and the IEC CO Technical Officer and the ISO CS Technical Program Manager giving the following information:

  - Organization is not-for-profit;

  - Organization is open to members worldwide or over a broad region;

  - Its activities and membership demonstrate that it has the competence and expertise to contribute to the development of International Standards or the authority to promote their implementation in the area of the technical committee or subcommittee concerned (Only relevant for category A liaisons);

- The name of the main contact person.

Note: Invariably the organization will have been in contact with the technical committee or subcommittee officers prior to submitting its application and in these cases the technical committee or subcommittee officers should ensure that the organization is aware of their obligations as given in clause 1.17.1 i.e., copyright, agreeing to the ISO/IEC procedures including IPR, and patent rights.

- The Office of the COE will confirm that the eligibility criteria have been fulfilled and then consult with the National Body where the organization making the application has its headquarters;

- In case of objections from the National Body where the organization making the application has its headquarters, the matter will be referred to the technical management board for decision;

- If there is no objection from the National Body where the organization making the application has its headquarters, the application will be sent to the technical committee or subcommittee secretary with a request to circulate it for vote;

- Approval criterion for category A or B liaisons is a 2/3rds majority of P-members voting to approve.

1.17.2.2 At the working group level (Category C liaisons)

The category of liaisons at the working group level (and in JTC 1 also the project level) is:

> **Category C**: Organizations that make a technical contribution to and participate actively in the work of a working group. This can include manufacturer associations, commercial associations, industrial consortia, user groups and professional and scientific societies. Liaison organizations shall be multinational (in their objectives and standards development activities) with individual, company or country membership and may be permanent or transient in nature.

1.17.3 Eligibility

1.17.3.1 At the technical committee/subcommittee level (Category A and B liaisons)

When an organization applies for a liaison with an ISO technical committee / subcommittee, the office of the CEO will check with the member body in the country in which the organization is located. If the member body does not agree that the eligibility criteria have been met, the matter will be referred to the TMB to define the eligibility.

The office of the CEO will also ensure that the organization meets the following eligibility criteria:

- it is not-for-profit;

- is a legal entity – the office of the CEO will request a copy of its statutes;

- it is membership-based and open to members worldwide or over a broad region;

- through its activities and membership demonstrates that it has the competence and expertise to contribute to the development of International Standards or the authority to promote their implementation; and

- has a process for stakeholder engagement and consensus decision-making to develop the input it provides (in ISO, see Guidance for ISO liaison organizations - Engaging stakeholders and building consensus http://www.iso.org/iso/guidance_liaison-organizations.pdf).

1.17.3.2 At the working group level (Category C liaisons)

When an organization applies for a liaison with a working group, the office of the CEO will check with the member body in the country in which the organization is located and will ensure that the organization meets the following eligibility criteria:

- it is not-for-profit;

- through its activities and membership demonstrates that it has the competence and expertise to contribute to the development of International Standards or the authority to promote their implementation; and

- has a process for stakeholder engagement and consensus decision-making to develop the input it provides (in ISO, see Guidance for ISO liaison organizations - Engaging stakeholders and building consensus http://www.iso.org/iso/guidance_liaison-organizations.pdf).

1.17.4 Acceptance (Category A, B and C liaisons)

Agreement to establish Category A, B and C liaisons requires approval of the application by two-thirds of the P-members voting.

Committees are urged to seek out the participation of all parties at the beginning of the development of a work item. Where a request for category C liaison is submitted late in the development stage of a particular work item, the P-members will consider the value that can be added by the organization in question despite its late involvement in the working group.

In JTC 1, Category C liaisons are proposed by JTC 1 to the ITTF, after receiving a recommendation from the appropriate JTC 1 subsidiary body, i.e., an SC (or a WG reporting directly to JTC 1). Each request for liaison status forwarded to JTC 1, from an appropriate JTC 1 subsidiary body shall contain a statement of expected benefits and responsibilities accepted by both the JTC 1 organization and the organization requesting liaison status.

In JTC 1, the ITTF shall reaffirm the liaison status of the organization if there is continued evidence of active participation in the work of the WG or project and appropriate National Body participation exists. If a request for liaison is considered by JTC 1 in the first instance, and Category C liaison is thought to be applicable, JTC 1 may request the appropriate JTC 1 subsidiary body or bodies to consider the request and apply the above procedure.

1.17.5 Rights and obligations

1.17.5.1 At the technical committee/subcommittee level (Category A and B liaisons)

Technical committees and subcommittees shall seek the full and, if possible, formal backing of the organizations having liaison status for each document in which the latter is interested. Any comments from liaison organizations should be given the same treatment as comments from member bodies. It should not be assumed that refusal by a liaison organization to provide its full backing is a sustained opposition. Where such objections are considered sustained oppositions, committees are invited to refer to clause 2.5.6 for further guidance.

1.17.5.2 At the working group level (Category C liaisons)

Category C liaison organizations have the right to participate as full members in a working group, maintenance team or project team (see 1.12.1) but not as project leaders or convenors. Category C liaison experts act as the official representative of the organization by which they are appointed. They may only attend committee plenary meetings if expressly invited by the committee to attend. If they are invited by the committee to attend, they may only attend as observers.

In JTC 1, representatives shall have the right to participate in the meetings of the subcommittee or working group whose work they have been designated to follow but shall not have the right to vote. They may contribute to the discussion in meetings, including the submission of written contributions, on matters within the competence of their organization.

In JTC 1, JTC 1 will work towards eliminating barriers to accessing or participating in JTC 1 activities and its body of work, especially for people with disabilities and older users.

1.17.6 Carrying over liaisons when a project committee is converted into a technical committee or a subcommittee

When a project committee is converted to a technical committee or a subcommittee, the new technical committee or subcommittee shall pass a resolution confirming which category A and B liaisons are carried over. Approval of the resolution requires a 2/3 majority of P-members voting.

Table 1 – Liaison categories

Category	A	B	C
Purpose	To make an effective contribution to the work of the committee.	To be kept informed of the work of the committee.	To make a technical contribution to drafting standards in a Working Group.
Eligibility	• Not for profit • Legal entity • Membership based (worldwide or over a broad region) • Relevant competence and expertise • Process for stakeholder engagement and consensus decision-making (See clause 1.17.3.1 for full details)	<u>Intergovernmental Organizations only</u> • Not for profit • Legal entity • Membership based (worldwide or over a broad region) • Relevant competence and expertise • Process for stakeholder engagement and consensus decision-making (See clause 1.17.3.1 for full details)	• Not for profit • Relevant competence and expertise • Process for stakeholder engagement and consensus decision-making (See clause 1.17.3.2 for full details)
Level	TC/SC	TC/SC	Working Group
Participation	Participate in TC/SC meetings, access to documents, may appoint experts to WGs and these experts may serve	To be kept informed of the work only (access to documents).	Full participation as a member of the WG (but cannot be convenor or Project Leader).

	as convenors or Project Leaders.		
Rights and obligations	No voting rights, but can comment (comments are given the same treatment as comments from member bodies). Can propose new work items (see clause 2.3.2).	No voting rights, but can comment (comments are given the same treatment as comments from member bodies). Cannot propose new work items.	Experts can attend committee meetings if expressly invited by the committee, but only as observers. Cannot propose new work items.

1.17.7 Category A Liaison with ITU-T

In JTC 1, a unique Category A liaison with the ITU-T is maintained. See Annex JB and the JTC 1 Standing Document 3 on "Guide for ITU-T and ISO/IEC JTC 1 Cooperation".

1.17.7.1 Liaison with ITU-T

All contributions to ITU-T should be subject to ITU-T Recommendations A.1 and A.2, and other ITU-T requirements as may be imposed. Specifically,

- each contribution should identify which, if any, prior contributions it supersedes;
- each contribution should be addressed to only one study group. However, other study groups which may be interested in the contribution may also be identified.

1.17.7.2 Collaborative Relationship with ITU-T

Two modes of collaboration with ITU-T are defined in Standing Document 3 "Guide for ITU-T and ISO/IEC JTC 1 Cooperation" collaborative interchange and collaborative team. A JTC 1 SC, in agreement with the corresponding ITU-T study group, may establish either of these two modes of collaboration as appropriate. JTC 1 shall make considered decisions when it comes to collaboration with ITU-T, evaluating each proposed project on a case-by-case basis.

JTC 1 shall consider at least the following criteria for each proposal:

1. Taking account of scarce technical resources;
2. Taking account of the JTC 1 scope;
3. Maximizing the efficiency of the standards development process;
4. Enhancing time to market of standards implementations;
5. Considering the impact of possible duplicative standards, and
6. Recognizing collaborative work with ITU-T in the specific area of technology related to the proposal.

When collaboration is planned from the onset of a new work item (NWI), the rationale (such as recognition that expertise missing in the JTC 1 SC is present in an ITU-T study group with applicable scope of work) and terms of reference for the collaborative project shall be included in the NP documentation, ensuring wide visibility of this proposed collaboration within JTC 1.

When collaboration is considered after the start of a JTC 1 project, any addition of a collaborative project can be considered a modification of the SC's Program of Work and treated as prescribed by the JTC 1 Consolidated Supplement by a default ballot (see 2.1.5.7 and JA 1.4). The rational and proposed terms of reference for the collaborative project shall accompany the default ballot.

Procedures for the operation of the two modes of collaboration are defined in Standing Document 3 "Guide for ITU-T and ISO/IEC JTC 1 Cooperation". These procedures deal primarily with the synchronisation of approval actions by JTC 1 and ITU-T and are intended to supplement, not modify JTC 1 approval requirements.

2 Development of International Standards

2.1 The project approach

2.1.1 General

The primary duty of a technical committee or subcommittee is the development and maintenance of International Standards. However, committees are also strongly encouraged to consider publication of intermediate deliverables as described in Clause 3.

International Standards shall be developed on the basis of a project approach as described below.

2.1.2 Strategic business plan

Each technical committee shall prepare a strategic business plan for its own specific field of activity,

a) taking into account the business environment in which it is developing its work programme;

b) indicating those areas of the work programme which are expanding, those which have been completed, and those nearing completion or in steady progress, and those which have not progressed and should be cancelled (see also 2.1.9);

c) evaluating revision work needed (see also the respective Supplements to the ISO/IEC Directives);

d) giving a prospective view on emerging needs.

The strategic business plan shall be formally agreed upon by the technical committee and be included in its report for review and approval by the technical management board on a regular basis.

2.1.3 Project stages

2.1.3.1 Table 2 shows the sequence of project stages through which the technical work is developed, and gives the name of the document associated with each project stage. The development of Technical Specifications, Technical Reports and Publicly Available Specifications is described in Clause 3. In JTC 1, the JTC 1 PAS (Publicly Available Specification) Transposition process is a different process from the one that results in PAS deliverables in ISO and IEC (see Annex F).

Table 2 — Project stages and associated documents

Project stage	Associated document	
	Name	Abbreviation
Preliminary stage	Preliminary work item	PWI
Proposal stage	New work item proposal [a]	NP
Preparatory stage	Working draft(s) [a]	WD
Committee stage	Committee draft(s) [a]	CD
Enquiry stage	Enquiry draft [b]	ISO/DIS IEC/CDV
Approval stage	final draft International Standard [c]	FDIS
Publication stage	International Standard	ISO, IEC or ISO/IEC

[a] These stages may be omitted as described in Annex F.
[b] Draft International Standard in ISO, committee draft for vote in IEC. In JTC 1, the enquiry draft is the DIS.
[c] May be omitted (see 2.6.4).

2.1.3.2   F.1 illustrates the steps leading to publication of an International Standard.

2.1.3.3 The ISO and IEC Supplements to the ISO/IEC Directives give a matrix presentation of the project stages, with a numerical designation of associated sub-stages. In JTC 1, Annex JD is used. In JTC 1, Standing Document 11 on "Progression of JTC 1 Projects" clause 1.1, Table 1 provide JTC 1 specific requirements for timeframes.

2.1.4 Project description and acceptance

A project is any work intended to lead to the issue of a new, amended or revised International Standard. A project may subsequently be subdivided (see also 2.1.5.4).

A project shall be undertaken only if a proposal has been accepted in accordance with the relevant procedures (see 2.3 for proposals for new work items, and the respective Supplements to the ISO/IEC Directives for review and maintenance of existing International Standards).

2.1.5 Programme of work

2.1.5.1 The programme of work of a technical committee or subcommittee comprises all projects allocated to that technical committee or subcommittee, including maintenance of published standards.

2.1.5.2 In establishing its programme of work, each technical committee or subcommittee shall consider sectoral planning requirements as well as requests for International Standards initiated by sources outside the technical committee, i.e. other technical committees, advisory groups of the technical management board, policy level committees and organizations outside ISO and IEC. (See also 2.1.2.)

2.1.5.3 Projects shall be within the agreed scope of the technical committee. Their selection shall be subject to close scrutiny in accordance with the policy objectives and resources of ISO and IEC. (See also Annex C.)

2.1.5.4 Each project in the programme of work shall be given a number (see IEC Supplements to the ISO/IEC Directives for document numbering at the IEC) and shall be retained in the programme of work under that number until the work on that project is completed or its cancellation has been agreed upon. The technical committee or subcommittee may subdivide a number if it is subsequently found necessary to subdivide the project itself. The subdivisions of the work shall lie fully within the scope of the original project; otherwise, a new work item proposal shall be made.

In JTC 1, to avoid undue delays in authorizing subdivisions of projects or minor enhancements of existing work, where the changes are not outside the scope of the original item, the subcommittee may proceed with such work if approved by a vote of its P-members. The change(s), however, shall be submitted to JTC 1 for endorsement and, if JTC 1 does not approve, the work shall cease.

2.1.5.5 The programme of work shall indicate, if appropriate, the subcommittee and/or working group to which each project is allocated.

2.1.5.6 The agreed programme of work of a new technical committee shall be submitted to the technical management board for approval.

> 2.1.5.7 In JTC 1, following its plenary meeting, a subcommittee shall submit to the JTC 1 secretariat as a single document the subcommittee's modified programme of work, including all proposed subdivisions of projects and minor enhancements of existing work, exclusive of proposals for new work. This document shall be considered using the Default Ballot process (see 1.4 of Annex JA on Voting).

2.1.6 Target dates

The technical committee or subcommittee shall establish, for each project on its programme of work, target dates for the completion of each of the following steps:

— completion of the first working draft (in the event that only an outline of a working document has been provided by the proposer of the new work item proposal – see 2.3);

— circulation of the first committee draft;

— circulation of the enquiry draft;

— circulation of the final draft International Standard (in agreement with the office of the CEO);

— publication of the International Standard (in agreement with the office of the CEO).

> Note: In JTC 1, the Final Draft International Standard (FDIS) shall be skipped if no technical changes are to be included.

These target dates shall correspond to the shortest possible development times, to produce International Standards rapidly and shall be reported to the office of the CEO, which distributes the information to all National Bodies. For establishment of target dates, see the respective Supplements to the ISO/IEC Directives.

In establishing target dates, the relationships between projects shall be taken into account. Priority shall be given to those projects intended to lead to International Standards upon which other International Standards will depend for their implementation. The highest priority shall be given to those projects having a significant effect on international trade and recognized as such by the technical management board.

The technical management board may also instruct the secretariat of the technical committee or subcommittee concerned to submit the latest available draft to the office of the CEO for publication as a Technical Specification (see 3.1).

All target dates shall be kept under continuous review and amended as necessary, and shall be clearly indicated in the programme of work. Revised target dates shall be notified to the technical management board. The technical management board will cancel all work items which have been on the work programme for more than 5 years and have not reached the approval stage (see 2.7).

~~In JTC 1, Standing Document 11 on "Progression of JTC 1 Projects" clause 1.1, provide JTC 1 specific requirements for timeframes.~~

2.1.6.1 General

In JTC 1, when a proposed new project is approved (whether for a new deliverable or for the revision of an existing deliverable), when submitting the results to the ISO Central Secretariat the committee secretariat shall also indicate the selected standards development track, as follows (all target dates are calculated from the date of adoption as an approved project, AWI (approved work item), stage 10.99):

NOTE   The deadlines for the various stages within the development tracks shall be established on a case-by-case basis.

SDT 18 standards development track — 18 months to publication

SDT 24 standards development track — 24 months to publication

SDT 36 standards development track — 36 months to publication

SDT 48 standards development track — 48 months to publication

NOTE   Projects using the 18-month development track shall be eligible for the 'Direct publication process' offered by ISO/CS if they successfully complete the DIS ballot within 13 months of the project's registration. This process reduces publication processing time by approximately one third.

Committee secretariats are reminded to perform risk assessments during project planning in order to identify potential problems in advance and set the target dates accordingly. The target dates shall be kept under continuous review by committee secretariats which shall ensure that they are reviewed and either confirmed or revised at each committee meeting. Such reviews shall also seek to confirm that projects are still market relevant and in cases in which they are found to be no longer required, or if the likely completion date is going to be too late, thus causing market players to adopt an alternative solution, the projects shall be cancelled or transformed into another deliverable (see 2.1.6.2).

NOTE   Time spent on round-robin testing during the development of a standard shall not be counted in the overall development time. The standards development track is paused on request from the secretariat to ISO/CS during round-robin testing in accordance with 2.6.4.

2.1.6.2 Automatic cancellation of projects (and their reinstatement)

In JTC 1, if the target date for DIS (stage 40.00) or publication (stage 60.60) is exceeded, the committee shall decide within 6 months on one of the following actions:

a) projects at the preparatory or committee stages: submission of a DIS - if the technical content is acceptable and mature;

b) projects at the enquiry stage: submission of a second DIS or FDIS - if the technical content is acceptable and mature;

c) publication of a TS - if the technical content is acceptable but unlikely sufficiently mature for a future International Standard;

d) publication of a TR - if the technical content is not considered to be acceptable for publication as a TS or for a future International Standard but is nevertheless considered to be of interest to the public;

e) submission of a request for extension to JTC 1 and ISO/TMB and IEC/SMB - if no consensus can be reached but there is strong interest from stakeholders to continue - a committee may be granted one extension of up to 9 months for the total project duration but the publication of intermediary deliverables (such as TS) is recommended;

f) cancellation of the work item - if the committee is unable to find a solution.

If, at the end of the six-month period, none of the above actions has been taken, the project shall be automatically cancelled by the ITTF. Such projects may only be reinstated with the approval of the ISO Technical Management Board and the IEC Standardization Management Board.

2.1.7 Project management

The secretariat of the technical committee or subcommittee is responsible for the management of all projects in the programme of work of that technical committee or subcommittee, including monitoring of their progress against the agreed target dates.

If target dates (see 2.1.6) are not met and there is insufficient support for the work (that is, the acceptance requirements for new work given in 2.3.5 are no longer met), the committee responsible shall cancel the work item.

2.1.8 Project leader

For the development of each project, a project leader (the WG convenor, a designated expert or, if appropriate, the secretary) shall be appointed by the technical committee or subcommittee, taking into account the project leader nomination made by the proposer of the new work item proposal (see 2.3.4). It shall be ascertained that the project leader will have access to appropriate resources for carrying out the development work. The project leader shall act in a purely international capacity, divesting him- or herself of a national point of view. The project leader should be prepared to act as consultant, when required, regarding technical matters arising at the proposal stage through to the publication stage (see 2.5 to 2.8).

The secretariat shall communicate the name and address of the project leader, with identification of the project concerned, to the office of the CEO.

In JTC 1, there are no project leaders. Working groups are led by a convenor, and projects shall be assigned project editors.

2.1.9 Progress control

~~In JTC 1, Standing Document 11 on "Progression of JTC 1 Projects" clause 2.3 provides JTC 1 specific reporting requirements.~~

Periodical progress reports to the technical committee shall be made by its subcommittees and working groups (see also ISO and IEC Supplements to the ISO/IEC Directives). Meetings between their secretariats will assist in controlling the progress.

The office of the CEO shall monitor the progress of all work and shall report periodically to the technical management board. For this purpose, the office of the CEO shall receive copies of documents as indicated in the ISO and IEC Supplements to the ISO/IEC Directives.

2.2 Preliminary stage

2.2.1 Technical committees or subcommittees may introduce into their work programmes, by a simple majority vote of their P-members, preliminary work items (for example, corresponding to subjects dealing with emerging technologies), which are not yet sufficiently mature for processing to further stages and for which no target dates can be established.

Such items may include, for example, those listed in the strategic business plan, particularly as given under 2.1.2 d) giving a prospective view on emerging needs.

2.2.2 All preliminary work items shall be registered into the programme of work.

2.2.3 All preliminary work items shall be subject to regular review by the committee. The committee shall evaluate the market relevance and resources required for all such items.

All preliminary work items that have not progressed to the proposal stage in the IEC by the expiration date given by the TC/SC, and in ISO within 3 years will be automatically cancelled.

2.2.4 This stage can be used for the elaboration of a new work item proposal (see 2.3) and the development of an initial draft.

2.2.5 Before progressing to the preparatory stage, all such items shall be subject to approval in accordance with the procedures described in 2.3.

2.3 Proposal stage

In the case of proposals to prepare management system deliverables, see Annex L~~J~~C.

2.3.1 ~~In JTC 1, a~~A new work item proposal (NP) is a proposal for:

- a new standard;
- a new part of an existing standard
- a Technical Specification (see 3.1)
-

~~a new standard;~~

~~a new part of an existing standard or Technical Specification (see also 2.1.5.4 for subdivision)~~

~~a new Technical Specification (see 3.1);~~

In JTC 1, the NP ~~ballot~~ stage (clause 2.3) is not required for:

- the revision or amendment of an existing standard or Technical Specification,

- the conversion of a TS to an IS

However, the committee shall pass a resolution containing the following elements:

1) target dates;
2) confirmation that the scope will not be expanded; and
3) project editor(s) if already assigned.

The committee shall also launch a call for experts (Form 4 is not required).

For the conversion of a TS to an IS, a two-thirds majority resolution is required.

If the revision or amendment results in an expanded scope, an NP ballot shall be initiated. and Form 4 is required.

2.3.2   A new work item proposal within the scope of an existing technical committee or subcommittee may be made in the respective organization by

— a National Body;

— the secretariat of that technical committee or subcommittee;

— another technical committee or subcommittee;

— an organization in category A liaison;

Note: in JTC 1, only JTC 1 Category A Liaisons;

— the technical management board or one of its advisory groups;

— the Chief Executive Officer.

2.3.3   Where both an ISO and an IEC technical committee are concerned, the Chief Executive Officers shall arrange for the necessary coordination. (See also Annex B.)

2.3.4   Each new work item proposal shall be presented using the appropriate form, and shall be fully justified and properly documented (see Annex C).

The proposers of the new work item proposal shall

— make every effort to provide a first working draft for discussion, or shall at least provide an outline of such a working draft;

— nominate a project leader. In JTC 1, there are no project leaders. In JTC1 the proposer should nominate a "project editor" if the New Work Item Proposal (NP) is to be allocated to an existing WG.

— In JTC 1, provide relevant details regarding any accessibility issues in the "supplemental information related to the proposal" field under "other" on the Form 4: New Work Item Proposal

The form shall be submitted to the office of the CEO or to the secretariat of the relevant committee for proposals within the scope of an existing committee.

The office of the CEO or the relevant committee chair and secretariat shall ensure that the proposal is properly developed in accordance with ISO and IEC requirements (see Annex C) and provides sufficient information to support informed decision making by National Bodies.

The office of the CEO or the relevant committee chair and secretariat shall also assess the relationship of the proposal to existing work, and may consult interested parties, including the technical management board or committees conducting related existing work. If necessary, an ad hoc group may be established to examine the proposal. Any review of proposals should not exceed 2 weeks.

In all cases, the office of the CEO or the relevant committee chair and secretariat may also add comments and recommendations to the proposal form. See Annex K for new work item proposals for project committees.

Copies of the completed form shall be circulated to the members of the technical committee or subcommittee for P-member ballot and to the O-members and liaison members for information.

The proposed date of availability of the publication shall be indicated on the form.

A decision upon a new work item proposal shall be taken by correspondence.

Votes shall be returned within 12 weeks. The committee may decide on a case-by-case basis by way of a resolution to shorten the voting period for new work item proposals to 8 weeks.

When completing the ballot form, National Bodies shall provide a statement justifying their decision for negative votes ("justification statement"). If no such statement is provided, the negative vote of a National Body will not be registered and considered.

2.3.5 Acceptance requires

a) approval of the work item by a 2/3 majority of the P-members of the technical committees or subcommittees voting – abstentions are excluded when the votes are counted; and

b) a commitment to participate actively in the development of the project, i.e. to make an effective contribution at the preparatory stage, by nominating technical experts and by commenting on working drafts, by at least 4 P-members in committees with 16 or less P-members, and at least 5 P-members in committees with 17 or more P-members;

Only P-members having also approved the inclusion of the work item in the programme of work [see a)] will be taken into account when making this tally. If experts are not nominated on the form accompanying an approval vote, then the National Body's commitment to active participation will not be registered and considered when determining if the approval criteria have been met on this ballot.

In JTC 1, if in the context of an NP, a P-member does not provide a clear justification statement for why it voted "no", the committee secretariat should go back to the P-member and give it two (2) weeks to provide an explanation.

If the P-member does not provide a response within that 2-week period, the vote will not be counted in the result.

Secretariats shall not make value judgments about the justification and shall ask the P-member in case of doubt.

If P-members do not name an expert in the Form, they have two (2) weeks following the result of the vote to name their expert. If this delay is not respected, the P-member's participation will not be counted, thereby affecting the approval requirement for (b) above.

Individual committees may increase this minimum requirement of nominated experts.

In cases, where it can be documented that the industry and/or technical knowledge exists only with a very small number of P-members, then the committee may request permission from the technical management board to proceed with fewer than 4 or 5 nominated technical experts.

In JTC 1, additional voting rules apply; see Annex JA.1 and JA.2.

2.3.6 Once a new work item proposal is accepted, it shall be registered in the programme of work of the relevant technical committee or subcommittee as a new project with the appropriate priority. The agreed target dates (see 2.1.6) shall be indicated on the appropriate form.

The voting results will be reported to the ISO Central Secretariat (using Form 6) or the IEC Central Office (using Form RVN) within 4 weeks after the close of the ballot.

2.3.7 The inclusion of the project in the programme of work concludes the proposal stage.

2.4 Preparatory Stage

In JTC 1, specific JTC 1 requirements for preparatory stage are contained in Standing Document 11 on "Progression of JTC 1 Projects" Clause 3 "Preparatory Stage Considerations".

2.4.1 The preparatory stage covers the preparation of a working draft (WD) conforming to the ISO/IEC Directives, Part 2.

2.4.2 When a new project is accepted the project leader shall work with the experts nominated by the P-members during the approval (see 2.3.5a). In JTC 1, there are no project leaders.

2.4.3 The secretariat may propose to the technical committee or subcommittee, either at a meeting or by correspondence, to create a working group the convenor of which will normally be the project leader.

Such a working group shall be set up by the technical committee or subcommittee, which shall define the task(s) and set the target date(s) for submission of draft(s) to the technical committee or subcommittee (see also 1.12). The working group convenor shall ensure that the work undertaken remains within the scope of the balloted work item.

2.4.4 In responding to the proposal to set up a working group those P-members having agreed to participate actively (see 2.3.5a)) shall each confirm their technical expert(s). Other P-members or A- or C- liaison organizations may also nominate expert(s).

2.4.5 The project leader is responsible for the development of the project and will normally convene and chair any meetings of the working group. S/he may invite a member of the working group to act as its secretary.

In JTC 1, a project editor should be identified as there are no project leaders (see 2.1.8). The working group develops one or more working drafts of the standard. Usually, a working draft undergoes several revisions before the working group recommends that it will be progressed to the Committee Stage. As decisions are made regarding the content of the working draft, the convenor should take care to assure consensus.

2.4.6 Every possible effort shall be made to prepare both a French and an English version of the text in order to avoid delays in the later stages of the development of the project.

If a trilingual (English — French — Russian) standard is to be prepared, this provision should include the Russian version.

In JTC 1, texts are only required to be prepared in English, except in exceptional instances.

2.4.7 For time limits relating to this stage, see 2.1.6.

2.4.8 The preparatory stage ends when a working draft is available for circulation to the members of the technical committee or subcommittee as a first committee draft (CD) and is registered by the office of the CEO. The committee may also decide to publish the final working draft as a PAS (see 3.2) to respond particular market needs.

2.5 Committee stage

In JTC 1, specific JTC 1 requirements for committee stage are contained in Standing Document 11 Progression of JTC 1 Projects" clause 4 "Committee Stage Considerations".

2.5.1 The committee stage is the principal stage at which comments from National Bodies are taken into consideration, with a view to reaching consensus on the technical content. National Bodies shall therefore carefully study the texts of committee drafts and submit all pertinent comments at this stage.

In JTC 1, any graphical symbol shall be submitted to the relevant ISO committee and/or IEC committee (as applicable) responsible for the registration of graphical symbols (see Annex JE).

2.5.2 As soon as it is available, a committee draft shall be circulated to all P-members and O-members of the technical committee or subcommittee for consideration, with a clear indication of the latest date for submission of replies. In JTC 1, organizations in liaison are asked to submit their comments.

A period of 8, 12 or 16 weeks as agreed by the technical committee or subcommittee shall be available for National Bodies to comment.

In JTC 1, the default for CD/CPDAM/PDTS/PDTR circulation is 8 weeks.

Comments shall be sent for preparation of the compilation of comments, in accordance with the instructions given.

National bodies shall fully brief their delegates on the national position before meetings.

2.5.3 No more than 4 weeks after the closing date for submission of replies, the secretariat shall prepare the compilation of comments and arrange for its circulation to all P-members and O-members of the technical committee or subcommittee. When preparing this compilation, the secretariat shall indicate its proposal, made in consultation with the chair of the technical committee or subcommittee and, if necessary, the project leader, for proceeding with the project, either

a) to discuss the committee draft and comments at the next meeting, or

b) to circulate a revised committee draft for consideration, or

c) to register the committee draft for the enquiry stage (see 2.6).

In the case of b) and c), the secretariat shall indicate in the compilation of comments the action taken on each of the comments received. This shall be made available to all P-members, if necessary by the circulation of a revised compilation of comments, no later than in parallel with the submission of a revised CD for consideration by the committee (case b) or simultaneously with the submission of the finalized version of the draft to the office of the CEO for registration for the enquiry stage (case c).

Committees are required to respond to all comments received. If, within 8 weeks from the date of dispatch, 2 or more P-members disagree with proposal b) or c) of the secretariat, the committee draft shall be discussed at a meeting (see 4.2.1.3).

In JTC 1, responsibility for the preparation of a revised CD text, disposition of comments report, and a recommendation on further processing may be delegated to a WG, ad hoc group, or Project Editor who reports back to the parent committee.

2.5.4 If a committee draft is considered at a meeting but agreement on it is not reached on that occasion, a further committee draft incorporating decisions taken at the meeting shall be distributed within 12 weeks for consideration. A period of 8, 12 or 16 weeks as agreed by the technical committee or subcommittee shall be available to National Bodies to comment on the draft and on any subsequent versions.

In JTC 1, the default for CD/CPDAM/PDTS/PDTR circulation is 8 weeks.

2.5.5 Consideration of successive drafts shall continue until consensus of the P-members of the technical committee or subcommittee has been obtained or a decision to abandon or defer the project has been made.

2.5.6 The decision to circulate an enquiry draft (see 2.6.1) shall be taken on the basis of the consensus principle.

It is the responsibility of the chair of the technical committee or subcommittee, in consultation with the secretary of his/her committee and, if necessary, the project leader, to judge whether there is sufficient support bearing in mind the definition of consensus given in ISO/IEC Guide 2:2004.

> "**consensus**: General agreement, characterized by the absence of sustained opposition to substantial issues by any important part of the concerned interests and by a process that involves seeking to take into account the views of all parties concerned and to reconcile any conflicting arguments.
>
> NOTE   Consensus need not imply unanimity."

The following applies to the definition of consensus:

In the process of reaching consensus, many different points of views will be expressed and addressed as the document evolves. However, "sustained oppositions" are views expressed at minuted meetings of committee, working group (WG) or other groups (e.g. task forces, advisory groups, etc.) and which are maintained by an important part of the concerned interest and which are incompatible with the committee consensus. The notion of "concerned interest(s)" will vary depending on the dynamics of the committee and shall therefore be determined by the committee leadership on a case by case basis. The concept of sustained opposition is not applicable in the context of member body votes on CD, DIS or FDIS since these are subject to the applicable voting rules.

Those expressing sustained oppositions have a right to be heard and the following approach is recommended when a sustained opposition is declared:

— The leadership shall first assess whether the opposition can be considered a "sustained opposition", i.e. whether it has been sustained by an important part of the concerned interest. If this is not the case, the leadership will register the opposition (i.e. in the minutes, records, etc.) and continue to lead the work on the document.

— If the leadership determines that there is a sustained opposition, it is required to try and resolve it in good faith. However, a sustained opposition is not akin to a right to veto. The obligation to address the sustained oppositions does not imply an obligation to successfully resolve them.

The responsibility for assessing whether or not consensus has been reached rests entirely with the leadership. This includes assessing whether there is sustained opposition or whether any sustained opposition can be resolved without compromising the existing level of consensus on the rest of the document. In such cases, the leadership will register the opposition and continue the work.

Those parties with sustained oppositions may avail themselves of appeals mechanisms as detailed in Clause 5.

In case of doubt concerning consensus, approval by a two-thirds majority of the P-members of the technical committee or subcommittee voting may be deemed to be sufficient for the committee draft to be accepted for registration as an enquiry draft; however, every attempt shall be made to resolve negative votes.

In JTC 1, abstentions are excluded when the votes are counted, as well as negative votes not accompanied to technical reasons.

The secretariat of the technical committee or subcommittee responsible for the committee draft shall ensure that the enquiry draft fully embodies decisions taken either at meetings or by correspondence.

2.5.7 When consensus has been reached in a technical committee or subcommittee, its secretariat shall submit the finalized version of the draft in electronic form suitable for distribution to the national members for enquiry (2.6.1), to the office of the CEO (with a copy to the technical committee secretariat in the case of a subcommittee) within a maximum of 16 weeks.

2.5.8 For time limits relating to this stage, see 2.1.6.

2.5.9 The committee stage ends when all technical issues have been resolved and a committee draft is accepted for circulation as an enquiry draft and is registered by the office of the CEO.

Texts that do not conform to the ISO/IEC Directives, Part 2 shall be returned to the secretariat with a request for correction before they are registered.

2.5.10 If the technical issues cannot all be resolved within the appropriate time limits, technical committees and subcommittees may wish to consider publishing an intermediate deliverable in the form of a Technical Specification (see 3.1) pending agreement on an International Standard.

2.6 Enquiry stage

2.6.1 At the enquiry stage, the enquiry draft (DIS in ISO, CDV in IEC) shall be circulated by the office of the CEO to all National Bodies for a 12-week vote. In JTC 1, the enquiry draft is a DIS or DAM. In JTC 1, the DIS or DAM shall be circulated for a 12-week vote, following a translation period of 8 weeks.

For policy on the use of languages, see Annex E. In JTC 1, texts are only required to be prepared in English, except in exceptional instances.

National bodies shall be advised of the date by which completed ballots are to be received by the office of the CEO.

At the end of the voting period, the Chief Executive Officer shall send within 4 weeks to the chair and secretariat of the technical committee or subcommittee the results of the voting together with any comments received, for further speedy action.

2.6.2 Votes submitted by National Bodies shall be explicit: positive, negative, or abstention.

A positive vote may be accompanied by editorial or technical comments, on the understanding that the secretary, in consultation with the chair of the technical committee or subcommittee and project leader, will decide how to deal with them.

If a National Body finds an enquiry draft unacceptable, it shall vote negatively and state the technical reasons. It may indicate that the acceptance of specified technical modifications will change its negative vote to one of approval, but it shall not cast an affirmative vote which is conditional on the acceptance of modifications.

In JTC 1, in the case where a National Body has voted negatively without submitting a justification, the vote will not be counted.

In JTC 1, in the case where a National Body has voted negatively and has submitted comments that are not clearly of a technical nature, the committee ~~secretary~~ manager shall contact the ~~ISO/CS~~ Technical Programme Manager within 2 weeks of the ballot closure.

In JTC 1, there are no constraints on the types of comments (technical, editorial, or general) National Bodies can submit with their votes; however, in the case of negative votes on enquiry drafts, National Bodies are required to describe their technical reasons.

2.6.3 An enquiry draft is approved if

a) a two-thirds majority of the votes cast by the P-members of the technical committee or subcommittee are in favour, and

b) not more than one-quarter of the total number of votes cast are negative.

Abstentions are excluded when the votes are counted, as well as negative votes not accompanied by technical reasons.

Comments received after the normal voting period are submitted to the technical committee or subcommittee secretariat for consideration at the time of the next review of the International Standard.

In JTC 1, additional voting rules apply; see Annex JA.1

2.6.4 On receipt of the results of the voting and any comments, the chair of the technical committee or subcommittee, in cooperation with its secretariat and the project leader, and in consultation with the office of the CEO, shall take one of the following courses of action:

a) when the approval criteria of 2.6.3 are met, and no technical changes are to be included to proceed to publication (see 2.8).

b) When the approval criteria of 2.6.3 are met, but technical changes are to be included, to register the enquiry draft, as modified, as a final draft international standard.

c) when the approval criteria of 2.6.3 are not met;

   1) to circulate a revised enquiry draft for voting (see 2.6.1), or

   NOTE   A revised enquiry draft will be circulated for a voting period of 8 weeks, which may be extended up to 12 weeks at the request of one or more P-members of the committee concerned. In JTC 1, a revised enquiry draft circulation period may be extended up to 12 weeks.

   2) to circulate a revised committee draft for comments, or

   3) to discuss the enquiry draft and comments at the next meeting. In JTC 1, a comment resolution meeting may be held by teleconference or using electronic means.

2.6.5 Not later than 12 weeks after the end of the voting period, a full report shall be prepared by the secretariat of the technical committee or subcommittee and circulated by the office of the CEO to the National Bodies. The report shall

a) show the result of the voting;

b) state the decision of the chair of the technical committee or subcommittee;

c) reproduce the text of the comments received; and

d) include the observations of the secretariat of the technical committee or subcommittee on each of the comments submitted.

Every attempt shall be made to resolve negative votes.

If, within 8 weeks from the date of dispatch, two or more P-members disagree with decision 2.6.4 c)1) or 2.6.4 c)2) of the chair, the draft shall be discussed at a meeting (see 4.2.1.3).

Committees are required to respond to all comments received.

2.6.6 When the chair has taken the decision to proceed to the approval stage (see 2.7) or publication stage (see 2.8), the secretariat of the technical committee or subcommittee shall prepare, within a maximum of 16 weeks after the end of the voting period and with the assistance of its editing committee, a final text and send it to the office of the CEO for preparation and circulation of the final draft International Standard.

The secretariat shall provide the office of the CEO with the text in a revisable electronic text and also in a format which permits validation of the revisable form.

Texts that do not conform to the ISO/IEC Directives, Part 2 shall be returned to the secretariat with a request for correction before they are registered.

2.6.7 For time limits relating to this stage, see 2.1.6.

2.6.8 The enquiry stage ends with the registration, by the office of the CEO, of the text for circulation as a final draft International Standard or publication as an International Standard, in the case of 2.6.4 a) and b).

2.7 Approval stage

2.7.1 At the approval stage, the final draft International Standard (FDIS) shall be distributed by the office of the CEO within 12 weeks to all National Bodies for an 8-week vote (6 weeks in IEC).

National Bodies shall be advised of the date by which ballots are to be received by the office of the CEO.

2.7.2 Votes submitted by National Bodies shall be explicit: positive, negative, or abstention.

A National Body may submit comments on any FDIS vote.

If a National Body finds a final draft International Standard unacceptable, it shall vote negatively and state the technical reasons. It shall not cast an affirmative vote that is conditional on the acceptance of modifications.

In JTC 1, in the case where a National Body has voted negatively without submitting a justification, the vote will not be counted.

In JTC 1, in the case where a National Body has voted negatively and has submitted comments that are not clearly of a technical nature, the secretariat shall contact the Technical Programme Manager within 2 weeks of the ballot closure.

2.7.3 A final draft International Standard having been circulated for voting is approved if

a) a two-thirds majority of the votes cast by the P-members of the technical committee or subcommittee are in favour, and

b) not more than one-quarter of the total number of votes cast are negative.

Abstentions are excluded when the votes are counted, as well as negative votes not accompanied by technical reasons.

In JTC 1, additional voting rules apply; see Annex JA.1.

2.7.4 The secretariat of the technical committee or subcommittee has the responsibility of bringing any errors that may have been introduced in the preparation of the draft to the attention of the office of the CEO by the end of the voting period; further editorial or technical amendments are not acceptable at this stage.

2.7.5 All comments received will be retained for the next review and will be recorded on the voting form as "noted for future consideration". However, the Secretary along with the office of the CEO may seek to resolve obvious editorial errors. Technical changes to an approved FDIS are not allowed.

Within 2 weeks after the end of the voting period, the office of the CEO shall circulate to all National Bodies a report showing the result of voting and indicating either the formal approval by National Bodies to issue the International Standard or formal rejection of the final draft International Standard.

2.7.6 If the final draft International Standard has been approved in accordance with the conditions of 2.7.3, it shall proceed to the publication stage (see 2.8).

2.7.7 If the final draft International Standard is not approved in accordance with the conditions in 2.7.3, the document shall be referred back to the technical committee or subcommittee concerned for reconsideration in the light of the technical reasons submitted in support of the negative votes.

The committee may decide to:

— resubmit a modified draft as a committee draft, enquiry draft or, in ISO and JTC 1, final draft International Standard;

— publish a Technical Specification (see 3.1);

— cancel the project.

2.7.8 The approval stage ends with the circulation of the voting report (see 2.7.5) stating that the FDIS has been approved for publication as an International Standard, with the publication of a Technical Specification (see 3.1.1.2), or with the document being referred back to the committee.

2.8 Publication stage

2.8.1 Within 6 weeks, the office of the CEO shall correct any errors indicated by the secretariat of the technical committee or subcommittee, and print and distribute the International Standard.

In JTC 1, the ITTF shall correct any errors indicated by the secretariat of the technical committee or subcommittee and print and distribute the International Standard within 8 weeks.

2.8.2 The publication stage ends with the publication of the International Standard.

2.9 Maintenance of Deliverables

2.9.1 Overview

The procedures for the maintenance of deliverables are given in the respective Supplements to the ISO/IEC Directives.

In JTC 1, the following procedures apply.
Additional procedures for defect correction of International Standards are found in the JTC 1 Standing Document 21 on "Defect Correction of International Standards".

See Standing Document 9 on "PAS Transposition Process" for additional requirements on the maintenance of documents approved through the PAS transposition process, maintained by the PAS submitter and administered by JTC 1.

See Standing Document 6 on "Technical Specifications and Technical Reports" for additional requirements on the maintenance of technical specifications and technical reports.

2.9.2 Systematic and Committee Reviews

2.9.2.1 General Principle

In JTC 1, every International Standard and International Technical Specification published jointly by ISO and IEC for JTC 1 shall be subject to systematic review in order to determine whether it should be confirmed, revised/amended, stabilized, or withdrawn, according to Table S1.

A committee may at any time between systematic reviews pass a resolution initiating a revision or amendment of a standard.

See clause 2.3.1 for the process for initiating a revision or amendment of an existing standard.

For minor changes, e.g., updating and editorial changes that do not impact technical content, a shortened procedure called "minor revision" may be applied. This is comprised only of the proposal for a minor revision by the committee (through a resolution), approval and publication stages (see 2.7 and 2.8). Subsequent to the resolution of the responsible technical committee and consultation with the responsible Technical Programme Manager, a final draft of the revised deliverable shall be circulated for an 8 week FDIS vote, and 12 weeks in the case of Vienna Agreement documents. The Foreword of the next edition of the deliverable shall indicate that it is a minor revision and list the updates and editorial changes made.

Table S1 — Timing of systematic reviews

Deliverable	Max. elapsed time before systematic review	Max. number of times deliverable may be confirmed	Max. life
International Standard	5 years	Not limited	Not limited
Technical Specification (see 3.1.3)	3 years	Once recommended	6 years recommended
Technical Report (see 3.3.3)	Not specified	Not specified	Not limited

While Technical Reports are not subject to systematic review (see Table S1), the responsible committee is still required to perform a review at intervals that should not exceed 5 years.

A systematic review will typically be initiated in the following circumstances:

— on the initiative and as a responsibility of the secretariat of the responsible committee, typically as the result of the elapse of the specified period since publication or the last confirmation of the document, or

— a default action by the ITTF, or

— at the request of one or more National Bodies or

— upon request of the committee ~~secretary~~ manager to ITTF.

The timing of a systematic review is normally based either on the year of publication or, where a document has already been confirmed, on the year in which it was last confirmed. However, it is not necessary to wait for the maximum period to elapse before a document is reviewed.

Where the relevant SC no longer exists, responsibility for the maintenance of such a standard may be given to a JTC 1 National Body or a JTC 1 Category A Liaison.

See clause 2.10.3 and 2.3.1 for the process for initiating a revision or amendment of an existing standard.

2.9.2.2 Systematic Review Requirements

In JTC 1, the systematic review ballot period is 20 weeks.

Before the systematic review ballot of National Bodies is initiated, the committee, by vote of the P-members of the responsible committee, either at plenary or by letter ballot, develops a recommendation for the disposition of standards and technical specifications to be included along with the systematic review ballot. ITTF submits the ballot for National Body approval using the systematic review ballot process.

After the closing of the systematic review ballot, the secretariat's proposal reflecting the voting results shall be circulated to the members of the technical committee or subcommittee. No more than 6 months after the closing of the systematic review ballot the committee shall take a final decision as to whether to revise, confirm, stabilize or withdraw the standard, following which the secretariat shall submit the committee's decision to ITTF.

> NOTE Systematic review ballots are administered electronically by the ITTF. P-members of a given committee have an obligation to vote on all systematic renew ballots for deliverables under the responsibilities of that committee. All P-members of ISO and IEC are invited to respond to such reviews. The purpose of the reviews has been extended to include obtaining information when National Bodies have needed to make modifications in order to make the deliverable suitable for national adoption. Such modifications need to be considered by committees in order to determine whether they need to be taken into account to improve the global relevance of a standard.

2.9.2.3 Interpretation of systematic review ballot results

In JTC 1, typically, a decision as to the appropriate action to take following a systematic review shall be based on a simple majority of ISO and IEC P-members voting for a specific action. However, in some cases a more detailed analysis of the results may indicate that another interpretation may be more appropriate. The committee decides upon a course of action and informs ITTF of the course of action.

> NOTE It is not feasible to provide concrete rules for all cases when interpreting the ballot results due to the variety of possible responses, degrees of implementation, and the relative importance of comments

In proposing future action, due account shall be taken of the maximum possible number of confirmations and specified maximum life of the deliverable concerned (see Table S1).

Where it has been verified that a standard or a technical specification is used, that it should continue to be made available, and that no technical changes are needed, the deliverable may be confirmed. The criteria are as follows:

— the standard or technical specification has been adopted with or without change or is used in at least five countries (when this criteria is not met, the deliverable should be withdrawn); and

- a simple majority of the P-members responding vote for confirmation.

If voting results on systematic review are not definitive for conformation, revision or withdrawal, and/or a decision is based on a determination of responses, the Secretariat shall invite approval of a proposed course of action within a specified time delay for example within eight weeks.

2.9.3 Revision or Amendment

In JTC 1, a committee may at any time pass a resolution initiating a revision or amendment of a deliverable.

Where it has been verified that a document is used, that it should continue to be made available, but that technical changes are needed, a deliverable may be proposed for amendment or revision. The criteria are as follows:

— the standard or technical specification has been adopted with or without change or is used in at least five countries (when this criteria is not met, the standard should be withdrawn); and

- a simple majority of the P-members of ISO and IEC voting considers there is a need for amendment or revision.

In that case, an item may be registered as an approved work item (stage 10.99).

A call for experts shall be launched. However, there is no minimum number of active P-members of the committee required.

Where an amendment or revision is not immediately started following approval by the committee, it is recommended that the project is first registered as a preliminary work item and that the standard is registered as confirmed. When it is eventually proposed for registration at stage 10.99, reference shall be made to the results of the preceding systematic review and the committee shall pass a resolution (see clause 2.3.1 for the process for initiating a revision or amendment of an existing standard).

Where it is decided that the International Standard or Technical Specification needs to be revised or amended, it becomes a new project and shall be added to the programme of work of the committee. The steps for revision or amendment are the same as those for preparation of a new deliverable (see the Consolidated JTC 1 Supplement, clauses 2.3 to 2.8), and include the establishment of target dates for the completion of the relevant stages.

For minor changes, e.g. updating and editorial changes, which do not impact the technical content, a shortened procedure called "minor revision" may be applied. This is comprised only of the proposal for a minor revision by the committee (through a resolution), approval and publication stages (see 2.7 and 2.8). Subsequent to the resolution of the committee and consultation with the ITTF, a final draft of the revised deliverable shall be circulated for a 8-week committee vote. The foreword of the next edition of the deliverable shall indicate that it is a minor revision and list the updates and editorial changes made.

2.9.3.1 Conversion to an International Standard (Technical Specifications only)

In addition to the three basic option of confirmation, amendment or revision, or withdrawal, in the cases of the systematic review of Technical Specifications a fourth option is its conversion to an International Standard.

To initiate conversion to an International Standard, a text, up-dated as appropriate, is submitted to the normal development procedures as specified for an International Standard (see clause 2.3.1).

The conversion procedure will typically start with a DIS vote. Where changes considered to be required are judged as being significant as to require a full review in the committee prior to DIS ballot, a revised version of the document shall be submitted for review and ballot as a CD.

2.9.4 Withdrawal

In JTC 1, in the case of the proposed withdrawal of an International Standard or a Technical Specification, the National Bodies shall be informed by the ITTF of the recommendation of the committee. ITTF shall hold a systematic review ballot so that National Bodies can vote on the recommendation.

2.9.5 Stabilization

In JTC 1, a standard may be a stabilized standard. A stabilized standard has on-going validity and effectiveness; is mature; and insofar as can be determined will not require further maintenance of any sort. A standard is in stabilized status that will no longer be subject to systematic review but is retained to provide for the continued viability of existing products or servicing of equipment that is expected to have a long working life.

To be designated a stabilized standard, at least one five-year review cycle shall pass after the last modification to the standard before it can be recommended for stabilization by the responsible committee.

A committee may recommend that a standard ~~it owns~~ for which it is responsible be put in stabilized status at the time of systematic review of that standard. In each case, the recommendation shall be accompanied by a statement of rationale and will result in a default ~~normal~~ systematic review ballot, as is done in the case of a reaffirmation recommendation.

Once a standard is given "stabilized" status, it will be recorded by ITTF on a master list of stabilized standards. This record will include the date of first addition to the list and the rationale provided as above. A stabilized standard is indicated as stabilized in the ISO Catalogue listing for that standard.

While stabilized standards are not subject to systematic review, the responsible committee should periodically (intervals that should not exceed 10 years) request its P-members to review the committee's stabilized standards to determine if stabilization of the standard is still relevant.

Where a committee or a P-member of JTC 1 or a P-member of a SC becomes aware that a standard in the stabilized state is

- no longer in use; or
- its use has been superseded; or
- it is now unsafe to continue to use the standard;

A default ballot concerning revision or withdrawal of the standard is to be initiated (see JA.1.3).

Note that Technical Specifications and Technical Reports cannot be stabilized.

2.9.6 Reinstatement of Withdrawn Standards

In JTC 1, if, following withdrawal of an International Standard, a committee determines that it is still needed, it may propose that the standard be reinstated. A ballot for reinstatement of the withdrawn standard shall be issued either as a draft International Standard, or an FDIS as initiated by the committee for voting by the ISO and IEC P-members. The balloting procedures of 2.6 and 2.7 shall apply. If approved, the standard shall be published as a new edition with a new date of publication. The foreword shall explain that the standard results from the reinstatement of the previous edition.

2.10 Corrections and amendments

2.10.1 General

A published International Standard may subsequently be modified by the publication of

— a technical corrigendum (in IEC only);

In JTC 1, the option to publish a technical corrigendum also applies.

— a corrected version

— an amendment; or

— a revision (as part of the maintenance procedure in 2.9).

In JTC 1, at the publication stage, the ITTF shall decide, in consultation with the Secretariat of JTC 1 or SC, and bearing in mind both the financial consequences to the organization and the interests of users of the IS, whether to publish an amendment or a new edition of the IS, incorporating the amendment.

[NOTE Where it is foreseen that there will be frequent additions to the provisions of an IS, the possibility should be borne in mind at the outset of developing these additions as a series of parts (see 5.5.1 of ISO/IEC Directives, Part 2)]

NOTE   In case of revision a new edition of the International Standard will be issued.

2.10.2 Corrections

In JTC 1, the term 'corrections' is a generic word referring to both technical corrigenda (COR) and corrected versions. In JTC 1, there remains the option to publish either a technical corrigendum or a corrected version.

A correction is only issued to correct an error or ambiguity inadvertently introduced either in drafting or in publishing and which could lead to incorrect or unsafe application of the publication.

Corrections are not issued to update information that has become outdated since publication.

Suspected errors shall be brought to the attention of the secretariat of the technical committee or subcommittee concerned. After confirmation by the secretariat and chair, if necessary in consultation with the project leader and P-members of the technical committee or subcommittee, the secretariat shall submit to the office of the CEO a proposal for correction, with an explanation of the need to do so.

The Chief Executive Officer shall decide, in consultation with the secretariat of the technical committee or subcommittee, and bearing in mind both the financial consequences to the organization and the interests of users of the publication, whether to publish a technical corrigendum (in IEC only) and / or a corrected version of the existing edition of the publication (see also 2.10.4). The secretariat of the committee will then inform the members of the committee of the outcome.

In JTC 1, the option to publish a technical corrigendum also applies.

In general, a correction will not be issued for a publication that is older than 3 years.

In JTC 1, ITTF shall decide, in consultation with the secretariat of the technical committee or subcommittee, and bearing in mind both the financial consequences to the organization and the interests of users of the publication, whether to publish a technical corrigendum or a corrected or updated reprint of the existing edition of the publication (see also 2.10.4). In general, a technical corrigendum will not be issued for an International Standard that is older than 3 years. In JTC 1 the procedures for developing and publishing Technical Corrigenda are given in JTC 1 Standing Document 21 on "Defect Correction of International Standards".

2.10.3 Amendments

An amendment alters and/or adds to previously agreed technical provisions in an existing International Standard. An amendment is considered as a partial revision: the rest of the International Standard is not open for comments.

An amendment is normally published as a separate document, the edition of the International Standard affected remaining in use.

The procedure for developing and publishing an amendment shall be as described in 2.3 (ISO and JTC 1), or the review and maintenance procedures (see IEC Supplement) and 2.4, 2.5, 2.6 (draft amendment, DAM), 2.7 (final draft amendment, FDAM), and 2.8. JTC 1 uses the same procedures as ISO.

In JTC 1, the default for CD/PDAMCDAM/DTS/DTR circulation is 8 weeks.

At the approval stage (2.7), the Chief Executive Officer shall decide, in consultation with the secretariat of the technical committee or subcommittee, and bearing in mind both the financial consequences to the organization and the interests of users of the International Standard, whether to publish an amendment or a new edition of the International Standard, incorporating the amendment. (See also 2.10.4.)

NOTE   Where it is foreseen that there will be frequent *additions* to the provisions of an International Standard, the possibility should be borne in mind at the outset of developing these additions as a series of parts (see ISO/IEC Directives, Part 2).

2.10.4 Avoidance of proliferation of modifications

No more than 2 separate documents in the form of technical corrigenda (in IEC only) or amendments shall be published modifying a current International Standard. The development of a third such document shall result in publication of a new edition of the International Standard.

In JTC 1, the option to publish a technical corrigendum also applies.

2.11 Maintenance agencies

When a technical committee or subcommittee has developed a standard that will require frequent modification, it may decide that a maintenance agency is required. Rules concerning the designation of maintenance agencies are given in Annex G.

2.12 Registration authorities

When a technical committee or subcommittee has developed a standard that includes registration provisions, a registration authority is required. Rules concerning the designation of registration authorities are given in Annex H.

In JTC 1, see also Annex JF, Registration Authority Policy and Standing Document 16 on "Registration Authorities" for additional requirements.

2.13   Copyright

The copyright for all drafts and International Standards and other publications belongs to ISO, IEC or ISO and IEC, respectively as represented by the office of the CEO.

The content of, for example, an International Standard may originate from a number of sources, including existing national standards, articles published in scientific or trade journals, original research and development work, descriptions of commercialized products, etc. These sources may be subject to one or more rights.

In ISO and IEC, there is an understanding that original material contributed to become a part of an ISO, IEC or ISO/IEC publication can be copied and distributed within the ISO and/or IEC systems (as relevant) as part of the consensus building process, this being without prejudice to the rights of the original copyright owner to exploit the original text elsewhere. Where material is already subject to copyright, the right should be granted to ISO and/or IEC to reproduce and circulate the material. This is frequently done without recourse to a written agreement, or at most to a simple written statement of acceptance. Where contributors wish a formal signed agreement concerning copyright of any submissions they make to ISO and/or IEC, such requests shall be addressed to ISO Central Secretariat or the IEC Central Office, respectively.

Attention is drawn to the fact that the respective members of ISO and IEC have the right to adopt and re-publish any respective ISO and/or IEC standard as their national standard. Similar forms of endorsement do or may exist (for example, with regional standardization organizations).

In JTC 1, the copyright for DISs/FDISs, International Standards, DAMs/FDAMs, amendments, technical corrigenda, technical specifications, and technical reports belongs to ISO and IEC.

For those standards requiring it, a register shall be published. The copyright for the register belongs to ISO and IEC which may license the copyright to the JTC 1 Registration Authority for as long as it functions in this capacity.

2.14 Reference to patented items (see also Annex I)

2.14.1 If, in exceptional situations, technical reasons justify such a step, there is no objection in principle to preparing an International Standard in terms which include the use of items covered by patent rights – defined as patents, utility models and other statutory rights based on inventions, including any published applications for any of the foregoing – even if the terms of the standard are such that there are no alternative means of compliance.

The rules given below shall be applied.

2.14.2 If technical reasons justify the preparation of a document in terms which include the use of items covered by patent rights, the following procedures shall be complied with:

a) The proposer of a proposal for a document shall draw the attention of the committee to any patent rights of which the proposer is aware and considers to cover any item of the proposal. Any party involved in the preparation of a document shall draw the attention of the committee to any patent rights of which it becomes aware during the development of the document.

b) If the proposal is accepted on technical grounds, the proposer shall ask any holder of such identified patent rights for a statement that the holder would be willing to negotiate worldwide licences under his/her rights with applicants throughout the world on reasonable and non-discriminatory terms and conditions. Such negotiations are left to the parties concerned and are performed outside ISO and/or IEC. A record of the right holder's statement shall be placed in the registry of the ISO Central Secretariat or IEC Central Office as appropriate, and shall be referred to in the introduction to the relevant document. If the right holder does not provide such a statement, the committee concerned shall not proceed with inclusion of an item covered by a patent right in the document without authorization from ISO Council or IEC Council Board as appropriate.

c) A document shall not be published until the statements of the holders of all identified patent rights have been received, unless the council board concerned gives authorization.

2.14.3 Should it be revealed after publication of a document that licences under patent rights, which appear to cover items included in the document, cannot be obtained under reasonable and non-discriminatory terms and conditions, the document shall be referred back to the relevant committee for further consideration.

3 Development of other deliverables

3.1 Technical Specifications

In JTC 1, See Standing Document 6 on "Technical Specifications and Technical Reports" for the JTC 1 specific requirements on this topic.

Technical Specifications may be prepared and published under the following circumstances and conditions.

3.1.1 When the subject in question is still under development or where for any other reason there is the future but not immediate possibility of an agreement to publish an International Standard, the technical committee or subcommittee may decide, by following the procedure set out in 2.3, that the publication of a Technical Specification would be appropriate. The procedure for preparation of such a Technical Specification shall be as set out in 2.4 and 2.5. The decision to publish the resulting document as a Technical Specification shall require a two-thirds majority vote of the P-members voting of the technical committee or subcommittee. A Technical Specification is a normative document.

When the required support cannot be obtained for a final draft International Standard to pass the approval stage (see 2.7), or in case of doubt concerning consensus, the technical committee or subcommittee may decide, by a two-thirds majority vote of P-members voting, that the document should be published in the form of a Technical Specification.

3.1.2 When the P-members of a technical committee or subcommittee have agreed upon the publication of a Technical Specification, the draft specification shall be submitted electronically by the secretariat of the technical committee or subcommittee to the office of the CEO within 16 weeks for publication. Competing technical specifications offering different technical solutions are possible provided that they do not conflict with existing International Standards.

3.1.3 Technical Specifications shall be subject to review by the technical committee or subcommittee not later than 3 years after their publication. The aim of such review shall be to re-examine the situation which resulted in the publication of a Technical Specification and if possible to achieve the agreement necessary for the publication of an International Standard to replace the Technical Specification. In IEC, the date for this review is based on the stability date which shall be agreed in advance of the publication of the Technical Specification (review date). In JTC 1, the IEC-specific procedures do not apply.

3.2 Publicly Available Specifications (PAS)

This section does not apply to JTC 1.

In JTC 1, the JTC 1 PAS (Publicly Available Specification) Transposition process is a different process from the one that results in PAS deliverables in ISO and IEC (see Annex F).

3.2.1 A PAS may be an intermediate specification, published prior to the development of a full International Standard, or, in IEC may be a "dual logo" publication published in collaboration with an external organization. It is a document not fulfilling the requirements for a standard.

A PAS is a normative document.

3.2.2 A proposal for submission of a PAS may be made by the Secretariat, an A-liaison or by any P-member of the committee. In IEC, a C-liaison may also submit a PAS (see 1.17).

3.2.3 The PAS is published after verification of the presentation and checking that there is no conflict with existing International Standards by the committee concerned and following simple majority approval of the P-members voting of the committee concerned. Competing PAS offering different technical solutions are possible provided that they do not conflict with existing International Standards.

3.2.4 A PAS shall remain valid for an initial maximum period of 3 years in ISO and 2 years in IEC. The validity may be extended for a single period up to a maximum of 3 years in ISO and 2 years in IEC, at the end of which it shall be transformed with or without change into another type of normative document, or shall be withdrawn.

3.3 Technical Reports

In JTC 1, See Standing Document 6 on "Technical Specifications and Technical Reports" for the JTC 1 specific requirements on this topic.

3.3.1 When a technical committee or subcommittee has collected data of a different kind from that which is normally published as an International Standard (this may include, for example, data obtained from a survey carried out among the National Bodies, data on work in other international organizations or data on the "state of the art" in relation to standards of National Bodies on a particular subject), the technical committee or subcommittee may decide, by a simple majority vote of P-members voting, to request the Chief Executive Officer to publish such data in the form of a Technical Report. The document shall be entirely informative in nature and shall not contain matter implying that it is normative. It shall clearly explain its relationship to normative aspects of the subject which are, or will be, dealt with in International Standards related to the subject. The Chief Executive Officer, if necessary in consultation with the technical management board, shall decide whether to publish the document as a Technical Report.

3.3.2 When the P-members of a technical committee or subcommittee have agreed upon the publication of a Technical Report, the draft report shall be submitted electronically by the secretariat of the technical committee or subcommittee to the Chief Executive Officer within 16 weeks for publication.

3.3.3 It is recommended that Technical Reports are regularly reviewed by the committee responsible, to ensure that they remain valid. Withdrawal of a Technical Report is decided by the technical committee or subcommittee responsible.

Technical Reports are not subject to systematic review.

4 Meetings

4.1 General

National Bodies are reminded that they are not permitted to charge delegates/experts any sort of participation fee, nor require accommodations at specific hotels or hotel rates for any meetings of technical committees, subcommittees, working groups, maintenance and project teams. The basic meeting facilities shall be funded entirely by resources from a National Body and/or voluntary sponsors. For more information in IEC, see Meeting Guide (http://www.iec.ch/members_experts/refdocs/iec/IEC_Meeting_Guide_2012.pdf) and for ISO, see Annex SF for further details. For JTC 1, see also JTC 1 Standing Document 19, "Meetings".

4.1.1 Technical committees and subcommittees shall use current electronic means to carry out their work (for example, e-mail, groupware and teleconferencing) wherever possible. A meeting of a technical committee or subcommittee should be convened only when it is necessary to discuss committee drafts (CD) or other matters of substance which cannot be settled by other means. In JTC 1, see also Standing Document 19 on "Meetings".

4.1.2 The technical committee secretariat should look ahead with a view to drawing up, in consultation with the office of the CEO, a planned minimum 2-year programme of meetings of the technical committee and its subcommittees and, if possible, its working groups, taking account of the programme of work.

In JTC 1, meetings of JTC 1 shall be convened by the JTC 1 secretariat at nominal ~~twelve~~six-month intervals and shall be of adequate duration to resolve all agenda items.

4.1.3   In planning meetings, account should be taken of the possible advantage of grouping meetings of technical committees and subcommittees dealing with related subjects, in order to improve communication and to limit the burden of attendance at meetings by delegates who participate in several technical committees or subcommittees.

In JTC 1, the possible advantage of grouping meetings applies also to working groups.

4.1.4   In planning meetings, account should also be taken of the advantages for the speedy preparation of drafts of holding a meeting of the editing committee immediately after the meeting of the technical committee or subcommittee and at the same place.

4.2   Procedure for calling a meeting

4.2.1   Technical committee and subcommittee meetings

In JTC 1, see also Standing Document 19 on "Meetings" for planning physical or electronic meetings.

4.2.1.1 The date and place of a meeting shall be subject to an agreement between the chair and the secretariat of the technical committee or subcommittee concerned, the Chief Executive Officer and the National Body acting as host. In the case of a subcommittee meeting, the subcommittee secretariat shall first consult with the secretariat of the parent technical committee in order to ensure coordination of meetings (see also 4.1.3).

4.2.1.2 A National Body wishing to act as host for a particular meeting shall contact the Chief Executive Officer and the technical committee or subcommittee secretariat concerned.

The National Body shall first ascertain that there are no restrictions imposed by its country to the entry of representatives of any P-member of the technical committee or subcommittee for the purpose of attending the meeting.

In JTC 1, in accrediting delegates to attend meetings, P-and O-members shall register them in the ISO Meetings application or the IEC Meeting Registration System (MRS), as appropriate.

The hosting National Body can access the list of delegates through the ISO Meetings application or the IEC Meeting Registration System (MRS) so that it can make appropriate arrangements for the meeting. It is the responsibility of the P-members and O-members with participants who need invitation letters to send the names of these participants directly to the hosting National Body.

The hosting organizations are advised to verify and provide information on access means to meeting facilities.

As per clause 4.2.1.3, a document describing logistics for the meeting shall be circulated. As well as location and transport information, it should provide details of the accessibility of meeting facilities.

During the planning process, there should be a request for notification of specific accessibility requirements. The hosting body should make its best efforts to satisfy these requirements.

In JTC 1, the hosting National Body is responsible for providing secretariat support and services for meetings unless alternative arrangements have been agreed with the responsible committee secretariat.

In JTC 1, the committee secretariat shall inform the hosting National Body of all accredited meeting attendees so that the latter can make appropriate arrangements for the meeting.

4.2.1.3 The secretariat shall ensure that arrangements are made for the agenda and logistical information to be circulated by the office of the CEO (in the IEC) or by the secretariat with a copy to the office of the CEO (in ISO) at the latest 16 weeks before the date of the meeting. In JTC 1 and its subcommittees, any comments on the agenda or proposals for the addition of new work item proposals should be sent to the committee secretariat by the members not later than 8 weeks before the meeting. The secretariat shall distribute such comments or proposals immediately in order to permit adequate preparation by delegates.

NOTE   All new work item proposals must be approved by correspondence (committee internal ballot – CIB) see 2.3.4.

Only those committee drafts for which the compilation of comments will be available at least 6 weeks before the meeting shall be included on the agenda and be eligible for discussion at the meeting.

Any other working documents, including compilations of comments on drafts to be discussed at the meeting, shall be distributed not less than 6 weeks in advance of the meeting.

The agenda shall clearly state the starting and estimated finishing times.

In the event of meetings over running the estimated finishing time, the Chair shall ensure that the P-members are willing to take voting decisions. However, if P-members leave, they may request the Chair not to take any further voting decisions.

In JTC 1, any decisions made after the estimated finishing time of the meeting and after any P-members have left shall be confirmed by correspondence after the meeting.

NOTE   Attendees should take the estimated meeting time into consideration when booking their travel.

In JTC 1 and its subcommittees, only those committee drafts for which the compilation of comments will be available at least four weeks before the meeting shall be included on the agenda and be eligible for discussion at the meeting.

4.2.2 Working group meetings

In JTC 1, see Standing Document 19 on "Meetings" for the JTC 1 specific requirements on WG meetings.

4.2.2.1 Working groups shall use current electronic means to carry out their work (for example, e-mail, groupware and teleconferencing) wherever possible. When a meeting needs to be held, notification by the convenor of the meetings of a working group shall be sent to its members and to the secretariat of the parent committee, at least 6 weeks in advance of the meeting.

In JTC 1, working group meeting agendas shall be distributed by the convenor or secretariat preferably four months but no less than 12 weeks in advance. Working group agendas shall be distributed to the members of the working group and the parent committee.

In JTC 1, see also Standing Document 19 on "Meetings" for requirement for planning Working Group physical or electronic meetings, in particular the requirements for posting a notice of the meeting, the agenda, and documents to be discussed.

Arrangements for meetings shall be made between the convenor and the member of the working group in whose country the meeting is to be held. The latter member shall be responsible for all practical working arrangements.

In JTC 1, as working groups may include a large number of participants, the meeting date and venue shall be agreed by the secretariat of the parent committee and the parent committee's National Body of the country in which the meeting is held.

4.2.2.2 If a working group meeting is to be held in conjunction with a meeting of the parent committee, the convenor shall coordinate arrangements with the secretariat of the parent committee. In particular, it shall be ensured that the working group members receive all general information for the meeting, which is sent to delegates to the meeting of the parent committee.

4.2.2.3 Either the WG (or PT/MT/AC in IEC) leader or the Secretary of the relevant committee shall notify National Body Secretariats of any WG (or PT/MT/AC in IEC) meeting held in their country.

4.3 Languages at meetings

While the official languages at meetings are English, French and Russian, meetings are conducted in English by default.

The National Body for the Russian Federation provides all interpretation and translation into or from the Russian language.

The chair and secretariat are responsible for dealing with the question of language at a meeting in a manner acceptable to the participants following the general rules of ISO or IEC, as appropriate. (See also Annex E.)

When at a meeting of JTC 1 or one of its subsidiary bodies a participant wishes, in view of exceptional circumstances, to speak in any other language, the chairman or convenor of the session shall be entitled to authorize this, for the session only, provided that a means of interpretation has been secured.

4.4 Cancellation of meetings

Every possible effort shall be made to avoid cancellation or postponement of a meeting once it has been convened. Nevertheless, if the agenda and basic documents are not available within the time required by 4.2.1.3, then the Chief Executive Officer has the right to cancel the meeting.

5 Appeals

5.1 General

5.1.1 National bodies have the right of appeal

a) to the parent technical committee on a decision of a subcommittee;

b) to the technical management board on a decision of a technical committee;

c) to the council board on a decision of the technical management board,

within 12 weeks in ISO and 8 weeks in IEC of the decision in question.

The decision of the council board on any case of appeal is final.

5.1.2 A P-member of a technical committee or subcommittee may appeal against any action, or inaction, on the part of the technical committee or subcommittee, when the P-member considers that such action or inaction is

a) not in accordance with

— the Statutes and Rules of Procedure;

— the ISO/IEC Directives; or

b) not in the best interests of international trade and commerce, or such public factors as safety, health or environment.

5.1.3 Matters under appeal may be either technical or administrative in nature.

Appeals on decisions concerning new work item proposals, committee drafts, enquiry drafts and final draft International Standards are only eligible for consideration if

— questions of principle are involved, or

— the contents of a draft may be detrimental to the reputation of ISO or IEC.

5.1.4 All appeals shall be fully documented to support the P-member's concern.

In JTC 1, all appeals shall be fully documented to support the JTC 1 National Body's concern. The appeal shall state the nature of the objection(s) including any direct and material adverse effects, the section(s) of these procedures or the standard that are at issue, actions or inactions that are at issue, and the specific remedial action(s) that would satisfy the appellant's concerns. Previous efforts to resolve the objection(s) and the outcome of each shall be noted.

5.2 Appeal against a subcommittee decision

5.2.1 The documented appeal shall be submitted by the P-member to the secretariat of the parent technical committee, with a copy to the Chief Executive Officer.

5.2.2 Upon receipt, the secretariat of the parent technical committee shall advise all its P-members of the appeal and take immediate action, by correspondence or at a meeting, to consider and decide on the appeal, consulting the Chief Executive Officer in the process.

5.2.3 If the technical committee supports its subcommittee, then the P-member which initiated the appeal may either

— accept the technical committee decision, or

— appeal against it.

5.3 Appeal against a technical committee decision

5.3.1 Appeals against a technical committee decision may be of 2 kinds:

— an appeal arising out of 5.2.3 above, or

— an appeal against an original decision of a technical committee.

5.3.2 The documented appeal shall, in all cases, be submitted to the Chief Executive Officer, with a copy to the chair and secretariat of the technical committee.

5.3.3 The Chief Executive Officer shall, following whatever consultations s/he deems appropriate, refer the appeal together with his/her comments to the technical management board within 4 weeks after receipt of the appeal.

5.3.4 The technical management board shall decide whether an appeal shall be further processed or not. If the decision is in favour of proceeding, the chair of the technical management board shall form a conciliation panel.

The conciliation panel shall hear the appeal within 12 weeks and attempt to resolve the difference of opinion as soon as practicable. The conciliation panel shall give a final report within 12 weeks. If the conciliation panel is unsuccessful in resolving the difference of opinion, this shall be reported to the Chief Executive Officer, together with recommendations on how the matter should be settled.

5.3.5 The Chief Executive Officer, on receipt of the report of the conciliation panel, shall inform the technical management board, which will make its decision.

5.4 Appeal against a technical management board decision

An appeal against a decision of the technical management board shall be submitted to the Chief Executive Officer with full documentation on all stages of the case.

The Chief Executive Officer shall refer the appeal together with his/her comments to the members of the council board within 4 weeks after receipt of the appeal.

The council board shall make its decision within 12 weeks.

5.5 Progress of work during an appeal process

When an appeal is against a decision respecting work in progress, the work shall be continued, up to and including the approval stage (see 2.7).

Annex A
(normative)

Guides

A.1 Introduction

In addition to International Standards, Technical Specifications, Publicly Available Specifications and Technical Reports prepared by technical committees, ISO and IEC publish Guides on matters related to international standardization. Guides shall be drafted in accordance with the ISO/IEC Directives, Part 2.

Guides shall not be prepared by technical committees and subcommittees. They may be prepared by an ISO Policy Development Committee, an IEC Advisory Committee or Strategic Group, an ISO group reporting to the ISO technical management board, or an ISO/IEC Joint Coordination Group. These bodies are referred to below as the "Committee or Group responsible for the project".

The procedure for preparation and publication of a Guide is as described below.

A.2 Proposal stage

The ISO and/or IEC technical management board will approve proposals for new Guides or revisions of Guides and decide on the secretariat and composition of the Committee or Group responsible for the project.

Once a project is approved by the ISO and/or IEC technical management board, the secretariat of the Committee or Group responsible for the project shall ensure that the appropriate interests in ISO and IEC are informed.

A.3 Preparatory stage

The Committee or Group responsible for the project shall ensure that the appropriate interests in ISO and IEC have the opportunity to be represented during the preparation of the working draft.

A.4 Committee stage

Once a working draft is available for circulation as a committee draft, the secretariat of the Committee or Group responsible for the project shall send it to the parent committee or ISO and/or IEC technical management board for vote, comments and to approve its advancement to the Enquiry stage.

A.5 Enquiry stage

A.5.1 The office of the CEOs shall circulate both the English and French texts of the revised draft Guide to all National Bodies for a 16-week vote.

A.5.2 The draft Guide is approved for publication as a Guide if not more than one-quarter of the votes cast are negative, abstentions being excluded when the votes are counted.

In the case of ISO/IEC Guides, the draft shall be submitted for approval to the National Bodies of both ISO and IEC. The National Bodies of both organizations need to approve the document if it is to be published as an ISO/IEC Guide.

If this condition is satisfied for only one of the organizations, ISO or IEC, the Guide may be published under the name of the approving organization only, unless the Committee or Group responsible for the project decides to apply the procedure set out in A.5.3.

A.5.3 If a draft Guide is not approved, or if it is approved with comments the acceptance of which would improve consensus, the chair of the Committee or Group responsible for the project may decide to submit an amended draft for an 8-week vote. The conditions for acceptance of the amended draft are the same as in A.5.2.

A.6 Publication stage

The publication stage shall be the responsibility of the office of the CEO of the organization to which the Committee or Group responsible for the project belongs.

In the case of a Joint ISO/IEC Group, the responsibility shall be decided by agreement between the Chief Executive Officers.

A.7 Withdrawal of a Guide

The Committee or Group responsible for the Guide shall be responsible for deciding if the Guide shall be withdrawn. The formal withdrawal shall be ratified by the technical management board (TMB) in accordance with its normal procedures.

Annex B
(normative)

ISO/IEC procedures for liaison and work allocation

B.1 Introduction

By the ISO/IEC Agreement of 1976 [1], ISO and IEC together form a system for international standardization as a whole. For this system to operate efficiently, the following procedures are agreed for coordination and allocation of work between the technical committees and subcommittees of both organizations.

B.2 General considerations

The allocation of work between ISO and IEC is based on the agreed principle that all questions relating to international standardization in the electrical and electronic engineering fields are reserved to IEC, the other fields being reserved to ISO and that allocation of responsibility for matters of international standardization where the relative contribution of electrical and non-electrical technologies is not immediately evident will be settled by mutual agreement between the organizations.

Questions of coordination and work allocation may arise when establishing a new ISO or IEC technical committee, or as a result of the activities of an existing technical committee.

The following levels of coordination and work allocation agreement are available. Matters should be raised at the next higher level only after all attempts to resolve them at the lower levels have failed.

a) **Formal liaisons** between ISO and IEC committees for normal inter-committee cooperation.

b) **Organizational consultations**, including technical experts and representatives of the Chief Executive Officers, for cases where technical coordination may have an effect on the future activities of the organizations in a larger sense than the point under consideration.

c) Decisions on work allocation

— by the technical management boards or, if necessary,

— the ISO/IEC Joint Technical Advisory Board (JTAB).

Questions affecting both ISO and IEC, on which it has not proved possible to obtain a common decision by the ISO Technical Management Board and the IEC Standardization Management Board, are referred to the ISO/IEC Joint Technical Advisory Board (JTAB) for decision (see 1.3.1).

---

[1] ISO Council resolutions 49/1976 and 50/1976 and IEC Administrative Circular No. 13/1977.

B.3 Establishing new technical committees

Whenever a proposal to establish a new technical committee is made to the National Bodies of ISO or of IEC respectively, the proposal shall also be submitted to the other organization requesting comment and/or agreement. As a result of these consultations, two cases may arise:

a) the opinion is unanimous that the work should be carried out in one of the organizations;

b) opinions are divided.

In case a), formal action may then be taken to establish the new technical committee according to the unanimous opinion.

In case b), a meeting of experts in the field concerned shall be arranged with representatives of the Chief Executive Officers with a view to reaching a satisfactory agreement for allocation of the work (i.e., organizational level). If agreement is reached at this level, formal action may be taken by the appropriate organization to implement the agreement.

In the case of disagreement after these consultations, the matter may be referred by either organization to the ISO/IEC Joint Technical Advisory Board (JTAB).

B.4 Coordinating and allocating work between ISO and IEC technical committees

B.4.1 Formal liaison at TC level

Most coordination needs arising between individual ISO and IEC committees are successfully dealt with through formal technical liaison arrangements. These arrangements, when requested by either organization, shall be honoured by the other organization. Requests for formal liaison arrangements are controlled by the offices of the CEOs. The requesting organization shall specify the type of liaison required, such as:

a) full or selective exchange of committee documents;

b) regular or selective attendance of liaison representatives at meetings;

c) participation in a standing coordination (or steering) committee for selected ISO and IEC technical committees;

d) setting up of a Joint Working Group (JWG).

B.4.2 Details of agreement

B.4.2.1 Continual efforts shall be made to minimize the overlap areas between IEC and ISO by entrusting areas of work to one of the two organizations.

For areas of work so entrusted, IEC and ISO shall agree through the JTAB on how the views and interests of the other organization are to be fully taken into account.

B.4.2.2 Five working modes of cooperation have been established, as follows:

Mode 1 – Informative relation

One organization is fully entrusted with a specific work area and keeps the other fully informed of all progress.

Mode 2 – Contributive relation

One organization should take the lead of the work and the other should make written contributions where considered appropriate during the progress of this work. This relation also includes the exchange of full information.

Mode 3 – Subcontracting relation

One organization is fully entrusted with the realization of the work on an identified item, but due to specialization of the other, a part of the work is subcontracted and that part is prepared under the responsibility of the second organization. Necessary arrangements shall be made to guarantee the correct integration of the resulting subcontracted work into the main part of the programme. To this end, the enquiry and approval stages are handled by the organization being the main contractor for the standardization task.

Mode 4 – Collaborative relation

One organization takes the lead in the activities, but the work sessions and meetings receive ~~delegates~~ liaison representatives from the other~~, who have observer status and who ensure the technical liaison with the other organization~~. Such ~~observers~~ liaison representatives should have the right to intervene in the debate but have no right to vote. The full flow of information is oriented through this liaison.

Mode 5 – Integrated liaison

Joint Working Groups and Joint Technical Committees ensure integrated meetings for handling together the realization of standards under a principle of total equality of participation.

Joint Working Groups between technical committees of the two organizations shall operate in accordance with 1.12.6.

B.4.2.3 The allocation of work between IEC and ISO for potentially overlapping areas will be set out as required in schedules or programmes which, when agreed by the relevant parties, will form addenda to this agreement.

A consequence of this agreement is that the parties agree to cross-refer to the relevant standards of the other in the respective competent fields of interest.

When the standard being referred to is updated, it is the responsibility of the body making the reference to take care of the updating of the reference where appropriate.

B.4.2.4 For work for which one organization has assumed responsibility and for which there will be subcontracting of work to the other, the fullest account shall be taken of the interests participating in the subcontracted work in defining the objectives of that work.

B.4.2.5 The necessary procedures for enquiry and approval shall be realized by the organization entrusted with a particular standardization task, except as otherwise agreed by the two technical management boards.

B.4.2.6 For standards developed under the Mode 5 – Integrated liaison, the committee, enquiry and approval stages shall be carried out in parallel in both ISO and IEC in accordance with the rules of the organization with the administrative lead. The committee/ organization with the administrative responsibility for the project shall submit drafts for the committee, enquiry and approval stages to the other organization two weeks prior to the circulation date.

B.4.2.7 When the enquiry draft has not fulfilled the approval criteria (see 2.6.3) in one of the organizations, then:

— the officers of the committees involved in the joint working group may select one of options given in 2.6.4 c) or

— in exceptional circumstances, if agreed between the officers of the ISO and IEC committees involved in the joint working group and the offices of the CEO, the project may proceed as a single logo standard of the organization in which the enquiry draft was approved. The joint working group is automatically disbanded.

B.4.2.8 If the final draft International Standard is not approved in accordance of the conditions in 2.7.3 then:

— the committees involved in the joint working group may select one of the options given in 2.7.7, noting that in IEC the circulation of a second final draft International Standard is not allowed and will require a derogation of the TMB or

— in exceptional circumstances, if agreed between the officers of the ISO and IEC committees involved in the joint working group and the offices of the CEO, the standard may be published as a single logo standard of the organization in which the final draft International Standard was approved. The joint working group is automatically disbanded.

B.4.2.9 Standards developed under the Mode 5 – Integrated liaison via a joint working group between ISO and IEC are published by the organization of the committee having administrative responsibility. That organization assigns the reference number of the standard and owns the copyright of the standard. The standard carries the logo of the other organization and may be sold by both organizations. The foreword of the International Standard will identify all the committees responsible for the development. For those standards where the committee with the administrative responsibility is in the IEC, then the foreword will also give the ISO voting results. ISO-lead documents are assigned numbers from 1 to 59999, IEC-lead documents are assigned numbers from 60000 to 79999. In the case of multi-part standards, some parts being under ISO responsibility and some being under IEC responsibility, a number in the 80000 series is assigned (e.g. ISO 80000-1, IEC 80000-6).

B.4.2.10 The maintenance procedures to be used for standards developed under the Mode 5 – Integrated liaison will be those currently applied in the organization which has the committee with the administrative responsibility.

B.4.2.11 If there is a reason, during the development of the project, to change from one mode of operation to another, a recommendation shall be made by both technical committees concerned and submitted to the two technical management boards for information.

B.4.3 Cooperation of secretariats

The secretariats of the technical committees/subcommittees from the two organizations concerned shall cooperate on the implementation of this agreement. There shall be a complete information flow on on-going work and availability on demand to each other of working documents, in accordance with normal procedures.

Annex C
(normative)

Justification of proposals for the establishment of standards

C.1 General

C.1.1 Because of the large financial resources and manpower involved and the necessity to allocate these according to the needs, it is important that any standardization activity begin by identifying the needs, determining the aims of the standard(s) to be prepared and the interests that may be affected. This will, moreover, help to ensure that the standards produced will cover appropriately the aspects required and be market relevant for the affected sectors. Any new activity shall therefore be reasonably justified before it is begun.

C.1.2 It is understood that, whatever conclusions may be drawn on the basis of the annex, a prerequisite of any new work to be commenced would be a clear indication of the readiness of a sufficient number of relevant interested parties to allocate necessary manpower, funds and to take an active part in the work.

C.1.3 This annex sets out rules for proposing and justifying new work, so that proposals will offer to others the clearest possible idea of the purposes and extent of the work, in order to ensure that standardization resources are really allocated by the parties concerned and are used to the best effect.

C.1.4 This annex does not contain rules of procedure for implementing and monitoring the guidelines contained in it, nor does it deal with the administrative mechanism which should be established to this effect.

C.1.5 This annex is addressed primarily to the proposer of any kind of new work to be started but may serve as a tool for those who will analyse such a proposal or comment on it, as well as for the body responsible for taking a decision on the proposal.

C.2 Terms and definitions

C.2.1
proposal for new work
proposal for a new field of technical activity or for a new work item

C.2.2
proposal for a new field of technical activity
proposal for the preparation of (a) standard(s) in a field that is not covered by an existing committee (such as a technical committee, subcommittee or project committee) of the organization to which the proposal is made

C.2.3
proposal for a new work item
proposal for the preparation of a standard or a series of related standards in the field covered by an existing committee (such as a technical committee) of the organization to which the proposal is made

C.3 General principles

C.3.1 Any proposal for new work shall lie within the scope of the organization to which it is submitted.

NOTE For example, the objects of ISO are laid down in its Statutes and of IEC in Article 2 of its Statutes.

C.3.2 The documentation justifying new work in ISO and IEC shall make a substantial case for the market relevance of the proposal.

C.3.3 The documentation justifying new work in ISO and IEC shall provide solid information as a foundation for informed ISO or IEC National Body voting.

C.3.4 Within the ISO and IEC systems, the onus is considered to be placed on the proposer to provide the proper documentation to support principles C.3.2 and C.3.3 stated above.

C.4 Elements to be clarified when proposing a new field of technical activity or a new work item

C.4.1 Proposals for new fields of technical activity and new work items shall include the following fields of information (C.4.2 to C.4.13)

C.4.2 Title

The title shall indicate clearly yet concisely the new field of technical activity or the new work item which the proposal is intended to cover.

EXAMPLE 1   (proposal for a new technical activity) "Machine tools".

EXAMPLE 2   (proposal for a new work item) "Electrotechnical products – Basic environmental testing procedures".

C.4.3 Scope

In JTC 1, additional factors such as cultural and linguistic adaptability and accessibility are to be considered.

C.4.3.1 For new fields of technical activity

The scope shall precisely define the limits of the field of activity. Scopes shall not repeat general aims and principles governing the work of the organization but shall indicate the specific area concerned.

EXAMPLE   "Standardization of all machine tools for the working of metal, wood and plastics, operating by removal of material or by pressure".

C.4.3.2 For new work items

The scope shall give a clear indication of the coverage of the proposed new work item and, if necessary for clarity, exclusions shall be stated.

EXAMPLE 1

This standard lists a series of environmental test procedures, and their severities, designed to assess the ability of electrotechnical products to perform under expected conditions of service.

Although primarily intended for such applications, this standard may be used in other fields where desired.

Other environmental tests, specific to the individual types of specimen, may be included in the relevant specifications.

EXAMPLE 2

Standardization in the field of fisheries and aquaculture, including, but not limited to, terminology, technical specifications for equipment and for their operation, characterization of aquaculture sites and maintenance of appropriate physical, chemical and biological conditions, environmental monitoring, data reporting, traceability and waste disposal.

Excluded:

— methods of analysis of food products (covered by ISO/TC 34);

— personal protective clothing (covered by ISO/TC 94);

— environmental monitoring (covered by ISO/TC 207).

C.4.4 Programme of work (for proposals for new fields of technical activity only)

In JTC 1, a programme of work is established and maintained within the overall business plan.

C.4.4.1 The proposed programme of work shall correspond to and clearly reflect the aims of the standardization activities and shall, therefore, show the relationship between the subjects proposed.

C.4.4.2 Each item on the programme of work shall be defined by both the subject and aspect(s) to be standardized (for products, for example, the items would be the types of products, characteristics, other requirements, data to be supplied, test methods, etc.).

C.4.4.3 Supplementary justification may be combined with particular items in the programme of work.

C.4.4.4 The proposed programme of work shall also suggest priorities and target dates for new work items (when a series of standards is proposed, priorities shall be suggested).

C.4.5 Indication(s) of the preferred type or types of deliverable(s) to be produced

In the case of proposals for new fields of technical activity, this may be provided under C.4.4.

C.4.6 A listing of relevant existing documents at the international, regional and national levels

Any known relevant documents (such as standards and regulations) shall be listed, regardless of their source and should be accompanied by an indication of their significance.

C.4.7 Relation to and impact on existing work

C.4.7.1 A statement shall be provided regarding any relation or impact the proposed work may have on existing work, especially existing ISO and IEC deliverables. The proposer should explain how the work differs from apparently similar work, or explain how duplication and conflict will be minimized.

C.4.7.2　If seemingly similar or related work is already in the scope of other committees of the organization or in other organizations, the proposed scope shall distinguish between the proposed work and the other work.

C.4.7.3　The proposer shall indicate whether his or her proposal could be dealt with by widening the scope of an existing committee or by establishing a new committee.

C.4.8　Relevant country participation

C.4.8.1　For proposals for new fields of technical activity, a listing of relevant countries should be provided where the subject of the proposal is important to their national commercial interests.

C.4.8.2　For proposals for new work item within existing committees, a listing of relevant countries should be provided which are not already P-members of the committee, but for whom the subject of the proposal is important to their national commercial interests.

C.4.9　Cooperation and liaison

C.4.9.1　A list of relevant external international organizations or internal parties (other than ISO and/or IEC committees) to be engaged as liaisons in the development of the deliverable(s) shall be provided.

C.4.9.2　In order to avoid conflict with, or duplication of efforts of, other bodies, it is important to indicate all points of possible conflict or overlap.

C.4.9.3　The result of any communication with other interested bodies shall also be included.

C.4.10 Affected stakeholders

A simple and concise statement shall be provided identifying and describing relevant affected stakeholder categories (including small and medium sized enterprises) and how they will each benefit from or be impacted by the proposed deliverable(s).

C.4.11 Base document (for proposals for new work items only)

C.4.11.1　When the proposer considers that an existing well-established document may be acceptable as a standard (with or without amendments) this shall be indicated with appropriate justification and a copy attached to the proposal.

C.4.11.2　All proposals for new work items shall include an attached existing document to serve as an initial basis for the ISO or IEC deliverable or a proposed outline or table of contents.

C.4.11.3　If an existing document is attached that is copyrighted or includes copyrighted content, the proposer shall ensure that appropriate permissions have been granted in writing for ISO or IEC to use that copyrighted content.

C.4.12 Leadership commitment

C.4.12.1　In the case of a proposal for a new field of technical activity, the proposer shall indicate whether his/her organization is prepared to undertake the secretariat work required.

C.4.12.2　In the case of a proposal for new work item, the proposer shall also nominate a project leader.

C.4.13 Purpose and justification

C.4.13.1 The purpose and justification of the standard to be prepared shall be made clear and the need for standardization of each aspect (such as characteristics) to be included in the standard shall be justified.

C.4.13.2 If a series of new work items is proposed the purpose and the justification of which is common, a common proposal may be drafted including all elements to be clarified and enumerating the titles and scopes of each individual item.

C.4.13.3 Please note that the items listed in the bullet points below represent a menu of suggestions or ideas for possible documentation to support the purpose and justification of proposals. Proposers should consider these suggestions, but they are not limited to them, nor are they required to comply strictly with them. What is most important is that proposers develop and provide purpose and justification information that is most relevant to their proposals and that makes a substantial business case for the market relevance and need of their proposals. Thorough, well-developed and robust purpose and justification documentation will lead to more informed consideration of proposals and ultimately their possible success in the ISO and IEC systems.

— A simple and concise statement describing the business, technological, societal or environmental issue that the proposal seeks to address, preferably linked to the Strategic Business Plan of the concerned ISO or IEC committee.

— Documentation on relevant global metrics that demonstrate the extent or magnitude of the economic, technological, societal or environmental issue, or the new market. This may include an estimate of the potential sales of the resulting standard(s) as an indicator of potential usage and global relevance.

— Technological benefit – a simple and concise statement describing the technological impact of the proposal to support coherence in systems and emerging technologies, convergence of merging technologies, interoperability, resolution of competing technologies, future innovation, etc.

— Economic benefit – a simple and concise statement describing the potential of the proposal to remove barriers to trade, improve international market access, support public procurement, improve business efficiency for a broad range of enterprises including small and medium sized ones, and/or result in a flexible, cost-effective means of complying with international and regional rules/conventions, etc. A simple cost/benefit analysis relating the cost of producing the deliverable(s) to the expected economic benefit to businesses worldwide may also be helpful.

— Societal benefit(s) – a simple and concise statement describing any societal benefits expected from the proposed deliverable(s).

— Environmental benefit(s) – a simple and concise statement describing any environmental or wider sustainability benefits expected from the proposed deliverable(s).

— A simple and concise statement clearly describing the intended use(s) of the proposed deliverable(s), for example, whether the deliverable is intended as requirements to support conformity assessment or only as guidance or recommended best practices; whether the deliverable is a management system standard; whether the deliverable is intended for use or

— reference in technical regulation; whether the deliverable is intended to be used to support legal cases in relation to international treaties and agreements.

— A simple and concise statement of metrics for the committee to track in order to assess the impact of the published standard over time to achieve the benefits to stakeholders documented under C.4.10 above.

— A statement assessing the prospect of the resulting deliverable(s) being compliant with, for the IEC, the IEC Global Relevance Policy: http://www.iec.ch/members_experts/refdocs/ac_cl/AC_200817e_AC.pdf and for ISO, with ISO's Global Relevance Policy http://www.iso.org/iso/standards_development/governance_of_technical_work/global_relevance_policy.htm and the ISO/TMB recommendations (see NOTE below) regarding sustainable development and sustainability, where relevant.

NOTE   For ISO, the ISO/TMB confirmed the following recommendations: 1) When a committee (in any sector) develops a standard dealing with sustainability/sustainable development the standard must remain within the context of the committee's scope of work; 2) The committee should also notify the TMB with the title and scope as early as possible; 3) The committee undertaking such work should clarify its intentions in the Introduction of the specific standard(s); 4) The most widely used definition of sustainable development is the one from the UN Brundtland committee on sustainable development: development that meets the needs of the present without compromising the ability of future generations to meet their own needs.

— A statement assessing the proposal's compliance with the Principles for developing ISO and IEC Standards related to or supporting public policy initiatives (for ISO see Annex SO in the Consolidated ISO Supplement and for IEC and ISO see *Using and referencing ISO and IEC standards for technical regulations*: http://www.iso.org/iso/standards_for_technical_regulations.pdf) and the possible relation of the resulting deliverable(s) to public policy, including a statement regarding the potential for easier market access due to conformity with appropriate legislation.

Annex D
(normative)

Resources of secretariats and qualifications of secretaries

D.1 Terms and definitions

D.1.1
secretariat
National Body to which has been assigned, by mutual agreement, the responsibility for providing technical and administrative services to a technical committee or subcommittee

D.1.2
secretary
individual appointed by the secretariat to manage the technical and administrative services provided

D.2 Resources of a secretariat

A National Body to which a secretariat has been assigned shall recognize that, no matter what arrangements it makes in its country to provide the required services, it is the National Body itself that is ultimately responsible for the proper functioning of the secretariat. National bodies undertaking secretariat functions shall become party to the ISO Service Agreement or IEC Basic Agreement, as appropriate.

The secretariat shall therefore have adequate administrative and financial means or backing to ensure:

a) facilities for word-processing in English and/or French, for providing texts electronically, and for any necessary reproduction of documents; In JTC 1, facilities for word-processing in English, for providing texts in machine-readable-form, and for any necessary reproduction of documents;

b) preparation of adequate technical illustrations;

c) identification and use, with translation where necessary, of documents received in the official languages;

d) updating and continuous supervision of the structure of the committee and its subsidiary bodies, if any;

e) reception and prompt dispatch of correspondence and documents;

f) adequate communication facilities by telephone, telefax and electronic mail;

g) access to the Internet;

h) arrangements and facilities for translation, interpretation and services during meetings, in collaboration with the host National Body, as required; In JTC 1, arrangements and facilities for translation, interpretation, and services are not required except as specified in 4.3;

i) attendance of the secretary at any meetings requiring his/her presence, including technical committee and/or subcommittee meetings, editing committee meetings, working group meetings, and consultations with the chair when necessary;

j) access by the secretary to basic International Standards (see the ISO/IEC Directives, Part 2, ~~2018, Annex D~~)on Referencing Documents and sources for drafting) and to International Standards, national standards and/or related documents in the field under consideration;

k) access by the secretary, when necessary, to experts capable of advising on technical issues in the field of the committee;

l) In JTC 1, the ability to fulfil the secretariat's electronic document distribution responsibilities as defined in the JTC 1 Standing Document 23, Access Control to JTC 1 Documents: open and restricted access.

Whilst the Chief Executive Officer endeavours to send his/her representative to the first meeting of a technical committee, to meetings of technical committees with new secretariats, and to any technical committee or subcommittee meeting where such presence is desirable for solving problems, the office of the CEO cannot undertake to carry out the work for a secretariat, on a permanent or temporary basis.

D.3 Requirements of a secretary

The individual appointed as secretary shall

a) have sufficient knowledge of English and/or French; In JTC 1, have sufficient knowledge of English;

b) be familiar with the *Statutes and rules of procedure*, as appropriate, and with the ISO/IEC Directives (see the respective Supplements to the ISO/IEC Directives); In JTC 1, also be familiar with the *Consolidated JTC 1 Supplement* and the JTC 1 Standing Documents and JTC 1 resolutions;

c) be in a position to advise the committee and any subsidiary bodies on any point of procedure or drafting, after consultation with the office of the CEO if necessary;

d) be aware of any council board or technical management board decision regarding the activities of the technical committees in general and of the committee for which s/he is responsible in particular;

e) be a good organizer and have training in and ability for technical and administrative work, in order to organize and conduct the work of the committee and to promote active participation on the part of committee members and subsidiary bodies, if any;

f) be familiar with the documentation supplied by the offices of CEO, in particular the use of electronic tools and services.

It is recommended that newly appointed secretaries of technical committees should make an early visit to the office of the CEO in Geneva in order to discuss procedures and working methods with the staff concerned.

Annex E
(normative)

General policy on the use of languages

E.1 Expressing and communicating ideas in an international environment

At the international level, it is common practice to publish deliverables in at least two languages. There are a number of reasons why it is advantageous to use two languages, for example:

— greater clarity and accuracy of meaning can be achieved by expressing a given concept in two languages which have different grammar and syntax;

— if consensus is reached on the basis of a text drafted in only one language, difficulties may arise when it comes to putting that text into another language. Some questions may have to be rediscussed, and this can cause delay if the text originally agreed upon has to be altered. Subsequent drafting into a second language of a text already approved in the first language often brings to light difficulties of expression that could have been avoided if both versions had been prepared at the same time and then amended together;

— to ensure that international meetings will be as productive as possible, it is important for the agreements reached to be utterly devoid of ambiguity, and there has to be no risk that these agreements can be called back into question because of misunderstandings of a linguistic nature;

— the use of two languages chosen from two linguistic groups widens the number of prospective delegates who might be appointed to attend the meetings;

— it becomes easier to express a concept properly in other languages if there are already two perfectly harmonized versions.

E.2 The use of languages in the technical work

The official languages are English, French and Russian.

The work of the technical committees and the correspondence are in English by default.

For the purposes of the above, the National Body for the Russian Federation provides all interpretation and translation into and from the Russian language.

In IEC, a definitive language of development for each deliverable shall be designated in the Foreword. Specific exceptions apply to the IEV and/or database standards.

E.3 International Standards

International Standards are published by the ISO and IEC in English and in French (and sometimes in multilingual editions also including Russian and other languages, especially in cases of terminology). These versions of a given International Standard are equivalent, and each is regarded as being an original-language version.

It is advantageous for the technical content of a standard to be expressed in both English and French from the outset of the drafting procedure, so that these two versions will be studied, amended and adopted at the same time and their linguistic equivalence will be ensured at all times. (See also the ISO/IEC Directives, Part 2, 2018clause on "Language versions")

This may be done

— by the secretariat or, under the latter's responsibility, with outside assistance, or

— by the editing committee of the responsible technical committee or subcommittee, or

— by National Bodies whose national language is English or French and under an agreement concluded between those National Bodies and the secretariat concerned.

When it is decided to publish a multilingual International Standard (a vocabulary, for example), the National Body for the Russian Federation takes charge of the Russian portion of the text; similarly, when it is decided to publish an International Standard containing terms or material in languages other than the official languages, the National Bodies whose national languages are involved are responsible for selecting the terms or for drafting the portions of text which are to be in those languages.

E.4 Other publications developed by technical committees

Other publications may be issued in one official language only.

E.5 Documents for technical committee and subcommittee meetings

E.5.1 Drafts and documents referred to the agenda

The documents prepared and circulated prior to a meeting are the following.

a) **Draft agendas**

Draft agendas are prepared and distributed in the language(s) of the meeting (English by default)in both English and French whenever possible by the responsible secretariats and are reproduced and distributed.

b) **Committee drafts referred to in the agenda**

It is desirable that versions of committee drafts both the English and the French versions of committee drafts referred to in the agenda will be available for the meeting in the language(s) of the meeting (English by default).

Enquiry drafts shall be available in English and French. The ISO Council or IEC Standardization Management Board guidelines shall be applied where one of the language versions is not available in due time.

Other documents (sundry proposals, comments, etc.) relating to agenda items may be prepared in only one language (English or French).

E.5.2 Documents prepared and circulated during a meeting

The documents prepared and circulated during a meeting are the following.

a) **Resolutions adopted during the meeting**

   An ad hoc drafting committee, formed at the beginning of each meeting and comprising the secretary and, whenever possible, one or more delegates of English and/or French mother tongue, edits each of the proposed resolutions.

b) **Brief minutes, if any, prepared after each session**

   If such minutes are prepared, they shall be drafted in English or French and preferably in both with, if necessary, the assistance of the ad hoc drafting committee.

E.5.3 Documents prepared and circulated after a meeting

After each technical committee or subcommittee meeting, the secretariat concerned shall draft a report of the meeting, which may be in only one language (English or French) and which includes, as annex, the full text of the resolutions adopted, preferably in both English and French.

E.6 Documents prepared in languages other than English or French

National bodies whose national language is neither English nor French may translate any documents circulated by secretariats into their own national language in order to facilitate the study of those documents by the experts of their country or to assist the delegates they have appointed to attend the meetings of the technical committees and subcommittees.

If one language is common to two or more National Bodies, one of them may at any time take the initiative of translating technical documents into that language and of providing copies to other National Bodies in the same linguistic group.

The terms of the above two paragraphs may be applied by the secretariats for their own needs.

E.7 Technical meetings

E.7.1 Purpose

The purpose of technical meetings is to achieve as full agreement as possible on the various agenda items and every effort shall be made to ensure that all delegates understand one another.

E.7.2 Interpretation of debates into English and French

Although the basic documents may be available in both English and French, it has to be determined according to the case whether interpretation of statements expressed in one language should be given in the other language

— by a volunteer delegate,

— by a staff member from the secretariat or host National Body, or

— by an adequately qualified interpreter. In JTC 1, the interpretation of debates into English and French is not applicable, except as specified in 4.3.

Care should also be taken that delegates who have neither English nor French as mother tongue can follow the meeting to a sufficient extent.

It is impractical to specify rules concerning the necessity of interpreting the debates at technical meetings. It is essential, of course that all delegates should be able to follow the discussions, but it may not be altogether essential to have a word-for-word interpretation of each statement made.

In view of the foregoing, and except in special cases where interpretation may not be necessary, the following practice is considered appropriate:

a) for meetings where procedural decisions are expected to be taken, brief interpretation may be provided by a member of the secretariat or a volunteer delegate;

b) at working group meetings, the members should, whenever possible, arrange between themselves for any necessary interpretation on the initiative and under the authority of the convenor of the working group.

To enable the secretariat responsible for a meeting to make any necessary arrangements for interpretation, the secretariat should be informed, at the same time as it is notified of attendance at the meeting, of the languages in which the delegates are able to express themselves and of any aid which delegates might be able to provide in the matter of interpretation.

In those cases where a meeting is conducted mainly in one language, the following practice should be adopted as far as is practicable in order to assist delegates having the other language:

a) the decision taken on one subject should be announced in both languages before passing to the next subject;

b) whenever a change to an existing text is approved in one language, time should be allowed for delegates to consider the effect of this change on the other language version;

c) a summary of what has been said should be provided in the other language if a delegate so requests.

E.7.3 Interpretation into English and French of statements made in other languages

When at a meeting of a technical committee or a subcommittee a participant wishes, in view of exceptional circumstances, to speak in any language other than English or French, the chair of the session shall be entitled to authorize this, for the session in question, provided that a means of interpretation has been secured.

In order to give all experts an equal opportunity to express their views at meetings of technical committees and subcommittees, a very flexible application of this provision is recommended.

Annex F (normative)

Options for development of a project

F.1 Simplified diagram of options

Project stage	Normal procedure	Draft submitted with proposal	"Fast-track procedure" [1]	Technical Specification [2]	Technical Report [3]	Publicly Available Specification [4]
Proposal stage (see 2.3)	Acceptance of proposal	Acceptance of proposal	Acceptance of proposal [1]	Acceptance of proposal		Acceptance of proposal [7]
Preparatory stage (see 2.4)	Preparation of working draft	*Study by working group* [5]		Preparation of draft		Approval of draft PAS
Committee stage (see 2.5)	Development and acceptance of committee draft	*Development and acceptance of committee draft* [5]		Acceptance of draft	Acceptance of draft	
Enquiry stage (see 2.6)	Development and acceptance of enquiry draft	Development and acceptance of enquiry draft	Acceptance of enquiry draft			
Approval stage (see 2.7)	Approval of FDIS [6]	Approval of FDIS [6]	Approval of FDIS [6]			
Publication stage (see 2.8)	Publication of International Standard	Publication of International Standard	Publication of International Standard	Publication of Technical Specification	Publication of Technical Report	Publication of PAS

Stages in *italics*, enclosed by dotted circles may be omitted.

1) See F.2.
2) See 3.1.
3) See 3.3.
4) See 3.2.
5) According to the result of the vote on the new work item proposal, both the preparatory stage and the committee stage may be omitted.
6) May be omitted if the enquiry draft was approved without negative votes.
7) See ISO and IEC Supplements for details on proposals for PAS.

In JTC 1, the following table is used.

Stage Name	Stage Description	Standard (see 2)	Fast Track IS (see F.2)	JTC 1 Publicly Available Specification (See F.3)	Technical Specification (see 3.1)	Technical Report (see 3.3)	Amendments (see 2.10.3)	Technical Corrigendum (see 2.10.2)
00 Preliminary stage (see 2.2)	Preparation of proposal	Preparation of NP	—	—	Preparation of NP		Preparation of NP	—
10 Proposal Stage (see 2.3)	Acceptance of proposal	Acceptance of NP	—	—	Acceptance of NP		Acceptance of NP	—
20 Preparatory stage (see 2.4)	Preparation of working draft	Preparation of WD	—	—	Preparation of WD		Preparation of WD	Preparation of Defect Report (DR)
30 Committee Stage (see 2.5)	Development and acceptance of committee draft	Development and acceptance of CD	—	—	Development and acceptance of PDTS	Development and acceptance of PDTR	Development and acceptance PDAMCDAM	Development and acceptance of DCOR
40 Enquiry Stage (see 2.6 and in IEC, see IEC Supplement E.3.1)	Development and acceptance of enquiry draft	Development and acceptance of DIS	Submission and acceptance of DIS	Submission and acceptance of DIS			Development and acceptance DAM	—
50 Approval Stage (see 2.7)	Approval of final draft	Approval of FDIS	Approval of FDIS	Approval of FDIS	—	—	Approval of FDAM	—
60 Publication Stage	Publication of document	Publication of IS	Publication of IS	Publication of IS	Publication of Technical Specification	Publication of Technical Report	Publication of Amendment	Publication of Technical Corrigendum

F.2 "Fast-track procedure"

F.2.1 Proposals to apply the fast-track procedure may be made as follows.

F.2.1.1 In JTC 1, only JTC 1 P-members and JTC 1 Category A Liaison organizations may propose Fast-Track submissions.

Any P-member or Category A liaison organization of a concerned technical committee or subcommittee may propose that an **existing standard from any source** be submitted for vote as an enquiry draft. The proposer shall obtain the agreement of the originating organization before making a proposal. The criteria for proposing an existing standard for the fast-track procedure are a matter for each proposer to decide.

In JTC 1, all fast-tracks are submitted to JTC 1, and the JTC1 secretariat submits them to the ITTF in accordance with F.4.1. The proposer of a fast-track document is encouraged to make a recommendation concerning the assignment of the document to a given subcommittee. The proposer of a fast-track document shall submit the name of an individual who has agreed to serve as project editor for the fast-track document. The proposer shall also submit an explanatory report similar to the PAS explanatory report (see F.3 below).

For its initial publication, the document is not required to be in ISO/IEC format, but can be published in its original format. The form of publication (e.g. reprint of original document or distribution of ISO/IEC cover page with reference) is to be determined by ITTF and the proposer as part of any publication agreements. However, subsequent revisions shall be in the format prescribed by the ISO/IEC Directives, Part 2.

In JTC 1, amendments to existing International Standards shall not be submitted via the fast-track procedure.

F.2.1.2 An international standardizing body recognized by the ISO or IEC council board may propose that a **standard developed by that body** be submitted for vote as a final draft International Standard.

F.2.1.3 An organization having entered into a formal technical agreement with ISO or IEC may propose, in agreement with the appropriate technical committee or subcommittee, that a **draft standard developed by that organization** be submitted for vote as an enquiry draft within that technical committee or subcommittee.

F.2.2 The proposal shall be received by the Chief Executive Officer, who shall take the following actions:

a) settle the copyright and/or trademark situation with the organization having originated the proposed document, so that it can be freely copied and distributed to National Bodies without restriction, and advise the organization that the ISO/IEC intellectual property policies shall apply to the proposed document, see in particular 2.13 and 2.14;

b) for cases F.2.1.1 and F.2.1.3, assess in consultation with the relevant secretariats which technical committee/subcommittee is competent for the subject covered by the proposed document; where no technical committee exists competent to deal with the subject of the document in question, the Chief Executive Officer shall refer the proposal to the technical management board, which may request the Chief Executive Officer to submit the document to the enquiry stage and to establish an ad hoc group to deal with matters subsequently arising;

c) ascertain that there is no evident contradiction with other International Standards;

d) distribute the proposed document as an enquiry draft (F.2.1.1 and F.2.1.3) in accordance with 2.6.1, or as a final draft International Standard (case F.2.1.2) in accordance with 2.7.1, indicating (in cases F.2.1.1 and F.2.1.3) the technical committee/subcommittee to the domain of which the proposed document belongs. In JTC 1, the subcommittee assignment recommendation and the name of the proposed project editor will also be distributed.

F.2.3 The period for voting and the conditions for approval shall be as specified in 2.6 for an enquiry draft and 2.7 for a final draft International Standard. In the case where no technical committee is involved, the condition for approval of a final draft International Standard is that not more than one-quarter of the total number of votes cast are negative.

In JTC 1, separately from their votes on the technical content of a standard, JTC 1 National Bodies shall be given the opportunity to comment on the specific subcommittee assignment of the project. However, comments on subcommittee assignments shall not influence the vote on technical content. In cases where subcommittee assignment is in question or where the fast-track document does not appear appropriate for any existing subcommittee, the JTC 1 secretariat may perform the duties normally assigned to the subcommittee secretariat until the final subcommittee assignment is determined.

In JTC 1, the proposer of the fast-track document has the right to withdraw the fast-track document from the fast-track process at any point prior to publication.

F.2.3.1 In JTC 1, a Ballot Resolution Meeting (see F.5 below) may be used to review the comments received on an enquiry draft (DIS) for fast-track ballots.

F.2.4 If, for an enquiry draft, the conditions of approval are met, the draft standard shall progress to the approval stage (2.7). If not, the proposal has failed and any further action shall be decided upon by the technical committee/subcommittee to which the document was attributed in accordance with F.2.2 b). If, for a final draft International Standard, the conditions of approval are met, the document shall progress to the publication stage (2.8). If not, the proposal has failed and any further action shall be decided upon by the technical committee/subcommittee to which the FDIS was attributed in accordance with F.2.2 b), or by discussion between the originating organization and the office of the CEO if no technical committee was involved.

In JTC 1, the committee leadership can decide whether or not to skip the FDIS vote, and go straight to publication — see 2.6.4.

In JTC 1, on receipt of the results of the voting and any comments, the chair of the technical committee or subcommittee, in cooperation with its secretariat and the project editor, and in consultation with the office of the CEO, shall take one of the following courses of action:

e) when the approval criteria of 2.6.3 are met, and no technical changes are to be included to proceed to publication (see 2.8).

f) When the approval criteria of 2.6.3 are met, but technical changes are to be included, to register the enquiry draft, as modified, as a final draft international standard.

g) when the approval criteria of 2.6.3 are not met:

  1) to circulate a revised enquiry draft for voting (see 2.6.1), or

  NOTE 2   A revised enquiry draft will be circulated for a voting period of 8 weeks, which may be extended up to 12 weeks at the request of one or more P-members of the committee concerned. In JTC 1, a revised enquiry draft circulation period may be extended up to 12 weeks.

> 2) to circulate a revised committee draft for comments, or
>
> 3) to discuss the enquiry draft and comments at the next meeting. A comment resolution meeting may be held by teleconference or using electronic means.

If the standard is published, its maintenance shall be handled by the technical committee/subcommittee to which the document was attributed in accordance with F.2.2 b), or, if no technical committee was involved, the approval procedure set out above shall be repeated if the originating organization decides that changes to the standard are required.

In JTC 1, the transposition and adoption process for a Fast-Track submission is described in F.4.

If the standard is published, its maintenance shall be handled by the technical committee/subcommittee to which the document was attributed in accordance with F.2.2 b), or, if no technical committee was involved, the approval procedure set out above shall be repeated if the originating organization decides that changes to the standard are required.

F.3 Preparation and Adoption of International Standards – JTC 1 PAS Transposition Process

JTC 1 provides Standing Document 9 on "Guide to the Transposition of Publically Available Specifications into International Standards", for potential PAS candidates.

F.3.1 Concepts

The JTC 1 PAS transposition process is based on the following key concepts:

Publicly Available Specification (PAS)

A technical specification is called a Publicly Available Specification (PAS) if it meets certain criteria making it suitable for possible processing as an International Standard. These criteria (see F.3.3 below) have been established in order to ensure a high level of quality, consensus, and proper treatment of Intellectual Property Rights (IPR) related matters.

PAS Mentor

An individual appointed by JTC 1 to assist a PAS Originator and/or Recognized PAS Submitter in creating and processing their submission(s), and to provide on-going advice.

PAS Originator

Any organization that has developed and hence owns a PAS which it considers proposing for transposition into an International Standard is called the PAS originator. There are no fundamental restrictions as to what form the organization should have, but constitutional characteristics of the organization are supposed to reflect the openness of the organization and the PAS development process. See Standing Document 9 on "Guide to the Transposition of Publically Available Specifications into International Standards", for the appropriate template.

Recognized PAS Submitter

A PAS originator shall apply to JTC 1 for recognition as a submitter of PAS(s) for transposition. Once approved, the status of a Recognized PAS submitter will remain valid for an initial period of two years, with the possibility of further extension (see F.3.4.1 below).

Explanatory Report

The submission of the PAS shall be accompanied by an explanatory report generated by the PAS originator. This report provides all information necessary to support the submission. In particular, it shall contain statements as to the extent that the PAS criteria are met by the specification. It should also clearly define the technical concepts used in the PAS. JTC 1 has developed a list of criteria to include in the explanatory report.

PAS Transposition Ballot

The PAS together with the corresponding explanatory report is submitted for ballot.

F.3.2 Applicability

These procedures apply to the transposition of a PAS into an International Standard. It is expected that these procedures will be used to process a broader class of documents from a more diverse set of sources than is currently served by the fast-track procedure (see F.2 above).

F.3.3 PAS Criteria

JTC 1 has established criteria that serve as a basis for the judgment as to whether a particular organization can be recognized and whether its specification can be accepted as a candidate for transposition into an International Standard. Such criteria may also be used by potential submitters to determine the level of suitability of their specification for the standardization process. The PAS criteria are broadly classified into two categories and address the following topics:

- Organization related criteria including:
    - Co-operative stance;
    - Characteristics of the organization;
    - Intellectual property rights.

- Document related criteria including:
    - Quality;
    - Consensus;
    - Alignment;
    - Maintenance.

Details can be found in the JTC 1 Standing Document 9 on "Guide to the Transposition of Publicly Available Specifications into International Standards".

F.3.4 Procedures

Based on the concepts provided in F.3.1 above, the PAS transposition process is described below. It is JTC 1's firm intention to provide full process transparency and the current status of any proposal from its web site (www.jtc1.org). Open dialogue (via the web site or any other available means) between the PAS Submitter and JTC 1 and JTC 1 National Bodies is strongly encouraged.

F.3.4.1 Recognition of PAS Submitter

A PAS originator interested in submitting an existing or forthcoming specification into the transposition process shall apply to the JTC 1 secretariat for recognition as a PAS submitter. Such application shall be accompanied by an identification of the initial PAS(s) which are planned to be submitted and by statements of the PAS originator regarding the organization related criteria (see below). The completed documentation shall be submitted to JTC 1 P-members 1 for a 12-week ballot. Approval as a Recognized PAS Submitter gives a PAS originator the right to submit specifications into the transposition process for a period of two years with the possibility of further extension of five year periods (see below). The recognition as a PAS submitter will terminate:

- In the absence of a successful ballot of JTC 1 P-members to confirm the status of the PAS submitter; or
- If the PAS originator fails to submit a specification to JTC 1 for transposition within the expected period (see F.3.4.2 below).

The initiative to submit an application for recognition shall come from a PAS Originator. Any JTC 1 National Body, a JTC 1 subcommittee, a JTC 1 Category A liaison, or a PAS Mentor may assist the PAS Originator in its interactions with JTC 1.

Since the ballot among JTC 1 P-members will take 12 weeks, the application for recognition should be submitted in time before the planned first submission of a PAS. While there are no particular requirements as to the format of the application, it should:

- Define the overall scope of the application;
- Identify the initial PAS(s) which are planned to be submitted, together with their scope;
- Address all mandatory elements of the organization acceptance criteria contained in the JTC 1 Standing Document 9 on "Guide to the Transposition of Publicly Available Specifications into International Standards".

The PAS submitter's expectation for maintenance of transposed PAS submissions is also stated in the application. JTC 1's intention for maintenance is to avoid any divergence between the current JTC 1 revision of a transposed PAS and the current revision of the original specification published by the PAS submitter. Therefore, the application should contain a description of how the submitting organization will work cooperatively with JTC 1 on maintenance of the standard. While JTC 1 is responsible for maintenance of the standard, this does not mean that JTC 1 itself shall perform the maintenance function. JTC 1 may approve the option of maintenance handled by the submitter as long as there is provision for participation of appropriate JTC 1 representatives, i.e. the submitters' group responsible for maintenance is designated as the JTC 1 maintenance group.

Six months prior to the expiration of an organization's status as an approved JTC 1 PAS submitter, the JTC 1 secretariat shall invite the submitter to review its future intentions as a PAS submitter and consider the following options with regard to its initial application for recognition as a JTC 1 PAS submitter:

- Revise (significant changes to the initial application, e.g. changes in scope, procedures);
- Withdraw (termination); or
- Reaffirm (extend current status with no significant changes).

If the PAS submitter chooses to revise, it shall submit a document to the JTC 1 secretariat stating the changes to the answers to the questions in the JTC 1 Standing Document 9 "Guide for the Transposition of Publicly Available Specifications" from its previous application. If the PAS submitter chooses to reaffirm, it shall identify subsequent PAS(s) intended for submission. In order to allow JTC 1 a timely reaction to the revision or affirmation, the necessary documentation should be submitted not later than 12 weeks prior to the expiration of its status as a PAS submitter. The JTC 1 secretariat shall issue a 12-week letter ballot on the request for either a revision or reaffirmation. Failure to respond to the secretariat's invitation for review of PAS submitter status will automatically result in termination of a PAS submitter's status at the conclusion of this term.

F.3.4.2 PAS Submission

Once a PAS originator has been recognized, a PAS submission to the JTC 1 secretariat may occur within the scope as identified on the application. When submitting a PAS to the JTC 1 secretariat, a Recognized PAS Submitter shall include an explanatory report and a statement that the conditions for recognition have not changed or an indication of the nature of changes that have occurred. The explanatory report shall address all mandatory elements of the document related criteria contained in the JTC 1 Standing Document 9 "Guide to the Transposition of Publicly Available Specifications" into International Standards.

If the recognized PAS submitter has received approval to perform maintenance functions, the PAS submitter should reconfirm their commitment to perform the duties of the JTC1 maintenance group in the explanatory report.

All submissions including the explanatory report shall occur in electronic form.

The first submission shall occur not later than six months after the initial recognition. On request by the PAS originator not later than six weeks before the end of this six-month period, the period may be extended for another six months, subject to approval by the JTC 1 chair and secretariat. Failure by the PAS originator to submit a specification within the expected period will result in the termination of its recognition status.

The format of the specification submitted is not regulated by JTC 1. Recognized PAS submitters are encouraged to apply, if flexibility still exists, a documentation style close to the ISO/IEC style in order to ease the later alignment process at the time of any revision.

The JTC 1 secretariat, after checking the recognition status of the submitter and the completeness of the application, shall forward the specification to the ITTF in accordance with F.4.1.

In view of the importance of the explanatory report for a successful transposition, the Recognized PAS submitter may request counsel and advice from a JTC 1 PAS Mentor, JTC 1 National Bodies, subcommittees or Category A liaison organizations during the generation of this report and throughout the transposition process. The counselling process could include a review of the submissions.

If the Recognized PAS submitter will not perform maintenance functions on the final International Standard, the Recognized PAS submitter is encouraged to make a recommendation concerning the assignment of the document to a given subcommittee. This recommendation (or in its absence, the JTC 1 secretariat's recommendation) shall be circulated to JTC 1 P-members together with the ballot, but the recommendation shall not influence the vote. In cases where the subcommittee assignment is in question or where the document does not appear appropriate for any existing subcommittee, the JTC 1 secretariat should perform the duties normally assigned to the subcommittee secretariat until the final subcommittee assignment is determined.

In cases where the Recognized PAS submitter has received approval to perform maintenance functions, on the final International Standard, the ballot is assigned to the JTC 1 Secretariat and the JTC 1 Secretariat shall perform all duties indicated in F.4.

F.4 Adoption of Submissions under the JTC 1 Fast-Track Procedure or JTC 1 PAS Transposition Process

F.4.1 The JTC 1 secretariat forwards the Fast-Track or PAS submission, together with the explanatory report and related documentation to ITTF.

F.4.2 The ITTF shall take the following actions:

- Settle the copyright or trademark situation, or both, with the PAS or Fast-Track submitter, so that the proposed text can be copied and distributed within ISO/IEC without restriction;
- Assess in consultation with the JTC 1 secretariat that JTC 1 is the competent committee for the subject covered in the proposed standard and ascertain that there is no evident contradiction with other ISO/IEC standards.
- Distribute the text of the proposed standard as a Draft International Standard (DIS), together with the explanatory report and related documentation, indicating that the standard falls within the scope of JTC 1.

F.4.3 The period for DIS voting shall be a 12-week ballot with a translation period of 8 weeks (see JA.6). In order to be accepted the DIS shall meet the conditions for approval as specified in 2.6.3. For JTC 1 Fast-Track and JTC 1 PAS, the "technical committee or subcommittee" of 2.6.3 a) is understood to be JTC 1.

F.4.4 Upon receipt of notification from the ITTF that a DIS has been registered, the JTC 1 secretariat shall inform the secretariat of the subcommittee recommended for assignment of the project of the DIS number, title, and ballot period dates, and shall send the subcommittee secretariat a copy of the DIS and its attached explanatory report. The JTC 1 secretariat shall also inform the ITTF of the subcommittee that will deal with the DIS ballot results, in order that the table of replies and any comments accompanying the votes may be sent by ITTF directly to the SC secretariat as well as to the JTC 1 secretariat.

F.4.5 Reflecting the importance of the Fast-Track process and the JTC 1 PAS Transposition Procedure, the JTC 1 secretariat shall also inform JTC 1 National Bodies and Liaison Organizations, and those organizations authorized to be Recognized PAS submitters, of the initiation of any Fast-Track or PAS ballot, the results of the ballot, and the identity of the JTC 1 subcommittee which will be responsible for any future work.

F.4.6 The Fast-Track or PAS submitter shall receive a copy of the ballot documentation.

F.4.7 Upon receipt of the notification from the JTC 1 secretariat that its subcommittee has been assigned the responsibility for dealing with the DIS, the subcommittee secretariat shall so inform the JTC 1 National Bodies that are members of the subcommittee, and shall make plans for a possible Ballot Resolution Meeting (see F.5 below).

F.4.8 Upon receipt of the DIS ballot results, and any comments, the SC secretariat shall distribute this material to the subcommittee, JTC 1 National Bodies and the PAS or Fast-Track Submitter.

The committee leadership can decide whether or not to skip the FDIS vote, and go straight to publication — see 2.6.4.

On receipt of the results of the voting and any comments, the chair of the technical committee or subcommittee, in cooperation with its secretariat and the project editor, and in consultation with the office of the CEO, shall take one of the following courses of action:

h) when the approval criteria of 2.6.3 are met, and no technical changes are to be included to proceed to publication (see 2.8).

i) When the approval criteria of 2.6.3 are met, but technical changes are to be included, to register the enquiry draft, as modified, as a final draft international standard.

j) when the approval criteria of 2.6.3 are not met;

   1) to circulate a revised enquiry draft for voting (see 2.6.1), or

   NOTE 2   A revised enquiry draft will be circulated for a voting period of 8 weeks, which may be extended up to 12 weeks at the request of one or more P-members of the committee concerned. In JTC 1, a revised enquiry draft circulation period may be extended up to 12 weeks.

   2) to circulate a revised committee draft for comments, or

   3) to discuss the enquiry draft and comments at the next meeting. A comment resolution meeting may be held by teleconference or using electronic means.

If the approval requirements in 2.6.3 for a DIS ballot have not been met, the proposal has failed. For JTC 1 Fast-Track and JTC 1 PAS, the "technical committee or subcommittee" of 2.6.3 a) is understood to be JTC 1. In this case, JTC 1 shall make known to the submitter the reasons which have led to the negative result. Based on this information, the submitter may choose to re-submit a modified specification as a new Fast-Track or PAS submission.

In the case where no negative votes and no comments other than editorial corrections and comments that cannot be accommodated by a textual change to the balloted document have been received, the text after incorporation of the editorial comments by the editor may proceed directly to publication and no ballot resolution meeting will be held.

In the case where one or more negative votes and no comments other than editorial corrections and comments that cannot be accommodated by a textual change to the balloted document have been received, the project editor shall incorporate the editorial corrections and send the corrected DIS to the ITTF for FDIS balloting. The ballot period for FDIS is 8 weeks.

If comments others than editorial corrections and comments that cannot be accommodated by a textual change to the balloted document have been received, a Ballot Resolution Meeting (BRM) is conducted (see F.5).

F.4.9 After the deliberations of a Ballot Resolution Meeting (if held following a successful DIS vote), the following cases may occur:

a) No changes other than editorial corrections have been made to the original DIS text, proceed to publication;

- ~~if no negative votes have been received, the text after incorporation of the editorial comments by the editor the text may proceed directly to publication; or~~
- ~~if negative votes have been received the project editor shall incorporate the editorial corrections and send the amended DIS to the subcommittee secretariat who shall forward it to the ITTF for FDIS balloting. The ballot period for FDIS is 8 weeks.~~

b) Changes other than editorial corrections have been agreed during the Ballot Resolution Meeting: in this case, the project editor shall prepare the amended DIS and send it to the committee secretariat who shall forward it to the ITTF for FDIS balloting. The ballot period for FDIS is 8 weeks.

F.4.10 Upon receipt of notification from the ITTF that a FDIS has been registered, the JTC 1 secretariat shall inform the secretariat of the SC recommended for assignment of the project of the FDIS number, title, and ballot period dates, and shall send the subcommittee secretariat a copy of the FDIS and the disposition of comments received on the DIS ballot, if any are received. The table of replies and any comments accompanying the votes will be sent by ITTF directly to the subcommittee secretariat as well as to the JTC 1 secretariat.

F.4.11 If the requirements of 2.7.3 are met, the text will be published by ITTF as an International Standard. For its initial publication, the document is not required to be in ISO/IEC format, but can be published in its original format. The form of publication (e.g. reprint of original document or distribution of ISO/IEC cover page with reference) is to be determined by ITTF and the PAS or Fast-Track submitter as part of any publication agreements. However, subsequent revisions shall be in the format prescribed by the ISO/IEC Directives, Part 2.

F.4.12 If it is impossible to agree to text meeting the FDIS approval requirements (see 2.7.3), the proposal has failed. In this case, JTC 1 shall make known to the submitter the reasons which have led to the negative result. Based on this information, the submitter may choose to re-submit a modified specification as a new Fast-Track or PAS submission.

F.4.13 The time period for post ballot activities by the respective responsible parties shall be as follows:

- immediately after the DIS and FDIS votes, the ITTF shall send the results of the vote to the JTC 1 secretariat and to the subcommittee secretariat, and the latter shall distribute the results without delay to JTC 1 National Bodies that are members of the subcommittee, to any National Bodies having voted that are not members of the subcommittee and to the proposer;
- as soon as possible after the distribution of the results of the vote to JTC 1 National Bodies that are members of the subcommittee, but in not less than two and one-half months the subcommittee secretariat shall convene a Ballot Resolution Meeting (BRM), if required;
- in not more than 4 weeks after the Ballot Resolution Meeting the subcommittee secretariat shall distribute the final report of the meeting and the amended DIS text.

F.4.14 If the proposed standard is accepted, it will be published following ISO and IEC standing copyright and other IPR policy. Its maintenance will be handled either by JTC 1 or by a JTC 1 designated maintenance group of the PAS submitter in accordance with JTC 1 rules.

It is at the discretion of the PAS or Fast-Track submitter to withdraw the document from the transposition process at any point prior to publication. It is also the right of the PAS or Fast-Track submitter to request that the document remain unchanged throughout the transposition process. Such a request should be clearly stated in the explanatory report.

F.5 JTC 1 PAS and Fast-Track Ballot Resolution Meetings

F.5.1 Ballot Resolution Meeting Purpose and Scope

In JTC 1, the purpose of a Ballot Resolution Meeting (BRM) is to review the comments received on an enquiry draft (DIS) for JTC 1 PAS or Fast-Track ballots (see F.4); further it shall formulate dispositions to those comments to receive the widest possible consensus. In some cases, the subcommittee secretariat may decide that a Ballot Resolution Meeting is unnecessary and assign the resolution of comments directly to the project editor.

F.5.2 Responsibilities of the assigned subcommittee for the Ballot Resolution Meeting

JTC 1 usually assigns an enquiry draft (DIS) to one of its subcommittees. Where the DIS is not assigned to a specific subcommittee, the JTC 1 secretariat will carry out the tasks assigned.

The assigned secretariat shall:

- schedule a Ballot Resolution Meeting, to be held not earlier than two and a half months after the distribution of the comments, to consider any comments on the DIS;
- appoint a convenor for the Ballot Resolution Meeting;
- notify the eligible attendees of the Ballot Resolution Meeting date(s), location, and convenor.

No later than 8 weeks before the start of the Ballot Resolution Meeting, the assigned secretariat shall send the logistical information and agenda together with the notification of the convenor to the JTC 1 secretariat for circulation to the recipients listed in F.5.4 below

F.5.3 Proposed dispositions of comments

The project editor assigned to the DIS shall prepare the Proposed Disposition of Comments (DoC) on the ISO template (final column).

No later than 4 weeks before the start of the Ballot Resolution Meeting, the assigned secretariat shall circulate the proposed Disposition of Comments document to the listed recipients in F.5.4 below.

F.5.4 Recipients and eligible attendance

The assigned secretariat shall make available the Proposed Disposition of Comments (DoC) via ITTF to the following who are eligible to attend or to nominate representatives to the Ballot Resolution Meeting:

- representatives of the National Bodies;
- representatives of the ISO and IEC Central Offices;
- the subcommittee chair;
- the subcommittee ~~secretary~~committee manager;
- the assigned project editor(s);
- the Ballot Resolution Meeting convenor;
- the Draft International Standard submitter; and
- JTC 1 Category A liaisons.

F.5.5 Meeting Procedures

The Ballot Resolution Meeting shall be convened as a separate meeting even if held in conjunction with other meetings of JTC 1 or the relevant subcommittee. The BRM may be held by teleconference or using electronic means or face-to-face (see Standing Document 19 on "Meetings").

The appointed convenor shall hold a roll-call.

The Ballot Resolution Meeting record shall list the Heads of Delegation (HoD), who represent their National Body positions, if needed in a vote, as well as all the other attendees and their roles.

The Ballot Resolution Meeting shall address and attempt as far as possible to resolve all comments raised during the Draft International Standard ballot to increase consensus on the resulting document.

For each of the comments, the project editor shall record the disposition on which the Ballot Resolution Meeting achieves consensus, or if that fails, the proposition that gets the majority support of those National Bodies that were present at the BRM and eligible to vote on the Draft International Standard ballot, in the final Disposition of Comments report.

When all DIS ballot comments have been addressed and the disposition of comments has been approved by the meeting, the Ballot Resolution Meeting goals have been met.

No longer than 4 weeks after the close of the meeting, or as permitted by ITTF, the subcommittee secretariat shall distribute

- a revision of the Draft International Standard balloted document that includes all changes agreed to at the Ballot Resolution Meeting;
- the disposition of comments report approved at the Ballot Resolution Meeting; and
- a Ballot Resolution Meeting report containing a list of attendees and their roles, referencing the final disposition of comments report and a recommendation for further processing of the draft International Standard.

These documents shall also be forwarded to ITTF for further circulation to the above listed recipients.

F.6 Progression of Fast-Track and PAS Submissions in JTC 1

In JTC 1, the following flow chart specifies the progression of Fast-Track and PAS Submissions as referenced in F.2, F.3, F.4, and F.5.

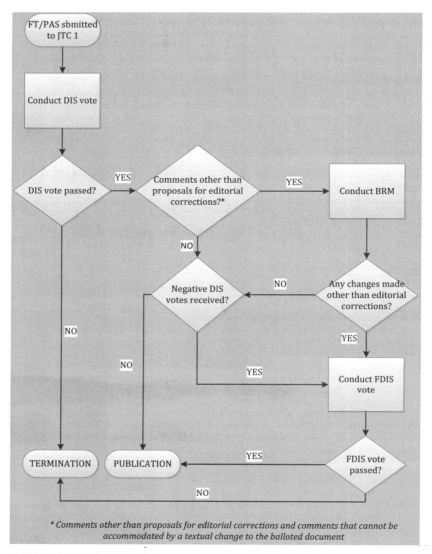

Figure F.1: Flowchart of Fast-Track and PAS Submissions for International Standards

Annex G
(normative)

Maintenance agencies

G.1 A technical committee or subcommittee developing an International Standard that will require a maintenance agency shall inform the Chief Executive Officer at an early stage in order that an ISO/TMB or IEC Council Board decision may be taken in advance of the publication of the International Standard.

G.2 The ISO/TMB or IEC Council Board designates maintenance agencies in connection with International Standards, including appointment of their members, on the proposal of the technical committee concerned.

G.3 The secretariat of a maintenance agency should be attributed wherever possible to the secretariat of the technical committee or subcommittee that has prepared the International Standard.

G.4 The Chief Executive Officer shall be responsible for contacts with external organizations associated with the work of a maintenance agency.

G.5 The rules of procedure of maintenance agencies shall be subject to ISO/TMB or IEC Council Board approval and any requested delegation of authority in connection with the updating of the International Standard or the issuing of amendments shall be specifically authorized by the ISO/TMB or IEC Council Board.

G.6 Any charges for services provided by a maintenance agency shall be authorized by the council board.

Annex H
(normative)

Registration authorities

H.1 A technical committee or subcommittee developing an International Standard that will require a registration authority shall inform the Chief Executive Officer at an early stage, in order to permit any necessary negotiations and to allow the technical management board to take a decision in advance of the publication of the International Standard.

In JTC 1, the group developing the standard which requires a Registration Authority shall develop the accompanying procedures which shall be approved by JTC 1 ballot.

H.2 The technical management board designates registration authorities in connection with International Standards on the proposal of the technical committee concerned.

H.3 Registration authorities should be qualified and internationally acceptable bodies; if there is no such organization available, such tasks may be conferred upon the office of the CEO by decision of the technical management board.

H.4 Registration authorities should be required to indicate clearly in their operations that they have been designated by ISO or IEC (for example, by including appropriate wording in the letterhead of the designated body).

H.5 Registration functions undertaken by the registration authority under the provisions of the relevant International Standard shall require no financial contribution from ISO or IEC or their members. This would not preclude, however, the charging for services provided by the registration authority if duly authorized by the council board.

In JTC 1, for further information on JTC 1 Registration Authorities, see the JTC 1 Standing Document 16 on "Registration Authorities".

Annex I
(normative)

Guideline for Implementation of the Common Patent Policy for ITU-T/ITU-R/ISO/IEC

The latest edition of the Guidelines for Implementation of the Common Patent Policy for ITU-T/ITU-R/ISO/IEC are available on the ISO website through the following link (including the forms in Word or Excel formats):

http://www.iso.org/iso/home/standards_development/governance_of_technical_work/patents.htm

They are also available on the IEC website through the following link:

http://www.iec.ch/members_experts/tools/patents/patent_policy.htm

Guidelines for Implementation of the Common Patent Policy for ITU-T/ITU-R/ISO/IEC

Summary

The Guidelines for Implementation of the Common Patent Policy for ITU-T/ITU-R/ISO/IEC are intended to clarify and facilitate implementation of the Patent Policy, a copy of which can be found in Annex 1 and also on the web site of each Organization.

The Patent Policy encourages the early disclosure and identification of Patents that may relate to Recommendations | Deliverables under development. In doing so, greater efficiency in standards development is possible and potential patent rights problems can be avoided.

History

Edition		
1.0	Published initial version	1 March 2007
2.0	Published first revision	23 April 2012
3.0	Published second revision	26 June 2015
4.0	Published third revision	2 November 2018

CONTENTS

Part I – Common guidelines
1 Purpose
2 Explanation of terms
3 Patent disclosure
4 Patent Statement and Licensing Declaration Form
    4.1 The purpose of the Declaration Form
    4.2 Contact information
5 Conduct of meetings
6 Patent Information database
7 Assignment or Transfer of Patent Rights

Part II – Organization-specific provisions
II.1 Specific provisions for ITU
II.2 Specific provisions for ISO and IEC

ANNEX 1: COMMON PATENT POLICY FOR ITU-T/ITU-R/ISO/IEC

ANNEX 2: PATENT STATEMENT AND LICENSING DECLARATION FORM FOR ITU-T OR ITU-R RECOMMENDATION | ISO OR IEC DELIVERABLE

ANNEX 3: GENERAL PATENT STATEMENT AND LICENSING DECLARATION FORM FOR ITU- T OR ITU-R RECOMMENDATION

Guidelines for Implementation of the Common Patent Policy for ITU-T/ITU-R/ISO/IEC

Revision 3, effective 2 November 2018

Part I – Common guidelines

1 Purpose

ITU, in its Telecommunication Standardization Sector (ITU-T) and its Radiocommunication Sector (ITU-R), ISO and IEC have had patent policies for many years, the purpose being to provide in simple words practical guidance to the participants in their Technical Bodies in case patent rights matters arise.

Considering that the technical experts are normally not familiar with the complex issue of patent law, the Common Patent Policy for ITU-T/ITU-R/ISO/IEC (hereafter referred to as the "Patent Policy") was drafted in its operative part as a checklist, covering the three different cases which may arise if a Recommendation | Deliverable requires licences for Patents to be practiced or implemented, fully or partly.

The Guidelines for Implementation of the Common Patent Policy for ITU-T/ITU-R/ISO/IEC (hereafter referred to as the "Guidelines") are intended to clarify and facilitate implementation of the Patent Policy, a copy of which can be found in Annex 1 and also on the web site of each Organization.

The Patent Policy encourages the early disclosure and identification of Patents that may relate to Recommendations | Deliverables under development. In doing so, greater efficiency in standards development is possible and potential patent rights problems can be avoided.

The Organizations should not be involved in evaluating patent relevance or essentiality with regards to Recommendations | Deliverables, interfere with licensing negotiations, or engage in settling disputes on Patents; this should be left - as in the past - to the parties concerned.

Organization-specific provisions are contained in Part II of this document. However, it is understood that those Organization-specific provisions shall contradict neither the Patent Policy nor the Guidelines.

2   Explanation of terms

Contribution: Any document submitted for consideration by a Technical Body.

Free of Charge: The words "Free of Charge" do not mean that the Patent Holder is waiving all of its rights with respect to the Patent. Rather, "Free of Charge" refers to the issue of monetary compensation; i.e., that the Patent Holder will not seek any monetary compensation as part of the licensing arrangement (whether such compensation is called a royalty, a one-time licensing fee, etc.). However, while the Patent Holder in this situation is committing to not charging any monetary amount, the Patent Holder is still entitled to require that the implementer of the relevant Recommendation | Deliverable sign a license agreement that contains other reasonable terms and conditions such as those relating to governing law, field of use, warranties, etc.

Organizations: ITU, ISO and IEC.

Patent: The word "Patent" means those claims contained in and identified by patents, utility models and other similar statutory rights based on inventions (including applications for any of these) solely to the extent that any such claims are essential to the implementation of a Recommendation | Deliverable. Essential patents are patents that would be required to implement a specific Recommendation | Deliverable.

Patent Holder: Person or entity that owns, controls and/or has the ability to license Patents.

Reciprocity: The word "Reciprocity" means that the Patent Holder shall only be required to license any prospective licensee if such prospective licensee will commit to license its Patent(s) for implementation of the same relevant Recommendation | Deliverable Free of Charge or under reasonable terms and conditions.

Recommendations | Deliverables: ITU-T and ITU-R Recommendations are referred to as "Recommendations", ISO deliverables and IEC deliverables are referred to as "Deliverables". The various types of Recommendation(s) | Deliverable(s) are referred to as "Document types" in the Patent Statement and Licensing Declaration Form (hereafter referred to as "Declaration Form") attached as Annex 2.

Technical Bodies: Study Groups, any subordinate groups and other groups of ITU-T and ITU-R and technical committees, subcommittees and working groups in ISO and IEC.

3 Patent disclosure

As mandated by the Patent Policy in its paragraph 1, any party participating[1] in the work of the Organizations should, from the outset, draw their attention to any known Patent or to any known pending Patent application, either its own or that of other organizations.

In this context, the words "from the outset" imply that such information should be disclosed as early as possible during the development of the Recommendation | Deliverable. This might not be possible when the first draft text appears since at this time, the text might be still too vague or subject to subsequent major modifications. Moreover, that information should be provided in good faith and on a best effort basis, but there is no requirement for patent searches.

In addition to the above, any party not participating in Technical Bodies may draw the attention of the Organizations to any known Patent, either their own and/or of any third-party.

When disclosing their own Patents, Patent Holders have to use the Patent Statement and Licensing Declaration Form (referred to as the "Declaration Form") as stated in Section 4 of these Guidelines.

---

[1] In the case of ISO and IEC, this includes any recipient of a draft standard at any stage in the standards development process.

Any communication drawing the attention to any third-party Patent should be addressed to the concerned Organization(s) in writing. The potential Patent Holder will then be requested by the Director/CEO of the relevant Organization(s) to submit a Declaration Form, if applicable.

The Patent Policy and these Guidelines also apply to any Patent disclosed or drawn to the attention of the Organizations subsequent to the approval of a Recommendation | Deliverable.

Whether the identification of the Patent took place before or after the approval of the Recommendation | Deliverable, if the Patent Holder is unwilling to license under paragraph 2.1 or

2.2 of the Patent Policy, the Organizations will promptly advise the Technical Bodies responsible for the affected Recommendation | Deliverable so that appropriate action can be taken. Such action will include, but may not be limited to, a review of the Recommendation | Deliverable or its draft in order to remove the potential conflict or to further examine and clarify the technical considerations causing the conflict.

4 Patent Statement and Licensing Declaration Form

4.1 The purpose of the Declaration Form

To provide clear information in the Patent Information databases of each Organization, Patent Holders have to use the Declaration Form, which is available on the web site of each Organization (the Declaration Form is included in Annex 2 for information purposes). They must be sent to the Organizations for the attention, for ITU, of the Directors of the TSB or the BR or, for ISO or IEC, of the CEOs. The purpose of the Declaration Form is to ensure a standardized submission to the respective Organizations of the declarations being made by Patent Holders.

The Declaration Form gives Patent Holders the means of making a licensing declaration relative to rights in Patents required for implementation of a specific Recommendation | Deliverable. Specifically, by submitting this Declaration Form the submitting party declares its willingness to license (by selecting option 1 or 2 on the Form) /or its unwillingness to license (by selecting option 3 on the Form), according to the Patent Policy, Patents held by it and whose licence would be required to practice or implement part(s) or all of a specific Recommendation | Deliverable.

If a Patent Holder has selected the licensing option 3 on the Declaration Form, then, for the referenced relevant ITU Recommendation and ISO or IEC Deliverable, the ITU, ISO and IEC require the Patent Holder to provide certain additional information permitting patent identification.

Multiple Declaration Forms are appropriate if the Patent Holder wishes to identify several Patents and classifies them in different options of the Declaration Form for the same Recommendation | Deliverable or if the Patent Holder classifies different claims of a complex patent in different options of the Declaration Form.

Information contained in a Declaration Form may be corrected in case of obvious errors, such as a typographical mistake in a standard or patent reference number. The licensing declaration contained in the Declaration Form remains in force unless it is superseded by another Declaration Form containing more favourable licensing terms and conditions from a licensee's perspective reflecting

(a) a change in commitment from option 3 to either option 1 or option 2, (b) a change in commitment from option 2 to option 1 or (c) un-checking one or more sub-options contained within option 1 or 2.

4.2 Contact information

In completing Declaration Forms, attention should be given to supplying contact information that will remain valid over time. Where possible, the "Name and Department" and e-mail address should be generic. Also it is preferable, where possible, that parties, particularly multinational organizations, indicate the same contact point on all Declaration Forms submitted.

With a view to maintaining up-to-date information in the Patent Information database of each Organization, it is requested that the Organizations be informed of any change or corrections to the Declaration Form submitted in the past, especially with regard to the contact person.

5 Conduct of meetings

Early disclosure of Patents contributes to the efficiency of the process by which Recommendations | Deliverables are established. Therefore, each Technical Body, in the course of the development of a proposed Recommendation | Deliverable, will request the disclosure of any known Patents essential to the proposed Recommendation | Deliverable.

Chairmen of Technical Bodies will, if appropriate, ask, at an appropriate time in each meeting, whether anyone has knowledge of patents, the use of which may be required to practice or implement the Recommendation | Deliverable being considered. The fact that the question was asked shall be recorded in the meeting report, along with any affirmative responses.

As long as the Organization concerned has received no indication of a Patent Holder selecting paragraph 2.3 of the Patent Policy, the Recommendation | Deliverable may be approved using the appropriate and respective rules of the Organization concerned. It is expected that discussions in Technical Bodies will include consideration of including patented material in a Recommendation | Deliverable, however the Technical Bodies may not take position regarding the essentiality, scope, validity or specific licensing terms of any claimed Patents.

6 Patent Information database

In order to facilitate both the standards-making process and the application of Recommendations | Deliverables, each Organization makes available to the public a Patent Information database composed of information that was communicated to the Organizations by the means of Declaration Forms. The Patent Information database may contain information on specific patents, or may contain no such information but rather a statement about compliance with the Patent Policy for a particular Recommendation | Deliverable.

The Patent Information databases are not certified to be either accurate or complete, but only reflect the information that has been communicated to the Organizations. As such, the Patent Information databases may be viewed as simply raising a flag to alert users that they may wish to contact the entities who have communicated Declaration Forms to the Organizations in order to determine if patent licenses must be obtained for use or implementation of a particular Recommendation | Deliverable.

7   Assignment or Transfer of Patent Rights

The rules governing the assignment or transfer of Patent rights are contained in the patent statement and licensing declaration forms (see Annexes 2 and 3). By complying with these rules, the Patent Holder has discharged in full all of its obligations and liability with regards to the licensing commitments after the transfer or assignment. These rules are not intended to place any duty on the Patent Holder to compel compliance with the licensing commitment by the assignee or transferee after the transfer occurs.

Part II – Organization specific provisions

II.1 Specific provisions for ITU

ITU-1 General Patent Statement and Licensing Declaration Form

Anyone may submit a General Patent Statement and Licensing Declaration Form which is available on the web sites of ITU-T and ITU-R (the form in Annex 3 is included for information purposes). The purpose of this form is to give Patent Holders the voluntary option of making a general licensing declaration relative to material protected by Patents contained in any of their Contributions. Specifically, by submitting its form, the Patent Holder declares its willingness to license its Patents owned by it in case part(s) or all of any proposals contained in its Contributions submitted to the Organization are included in Recommendation(s) and the included part(s) contain items for which Patents have been filed and whose licence would be required to practice or implement Recommendation(s).

The General Patent Statement and Licensing Declaration Form is not a replacement for the "individual" (see clause 4 of Part I) Declaration Form, which is made per Recommendation, but is expected to improve responsiveness and early disclosure of the Patent Holder's compliance with the Patent Policy. Therefore, in addition to its existing General Patent Statement and Licensing Declaration in respect of its Contributions, the Patent Holder should, when appropriate (e.g. if it becomes aware that it has a Patent for a specific Recommendation), also submit an "individual" Patent Statement and Licensing Declaration Form:

— for the Patents contained in any of its Contributions submitted to the Organization which are included in a Recommendation, any such "individual" Patent Statement and Licensing Declarations may contain either the same licensing terms and conditions as in the General Patent Statement and Licensing Declaration Form, or more favourable licensing terms and conditions from a licensee's perspective as defined in the "individual" (see clause 4.1 of Part I) Declaration Form; and

— for the Patents that the Patent Holder did not contribute to the Organization which are included in a Recommendation, any such "individual" Patent Statement and Licensing Declarations may contain any of the three options available on the Form (see clause 4.1 of Part I), regardless of the commitment in its existing General Patent Statement and Licensing Declaration.

The General Patent Statement and Licensing Declaration remains in force unless it is superseded by another General Patent Statement and Licensing Declaration form containing more favourable licensing terms and conditions from a licensee's perspective reflecting (a) a change in commitment from option 2 to option 1 or (b) un-checking one or more sub- options contained within option 1 or 2.

The ITU Patent Information database also contains a record of General Patent Statement and Licensing Declarations.

ITU-2 Notification

Text shall be added to the cover sheets of all new and revised ITU-T and ITU-R Recommendations, where appropriate, urging users to consult the ITU Patent Information database. The wording is:

"ITU draws attention to the possibility that the practice or implementation of this Recommendation may involve the use of a claimed Intellectual Property Right. ITU takes no position concerning the evidence, validity or applicability of claimed Intellectual Property Rights, whether asserted by ITU members or others outside of the Recommendation development process.

As of the date of approval of this Recommendation, ITU [had/had not] received notice of intellectual property, protected by patents, which may be required to implement this Recommendation. However, implementers are cautioned that this may not represent the latest information and are therefore strongly urged to consult the ITU Patent Information database."

II.2 Specific provisions for ISO and IEC

ISO/IEC-1 Consultations on draft Deliverables

All drafts submitted for comment shall include on the cover page the following text:

"Recipients of this draft are invited to submit, with their comments, notification of any relevant patent rights of which they are aware and to provide supporting documentation."

ISO/IEC-2 Notification

A published document for which no patent rights are identified during the preparation thereof shall contain the following notice in the foreword:

"Attention is drawn to the possibility that some of the elements of this document may be the subject of patent rights. ISO [and/or] IEC shall not be held responsible for identifying any or all such patent rights."

A published document for which patent rights have been identified during the preparation thereof shall include the following notice in the introduction:

"The International Organization for Standardization (ISO) [and/or] International Electrotechnical Commission (IEC) draws attention to the fact that it is claimed that compliance with this document may involve the use of a patent concerning (... subject matter ...) given in (... subclause ...).

ISO [and/or] IEC take[s] no position concerning the evidence, validity and scope of this patent right.

The holder of this patent right has assured the ISO [and/or] IEC that he/she is willing to negotiate licences under reasonable and non-discriminatory terms and conditions with applicants throughout the world. In this respect, the statement of the holder of this patent right is registered with ISO [and/or] IEC. Information may be obtained from:

name of holder of patent right ...

address ...

Attention is drawn to the possibility that some of the elements of this document may be the subject of patent rights other than those identified above. ISO [and/or] IEC shall not be held responsible for identifying any or all such patent rights."

ISO/IEC - 3 National Adoptions

Patent Declarations in ISO, IEC and ISO/IEC Deliverables apply only to the ISO and/or IEC documents indicated in the Declaration Forms. Declarations do not apply to documents that are altered (such as through national or regional adoption). However, implementations that conform to identical national and regional adoptions and the respective ISO and/or IEC Deliverables, may rely on Declarations submitted to ISO and/or IEC for such Deliverables.

ANNEX 1

COMMON PATENT POLICY FOR ITU-T/ITU-R/ISO/IEC

The following is a "code of practice" regarding patents covering, in varying degrees, the subject matters of ITU-T Recommendations, ITU-R Recommendations, ISO deliverables and IEC deliverables (for the purpose of this document, ITU-T and ITU-R Recommendations are referred to as "Recommendations", ISO deliverables and IEC deliverables are referred to as "Deliverables"). The rules of the "code of practice" are simple and straightforward. Recommendations |Deliverables are drawn up by technical and not patent experts; thus, they may not necessarily be very familiar with the complex international legal situation of intellectual property rights such as patents, etc.

Recommendations | Deliverables are non-binding; their objective is to ensure compatibility of technologies and systems on a worldwide basis. To meet this objective, which is in the common interests of all those participating, it must be ensured that Recommendations | Deliverables, their applications, use, etc. are accessible to everybody.

It follows, therefore, that a patent embodied fully or partly in a Recommendation | Deliverable must be accessible to everybody without undue constraints. To meet this requirement in general is the sole objective of the code of practice. The detailed arrangements arising from patents (licensing, royalties, etc.) are left to the parties concerned, as these arrangements might differ from case to case.

This code of practice may be summarized as follows:

1   The ITU Telecommunication Standardization Bureau (TSB), the ITU Radio-communication Bureau (BR) and the offices of the CEOs of ISO and IEC are not in a position to give authoritative or comprehensive information about evidence, validity or scope of patents or similar rights, but it is desirable that the fullest available information should be disclosed. Therefore, any party participating in the work of ITU, ISO or IEC should, from the outset, draw the attention of the Director of ITU-TSB, the Director of ITU-BR, or the offices of the CEOs of ISO or IEC, respectively, to any known patent or to any known pending patent application, either their own or of other organizations, although ITU, ISO or IEC are unable to verify the validity of any such information.

2   If a Recommendation | Deliverable is developed and such information as referred to in paragraph 1 has been disclosed, three different situations may arise:

2.1 The patent holder is willing to negotiate licences free of charge with other parties on a non-discriminatory basis on reasonable terms and conditions. Such negotiations are left to the parties concerned and are performed outside ITU-T/ITU-R/ISO/IEC.

2.2 The patent holder is willing to negotiate licences with other parties on a non-discriminatory basis on reasonable terms and conditions. Such negotiations are left to the parties concerned and are performed outside ITU-T/ITU-R/ISO/IEC.

2.3 The patent holder is not willing to comply with the provisions of either paragraph 2.1 or paragraph 2.2; in such case, the Recommendation | Deliverable shall not include provisions depending on the patent.

3   Whatever case applies (2.1, 2.2 or 2.3), the patent holder has to provide a written statement to be filed at ITU-TSB, ITU-BR or the offices of the CEOs of ISO or IEC, respectively, using the appropriate "Patent Statement and Licensing Declaration" Form. This statement must not include additional provisions, conditions, or any other exclusion clauses in excess of what is provided for each case in the corresponding boxes of the form.

ANNEX 2

PATENT STATEMENT AND LICENSING DECLARATION FORM FOR ITU-T OR ITU-R RECOMMENDATION | ISO OR IEC DELIVERABLE

5.2 Patent Statement and Licensing Declaration

for ITU-T or ITU-R Recommendation | ISO or IEC Deliverable

This declaration does not represent an actual grant of a license

Please return to the relevant organization(s) as instructed below per document type:

Director Telecommunication Standardization Bureau International Telecommunication Union Place des Nations CH-1211 Geneva 20, Switzerland Fax: +41 22 730 5853 Email: tsbdir@itu.int	Director Radiocommunication Bureau International Telecommunication Union Place des Nations CH-1211 Geneva 20, Switzerland Fax: +41 22 730 5785 Email: brmail@itu.int	Secretary-General International Organization for Standardization 8 Chemin de Blandonnet CP 401 1214 Vernier, Geneva Switzerland Fax: +41 22 733 3430 Email: patent.statements@iso.org	General Secretary International Electrotechnical Commission 3 rue de Varembé CH-1211 Geneva 20 Switzerland Fax: +41 22 919 0300 Email: inmail@iec.ch

Patent Holder: Legal Name
Contact for license application: Name & Department Address
Tel.
Fax
E-mail
URL (optional)
Document type: ☐ **ITU-T Rec. (*)** ☐ **ITU-R Rec. (*)** ☐ **ISO Deliverable (*)** ☐ **IEC Deliverable (*)** (please return the form to the relevant Organization) ☐ **Common text or twin text (ITU-T Rec.	ISO/IEC Deliverable (*))** (for common text or twin text, please return the form to each of the three Organizations: ITU-T, ISO, IEC) ☐ **ISO/IEC Deliverable (*)** (for ISO/IEC Deliverables, please return the form to both ISO and IEC)
(*)Number
(*)Title

Licensing declaration:

The Patent Holder believes that it holds granted and/or pending applications for Patents, the use of which would be required to implement the above document and hereby declares, in accordance with the Common Patent Policy for ITU-T/ITU-R/ISO/IEC, that (check <u>one</u> box only):

☐ 1. The Patent Holder is prepared to grant a <u>Free of Charge</u> license to an unrestricted number of applicants on a worldwide, non-discriminatory basis and under other reasonable terms and conditions to make, use, and sell implementations of the above document.
Negotiations are left to the parties concerned and are performed outside the ITU-T, ITU-R, ISO or IEC.

Also mark here___ if the Patent Holder's willingness to license is conditioned on <u>Reciprocity</u> for the above document.

Also mark here if the Patent Holder reserves the right to license on reasonable terms and conditions (but not <u>Free of Charge</u>) to applicants who are only willing to license their Patent, whose use would be required to implement the above document, on reasonable terms and conditions (but not Free of Charge).

☐ 2. The Patent Holder is prepared to grant a license to an unrestricted number of applicants on a worldwide, non-discriminatory basis and on reasonable terms and conditions to make, use and sell implementations of the above document.
Negotiations are left to the parties concerned and are performed outside the ITU-T, ITU-R, ISO, or IEC.

Also mark here if the Patent Holder's willingness to license is conditioned on <u>Reciprocity</u> for the above document.

☐ 3. The Patent Holder is unwilling to grant licenses in accordance with provisions of either 1 or 2 above.
In this case, the following information must be provided to ITU, ISO and/or IEC as part of this declaration:
- granted patent number or patent application number (if pending);
- an indication of which portions of the above document are affected;
- a description of the Patents covering the above document.

Free of Charge: The words "Free of Charge" do not mean that the Patent Holder is waiving all of its rights with respect to the Patent. Rather, "Free of Charge" refers to the issue of monetary compensation; *i.e.*, that the Patent Holder will not seek any monetary compensation as part of the licensing arrangement (whether such compensation is called a royalty, a one-time licensing fee, etc.). However, while the Patent Holder in this situation is committing to not charging any monetary amount, the Patent Holder is still entitled to require that the implementer of the same above document sign a license agreement that contains other reasonable terms and conditions such as those relating to governing law, field of use, warranties, etc.

Reciprocity: The word "Reciprocity" means that the Patent Holder shall only be required to license any prospective licensee if such prospective licensee will commit to license its Patent(s) for implementation of the same above document Free of Charge or under reasonable terms and conditions.

Patent: The word "Patent" means those claims contained in and identified by patents, utility models and other similar statutory rights based on inventions (including applications for any of these) solely to the extent that any such claims are essential to the implementation of the same above document. Essential patents are patents that would be required to implement a specific Recommendation | Deliverable.

Assignment/transfer of Patent rights: Licensing declarations made pursuant to Clause 2.1 or 2.2 of the Common Patent Policy for ITU-T/ITU-R/ISO/IEC shall be interpreted as encumbrances that bind all successors-in-interest as to the transferred Patents. Recognizing that this interpretation may not apply in all jurisdictions, any Patent Holder who has submitted a licensing declaration according to the Common Patent Policy - be it selected as option 1 or 2 on the Patent Declaration form - who transfers ownership of a Patent that is subject to such licensing declaration shall include appropriate provisions in the relevant transfer documents to ensure that, as to such transferred Patent, the licensing declaration is binding on the transferee and that the transferee will similarly include appropriate provisions in the event of future transfers with the goal of binding all successors-in-interest.

Patent Information (desired but not required for options 1 and 2; required in ITU, ISO and IEC for option 3 (NOTE))

No.	Status [granted/ pending]	Country	Granted Patent Number or Application Number (if pending)	Title
1				
2				
3				
4				
5				
6				
7				
8				
9				
10				

☐ Check here if additional patent information is provided on additional pages.

NOTE: For option 3, the additional minimum information that shall also be provided is listed in the option 3 box above.

Signature (include on final page only):
Patent Holder _____
Name of authorized person _____
Title of authorized person _____
Signature _____
Place, Date

FORM: 2 November 2018

ANNEX 3

GENERAL PATENT STATEMENT AND LICENSING DECLARATION FORM FOR ITU-T OR ITU-R RECOMMENDATION

General Patent Statement and Licensing Declaration for ITU-T or ITU-R Recommendation

This declaration does not represent an actual grant of a license

Please return to the relevant bureau:

Director
Telecommunication Standardization Bureau
International Telecommunication Union
Place des Nations
CH-1211 Geneva 20,
Switzerland
Fax: +41 22 730 5853
Email: tsbdir@itu.int

Director
Radiocommunication Bureau
International Telecommunication Union
Place des Nations
CH-1211 Geneva 20,
Switzerland
Fax: +41 22 730 5785
Email: brmail@itu.int

Patent Holder:
Legal Name
Contact for license application:
Name &
Department
Address
Tel.
Fax
E-mail
URL (optional)
Licensing declaration:
In case part(s) or all of any proposals contained in Contributions submitted by the Patent Holder above are included in ITU-T/ITU-R Recommendation(s) and the included part(s) contain items for which Patents have been filed and whose use would be required to implement ITU-T/ITU-R Recommendation(s), the above Patent Holder hereby declares, in accordance with the Common Patent Policy for ITU-T/ITU-R/ISO/IEC (check <u>one</u> box only):

☐ 1. The Patent Holder is prepared to grant a <u>Free of Charge</u> license to an unrestricted number of applicants on a worldwide, non-discriminatory basis and under other reasonable terms and conditions to make, use, and sell implementations of the relevant ITU-T/ITU-R Recommendation.

Negotiations are left to the parties concerned and are performed outside the ITU-T/ITU-R.

Also mark here___if the Patent Holder's willingness to license is conditioned on <u>Reciprocity</u> for the above ITU-T/ITU-R Recommendation.

Also mark here if the Patent Holder reserves the right to license on reasonable terms and conditions (but not <u>Free of Charge</u>) to applicants who are only willing to license their patent claims, whose use would be required to implement the above ITU-T/ITU-R Recommendation, on reasonable terms and conditions (but not Free of Charge).

☐	2.	The Patent Holder is prepared to grant a license to an unrestricted number of applicants on a worldwide, non-discriminatory basis and on reasonable terms and conditions to make, use and sell implementations of the relevant ITU-T/ITU-R Recommendation.
		Negotiations are left to the parties concerned and are performed outside the ITU-T/ITU-R.
		Also mark here if the Patent Holder's willingness to license is conditioned on <u>Reciprocity</u> for the above ITU-T/ITU-R Recommendation.

Free of Charge: The words "Free of Charge" do not mean that the Patent Holder is waiving all of its rights with respect to the Patent. Rather, "Free of Charge" refers to the issue of monetary compensation; *i.e.*, that the Patent Holder will not seek any monetary compensation as part of the licensing arrangement (whether such compensation is called a royalty, a one-time licensing fee, etc.). However, while the Patent Holder in this situation is committing to not charging any monetary amount, the Patent Holder is still entitled to require that the implementer of the relevant ITU-T/ITU-R Recommendation sign a license agreement that contains other reasonable terms and conditions such as those relating to governing law, field of use, warranties, etc.

Reciprocity: The word "Reciprocity" means that the Patent Holder shall only be required to license any prospective licensee if such prospective licensee will commit to license its Patent(s) for implementation of the relevant ITU-T/ITU-R Recommendation Free of Charge or under reasonable terms and conditions.

Patent: The word "Patent" means those claims contained in and identified by patents, utility models and other similar statutory rights based on inventions (including applications for any of these) solely to the extent that any such claims are essential to the implementation of the relevant Recommendation | Deliverable. Essential patents are patents that would be required to implement a specific Recommendation | Deliverable.

Assignment/transfer of Patent rights: Licensing declarations made pursuant to Clause 2.1 or 2.2 of the Common Patent Policy for ITU-T/ITU-R/ISO/IEC shall be interpreted as encumbrances that bind all successors-in-interest as to the transferred Patents. Recognizing that this interpretation may not apply in all jurisdictions, any Patent Holder who has submitted a licensing declaration according to the Common Patent Policy - be it selected as option 1 or 2 on the Patent Declaration form - who transfers ownership of a Patent that is subject to such licensing declaration shall include appropriate provisions in the relevant transfer documents to ensure that, as to such transferred Patent, the licensing declaration is binding on the transferee and that the transferee will similarly include appropriate provisions in the event of future transfers with the goal of binding all successors-in-interest.

Signature:
Patent Holder _____
Name of authorized person _____
Title of authorized person _____
Signature _____
Place, Date

FORM: 26 June 2015

Annex J
(normative)

Formulating scopes of technical committees and subcommittees

J.1 Introduction

The scope of a technical committee or subcommittee is a statement precisely defining the limits of the work of that committee. As such it has a number of functions:

— it assists those with queries and proposals relating to a field of work to locate the appropriate committee;

— it prevents overlapping the work programmes of two or more ISO and/or IEC committees.

— it can also help guard against moving outside the field of activities authorized by the parent committee.

J.2 Formulation of scopes

Basic rules for the formulation of scopes of technical committees and subcommittees are given in 1.5.10.

The order of the elements of a scope shall be:

— basic scope;

— in the ISO, horizontal functions, where applicable;

— in the IEC, horizontal and/or group safety functions where applicable;

— in JTC 1, horizontal functions where applicable;

— exclusions (if any);

— notes (if any).

J.3 Basic scope

Scopes of technical committees shall not refer to the general aims of international standardization or repeat the principles that govern the work of all technical committees.

In exceptional cases, explanatory material may be included if considered important to the understanding of the scope of the committee. Such material shall be in the form of "Notes".

J.4 Exclusions

Should it be necessary to specify that certain topics are outside the scope of the technical committee, these shall be listed and be introduced by the words "Excluded ..."

Exclusions shall be clearly specified.

Where the exclusions are within the scope of one or more other existing ISO or IEC technical committees, these committees shall also be identified.

EXAMPLE 1 "Excluded: Those ... covered by ISO/TC ...".

EXAMPLE 2 "Excluded: Standardization for specific items in the field of ... (ISO/TC ...), ... (IEC/TC ...), etc.".

It is *not* necessary to mention self-evident exclusions.

EXAMPLE 3 "Excluded: Products covered by other ISO or IEC technical committees".

EXAMPLE 4 "Excluded: ... Specifications for electrical equipment and apparatus, which fall within the scope of IEC committees".

J.5 Scopes of committees related to products

Scopes of committees related to products shall clearly *indicate the field, application area or market sector* which they intend to cover, in order to easily ascertain whether a particular product is, or is not, within that field, application area or market sector.

EXAMPLE 1 "Standardization of ... and ... used in ...".

EXAMPLE 2 "Standardization of materials, components and equipment for construction and operation of ... and ... as well as equipment used in the servicing and maintenance of ...".

The limits of the scope can be defined by *indicating the purpose* of the products, or by *characterizing* the products.

The scope *should not enumerate the types* of product covered by the committee since to do so might suggest that other types can be, or are, standardized by other committees. However, if this is the intention, then it is preferable to list those items which are excluded from the scope.

The *enumeration of aspects* such as terminology, technical requirements, methods of sampling, test methods, designation, marking, packaging, dimensions, etc. suggests a restriction in the scope to those particular aspects, and that other aspects may be standardized by other committees. The aspects of the products to be standardized should therefore not be included in the scope unless it is intended that the scope is limited to those particular aspects.

If the scope makes no mention of any aspect, this means that the subject *in its entirety* is covered by the committee.

NOTE The coverage does not necessarily mean the need for preparing a standard. It only means that standards on any aspect, if needed, will be prepared by that committee and no other.

An example of unnecessary enumeration of aspects is as follows:

EXAMPLE 3 "Standardization of classification, terminology, sampling, physical, chemical or other test methods, specifications, etc.".

Mention of priorities, whether referring to type of product or aspect, shall not appear in the scope since these will be indicated in the programme of work.

J.6 Scopes of committees not related to products

If the scope of a committee is intended to be limited to *certain aspects* which are unrelated, or only indirectly related to products, the scope shall only indicate the aspect to be covered (e.g. safety colours and signs, non-destructive testing, water quality).

The term *terminology* as a possible aspect of standardization should not be mentioned unless this aspect is the only task to be dealt with by the committee. If this is not the case, the mention of terminology is superfluous since this aspect is a logical part of any standardization activity.

Annex K
(normative)

Project committees

K.1 Proposal stage

A new work item proposal not falling within the scope of an existing technical committee shall be presented using the appropriate form and fully justified (see 2.3.4) by one of the bodies authorized to make new work item proposals (see 2.3.2).

The office of the CEO may decide to return the proposal to the proposer for further development before circulation for voting. In this case, the proposer shall make the changes suggested or provide justification for not making the changes. If the proposer does not make the changes and requests that its proposal be circulated for voting as originally presented, the technical management board will decide on appropriate action. This could include blocking the proposal until the changes are made or accepting that it be balloted as received.

In all cases, the office of the CEO may also include comments and recommendations to the proposal form.

For details relating to justification of the proposal, see Annex C.

In the case of a proposal to establish a project committee to prepare management systems standards, see Annex L.

It shall be submitted to the secretariat of the technical management board which shall arrange for it to be submitted to all National Bodies for voting.

Proposers are also encouraged to indicate the date of the first meeting of the project committee (see K.3).

If the proposal was not submitted by a National Body, the submission to the National Bodies shall include a call for offers to assume the secretariat of a project committee.

Votes shall be returned within 12 weeks.

Acceptance requires:

— approval by a 2/3 majority of the National Bodies voting;

— a commitment to participate actively by at least five National Bodies that approved the new work item proposal and nominated technical experts.

K.2 Establishment of a project committee

The technical management board shall review the results of voting on the new work item proposal and if the approval criteria are met, shall establish a project committee (the reference number shall be the next available number in the technical committee/ project committee sequence).

The secretariat of the project committee shall be allocated to the National Body that submitted the proposal, or the technical management board shall decide on the allocation amongst the offers received if the proposal did not originate from a National Body.

National Bodies that approved the new work item proposal and nominated (a) technical expert(s) shall be registered as P-members of the project committee. National Bodies that approved the new work item proposal but did not make a commitment to participate actively shall be registered as O-members. National bodies that voted negatively, but nevertheless indicated that they would participate actively if the new work item was approved, shall be registered as P-members. National Bodies voting negatively without indicating a wish to participate shall be registered as O-members.

The office of the CEO shall announce to the National Bodies the establishment of the project committee and its membership.

National Bodies will be invited to confirm/change their membership status by informing the office of the CEO.

The secretariat will contact any potential liaison organizations identified in the new work item proposal or in National Body comments thereon and will invite them to indicate whether they have an interest in the work and, if so, which category of liaison they would be interested in. Requests for liaison will be processed according to the existing procedures.

K.3 First meeting of a project committee

The procedure for calling a project committee meeting shall be carried out in accordance with Clause 4, with the exception that a six weeks' notice period may be used if the date of the first meeting was communicated at the time of submission of the proposal.

The chair of the project committee shall be the project leader nominated in the new work item proposal or shall be nominated by the secretariat if no project leader was nominated in the new work item proposal.

The first meeting shall confirm the scope of the new work item. In case revision is necessary (for purposes of clarification but not extension of the scope), the revised scope shall be submitted to the technical management board for approval. It shall also confirm the project plan and in ISO the development track and decide on any substructures needed to carry out the work.

If it is determined that the project needs to be subdivided to produce two or more publications, this is possible provided that the subdivisions of the work lie fully within the scope of the original new work item proposal. If not, a new work item will need to be prepared for consideration by the technical management board.

NOTE   Project committees are exempted from the requirement to establish a strategic business plan.

K.4 Preparatory stage

The preparatory stage shall be carried out in accordance with 2.4.

K.5 Committee, enquiry, approval and publication stages

The committee, enquiry, approval and publication stages shall be carried out in accordance with 2.5 to 2.8.

K.6 Disbanding of a project committee

Once the standard(s) is/are published, the project committee shall be disbanded.

K.7 Maintenance of standard(s) prepared by a project committee

The National Body which held the secretariat shall assume responsibility for the maintenance of the standard(s) according to the procedures given in 2.9 unless the project committee has been transformed into a technical committee (see 1.10) in which case the technical committee shall be given the responsibility for the maintenance of the standard.

Annex L
(normative)

Proposals for management system standards

L.1 General

Whenever a proposal is made to prepare a new management system standard (MSS), including sector-specific MSS, a justification study (JS) shall be carried out in accordance with Appendix 1 to this annex.

NOTE No JS is needed for the revision of an existing MSS whose development has already been approved and provided the scope is confirmed (unless it was not provided during its first development).

To the extent possible, the proposer shall endeavour to identify the full range of deliverables which will constitute the new or revised MSS family, and a JS shall be prepared for each of the deliverables.

L.2 Terms and definitions

For the purposes of this annex, the following terms and definitions apply.

L.2.1
management system
See definition contained in Appendix 2 (clause 3.4) of this annex.

L.2.2
management system standard
MSS
standard for *management systems* (L.5.1)

> Note 1 to entry: For the purposes of this document, this definition also applies to other ISO and IEC deliverables (e.g. TS, PAS).

L.2.3
generic MSS
MSS designed to be widely applicable across economic sectors, various types and sizes of organizations and diverse geographical, cultural and social conditions

L.2.4
sector specific MSS
MSS that provides additional requirements or guidance for the application of a generic MSS to a specific economic or business sector

L.2.5
Type A MSS
MSS providing requirements

EXAMPLES

— Management system requirements standards (specifications).

— Management system sector-specific requirements standards.

L.2.6
Type B MSS
MSS providing guidelines

EXAMPLES

— Guidance on the use of management system requirements standards.

— Guidance on the establishment of a management system.

— Guidance on the improvement/enhancement of a management system.

L.2.7
high level structure
HLS
outcome of the work of the ISO/TMB/JTCG "Joint technical Coordination Group on MSS" which refers to high level structure (HLS), identical subclause titles, identical text and common terms and core definitions. See Appendix 2 to this annex.

L.3 Obligation to submit a JS

All MSS [including sector-specific MSS (L.2.4), see Annex M] proposals and their JS shall be identified by the relevant TC/SC/PC (or SyC, in IEC) leadership and the JS shall be sent to the TMB (or its MSS task force) for evaluation and approval before the NP ballot takes place. It is the responsibility of the relevant TC/SC/PC secretariat to identify all MSS proposals, without exception, so that there will be no MSS proposals which fail (with knowledge or without knowledge) to carry out the JS or which fail to be sent to the ISO/TMB for evaluation.

No JS is required for a Type B MSS providing guidance on a specific Type A MSS for which a JS has already been submitted and approved.

EXAMPLE   ISO/IEC 27003:2010 (*Information technology — Security techniques — Information security management system implementation guidance*) does not need to have JS submitted as ISO/IEC 27001:2013 (*Information technology — Security techniques — Information security management systems — Requirements*) has already had a JS submitted and approved.

L.4 Cases where no JS have been submitted

MSS proposals which have not been submitted for TMB evaluation before the NP ballot will be sent to the TMB for evaluation and no new ballot should take place before the TMB decision (project on hold). It is considered good practice that the TC/SC/PC (and/or SyC, in IEC) members endorse the JS prior it being sent to the TMB.

NOTE   Already published MSS which did not have a JS submitted will be treated as new MSS at the time of revision, i.e. a JS is to be presented and approved before any work can begin.

L.5 Applicability of this annex

The above procedures apply to all ISO and IEC deliverables, including IWAs.

L.6 General principles

All projects for new MSS (or for MSS which are already published but for which no JS was completed) shall undergo a JS (see L.1 and Note to L.3). The following general principles provide guidance to assess the market relevance of proposed MSS and for the preparation of a JS. The justification criteria questions in Appendix 1 to this annex are based on these principles. The answers to the questions will form part of the JS. An MSS should be initiated, developed and maintained only when all of the following principles are observed.

1) **Market relevance** — Any MSS should meet the needs of, and add value for, the primary users and other affected parties.

2) **Compatibility** — Compatibility between various MSS and within an MSS family should be maintained.

3) **Topic coverage** — A generic MSS (L.5.3) should have sufficient application coverage to eliminate or minimize the need for sector-specific variances.

4) **Flexibility** — An MSS should be applicable to organizations in all relevant sectors and cultures and of every size. An MSS should not prevent organizations from competitively adding to or differentiating from others, or enhancing their management systems beyond the standard.

5) **Free trade** — An MSS should permit the free trade of goods and services in line with the principles included in the WTO Agreement on Technical Barriers to Trade.

6) **Applicability of conformity assessment** — The market need for first-, second- or third-party conformity assessment, or any combination thereof, should be assessed. The resulting MSS should clearly address the suitability of use for conformity assessment in its scope. An MSS should facilitate joint audits.

7) **Exclusions** — An MSS should not include directly related product (including services) specifications, test methods, performance levels (i.e. setting of limits) or other forms of standardization for products produced by the implementing organization.

8) **Ease of use** — It should be ensured that the user can easily implement one or more MSS. An MSS should be easily understood, unambiguous, free from cultural bias, easily translatable, and applicable to businesses in general.

L.7 Justification study process and criteria

L.7.1 General

This clause describes the justification study (JS) process for justifying and evaluating the market relevance of proposals for an MSS. Appendix 1 to this annex provides a set of questions to be addressed in the justification study.

L.7.2 Justification study process

The JS process applies to any MSS project and consists of the following:

a) the development of the JS by (or on behalf of) the proposer of an MSS project;

b) an approval of the JS by the TMB (or in ISO, the ISO/TMB MSS task force).

The JS process is followed by the normal ISO or IEC balloting procedure for new work item approval as appropriate.

L.7.3 Justification study criteria

Based on Annex C and the general principles stated above, a set of questions (see Appendix 1 to this annex) shall be used as criteria for justifying and assessing a proposed MSS project and shall be answered by the proposer. This list of questions is not exhaustive and any additional information that is relevant to the case should be provided. The JS should demonstrate that all questions have been considered. If it is decided that they are not relevant or appropriate to a particular situation, then the reasons for this decision should be clearly stated. The unique aspect of a particular MSS may require consideration of additional questions in order to assess objectively its market relevance.

L.8 Guidance on the development process and structure of an MSS

L.8.1 General

The development of an MSS will have effects in relation to

— the far-reaching impact of these standards on business practice,

— the importance of worldwide support for the standards,

— the practical possibility for involvement by many, if not all, National Bodies, and

— the market need for compatible and aligned MSS.

This clause provides guidance in addition to the procedures laid down in other clauses of the ISO/IEC Directives, in order to take these effects into account.

All MSS (whether they are Type A or Type B MSS, generic or sector-specific) shall, in principle, use consistent structure, common text and terminology so that they are easy to use and compatible with each other. The guidance and structure given in Appendix 2 to this annex shall, in principle, also be followed.

A Type B MSS which provides guidance on another MSS of the same MSS family should follow the same structure (i.e. clause numbering). Where MSS providing guidance (Type B MSS) are involved, it is important that their functions be clearly defined together with their relationship with the MSS providing requirements (Type A MSS), for example:

— guidance on the use of the requirements standard;

— guidance on the establishment/implementation of the management system;

— guidance on improvement/enhancement of the management system.

Where the proposed MSS is sector-specific:

— it should be compatible and aligned with the generic MSS;

— rules and principles specified in Annex M shall be followed;

— the relevant committee responsible for the generic MSS may have additional requirements to be met or procedures to be followed (see Annex M);

— other committees may need to be consulted, as well as ISO CASCO and IEC CAB on conformity assessment issues.

In the case of sector specific documents, their function and relationship with the generic MSS should be clearly defined (e.g. additional sector-specific requirements; elucidation; or both as appropriate).

Sector-specific documents should always show clearly (e.g. by using different typographical styles) the kind of sector-specific information being provided.

NOTE 1   Where the identical text or any of the requirements cannot be applied in a specific MSS, due to special circumstances, this should be reported to the TMB through the TMB Secretary at tmb@iso.org (see L.9.3) or the IEC/SMB Secretary.

L.8.2 MSS development process

L.8.2.1 General

In addition to the JS, the development of an MSS should follow the same requirements as other ISO and IEC deliverables (see Clause 2).

L.8.2.2 Design specification

To ensure that the intention of the standard, as demonstrated by the justification study, will be maintained, a design specification may be developed before a working draft is prepared.

The responsible committee will decide whether the design specification is needed and in case it is felt necessary, it will decide upon its format and content that is appropriate for the MSS and should set up the necessary organization to carry out the task.

The design specification should typically address the following.

User needs	The identification of the users of the standard and their associated needs, together with the costs and benefits for these users.
Scope	The scope and purpose of the standard, the title and the field of application.
Compatibility	How compatibility within this and with other MSS families will be achieved, including identification of the common elements with similar standards, and how these will be included in the recommended structure (see Appendix 2 to this annex).
Consistency	Consistency with other documents (to be) developed within the MSS family.

NOTE   Most, if not all of the information on user needs and scope will be available from the justification study.

The design specification should ensure that

a) the outputs of the justification study are translated correctly into requirements for the MSS,

b) the issues of compatibility and alignment with other MSS are identified and addressed,

c) a basis for verification of the final MSS exists at appropriate stages during the development process,

d) the approval of the design specification provides a basis for ownership throughout the project by the members of the TC/SC(s), and/or SyC in IEC,

e) account is taken of comments received through the NP ballot phase, and

f) any constraints are taken into account.

The Committee developing the MSS should monitor the development of the MSS against the design specification in order to ensure that no deviations happen in the course of the project.

L.8.2.3 Producing the deliverables

L.8.2.3.1 Monitoring output

In the drafting process, the output should be monitored for compatibility and ease of use with other MSS, by covering issues such as

— the high level structure (HLS), identical subclause titles, identical text and common terms and core definitions,

— the need for clarity (both in language and presentation), and

— avoiding overlap and contradiction.

L.8.2.4 Transparency of the MSS development process

MSS have a broader scope than most other types of standard. They cover a large field of human endeavour and have an impact on a wide range of user interests.

Committees preparing MSS should accordingly adopt a highly transparent approach to the development of the standards, ensuring that

— possibilities for participation in the process of developing standards are clearly identified, and

— the development processes being used are understood by all parties.

Committees should provide information on progress throughout the life cycle of the project, including

— the status of the project to date (including items under discussion),

— contact points for further information,

— communiqués and press releases on plenary meetings, and

— regular listings of frequently asked questions and answers.

In doing this, account needs to be taken of the distribution facilities available in the participating countries.

Where it may be expected that users of a Type A MSS are likely to demonstrate conformity to it, the MSS shall be so written that conformity can be assessed by a manufacturer or supplier (first party, or self-declaration), a user or purchaser (second party) or an independent body (third party, also known as certification or registration).

Maximum use should be made of the resources of the ISO Central Secretariat or IEC Central Office to facilitate the transparency of the project and the committee should, in addition, consider the establishment of a dedicated open-access website.

Committees should involve the national member bodies to build up a national awareness of the MSS project, providing drafts as appropriate for different interested and affected parties, including accreditation bodies, certification bodies, enterprises and the user community, together with additional specific information as needed.

The committee should ensure that technical information on the content of the MSS under development is readily available to participating members, especially those in developing countries.

L.8.2.5 Process for interpretation of a standard

The committee may establish a process to handle interpretation questions related to its standards from the users, and may make the resulting interpretations available to others in an expedient manner. Such a mechanism can effectively address possible misconceptions at an early stage and identify issues that may require improved wording of the standard during the next revision cycle. Such processes are considered, in ISO, to be "committee specific procedures" [see Foreword f)]. In IEC, the committee shall use the Interpretation Sheet process (see IEC Supplement).

L.9 High level structure, identical core text and common terms and core definitions for use in management systems standards

L.9.1 Introduction

The aim of this document is to enhance the consistency and alignment of MSS by providing a unifying and agreed upon high level structure, identical core text and common terms and core definitions. The aim is that all Type A MSS (and B where appropriate) are aligned and the compatibility of these standards is enhanced. It is envisaged that individual MSS will add additional "discipline-specific" requirements as required.

NOTE   In L.9.1 and L.9.4, "discipline-specific" is used to indicate specific subject(s) to which a management system standard refers, e.g. energy, quality, records, environment etc.

The intended audience for this document is Technical Committees (TC), Subcommittees (SC) and Project Committees (PC) (and in IEC, SyCs) and others that are involved in the development of MSS.

This common approach to new MSS and future revisions of existing standards will increase the value of such standards to users. It will be particularly useful for those organizations that choose to operate a single (sometimes called "integrated") management system that can meet the requirements of two or more MSS simultaneously.

Appendix 2 to this annex sets out the high level structure, identical core text and common terms and core definitions that form the nucleus of future and revised Type A MSS and Type B MSS when possible.

Appendix 3 to this annex sets out guidance to the use of Appendix 2 to this annex.

L.9.2 Use

MSS include the high level structure and identical core text as presented in Appendix 2 to this annex. The common terms and core definitions are either included or normatively referenced an international standard where they are included.

NOTE   The high level structure includes the main clauses (1 to 10) and their titles, in a fixed sequence. The identical core text includes numbered subclauses (and their titles) as well as text within the subclauses.

L.9.3 Non applicability

If due to exceptional circumstances the high level structure or any of the identical core text, common terms and core definitions cannot be applied in the management system standard, then the TC/PC/SC needs to explain its rationale for review by:

a)   providing an initial deviation report to ISO/CS or IEC/CO with the DIS submission;

b)   providing a final deviation report to TMB (through the ISO/TMB Secretary at tmb@iso.org) or IEC/SMB Secretary upon submission of the final text of the standard for publication.

The TC/PC/SC (or in IEC, SyC) shall use the ISO or IEC commenting template to provide its deviation reports.

NOTE 1   The final deviation report can be an updated version of the initial deviation report.

NOTE 2   The TC/PC/SC (or in IEC, SyC) strives to avoid any non-applicability of the high level structure or any of the identical core text, common terms and core definitions.

L.9.4 Using Appendix 2 to this annex

Discipline-specific text additions to Appendix 2 to this annex are managed as follows.

1. Discipline-specific additions are made by the individual TC, PC, SC (or in IEC, SyC) or other group that is developing the specific management system standard.

2. Discipline-specific text does not affect harmonization or contradict or undermine the intent of the high level structure, identical core text, common terms and core definitions.

3. Insert additional subclauses, or sub-subclauses (etc.) either ahead of an identical text subclause (or sub-subclause etc.), or after such a subclause (etc.) and renumbered accordingly.

NOTE 1   Hanging paragraphs are not permitted (see ISO/IEC Directives, Part 2).

NOTE 2   Attention is drawn to the need to check cross referencing.

4. Add or insert discipline-specific text within Appendix 2 to this annex. Examples of additions include:

   1) a) new bullet points;

   2) b) discipline-specific explanatory text (e.g. Notes or Examples), in order to clarify requirements;

   3) c) discipline-specific new paragraphs to subclauses (etc.) within the identical text;

   4) d) adding text that enhances the existing requirements in Appendix 2 to this annex.

5. Avoid repeating requirements between identical core text and discipline-specific text by adding text to the identical core text, taking account of point 2 above.

6. Distinguish between discipline-specific text and identical core text from the start of the drafting process. This aids identification of the different types of text during the development and balloting stages.

   NOTE 1   Distinguishing options include by colour, font, font size, italics, or by being boxed separately, etc.

   NOTE 2   Identification of distinguishing text is not necessarily carried into the published version.

7. Understanding of the concept of "risk" may be more specific than that given in the definition under 3.9 of Appendix 2 to this annex. In this case, a discipline-specific definition may be needed. The discipline-specific terms and definitions are differentiated from the core definitions, e.g. (XXX) risk.

   NOTE   The above can also apply to a number of other definitions.

8. Common terms and core definitions will be integrated into the listing of terms and definitions in the discipline-specific management system standard, consistent with the concept system of that standard.

L.9.5 Implementation

Follow the sequence, high level structure, identical core text, common terms and core definitions for any new management system standard and for any revisions to existing management system standards.

L.9.6 Guidance

Find supporting guidance in Appendix 3 to this annex.

Appendix 1

(normative)

Justification criteria questions

1. General

The list of questions to be addressed in the justification study are in line with the principles listed in L.6. This list is not exhaustive. Additional information not covered by the questions should be provided if it is relevant to the case.

Each general principle should be given due consideration and, ideally, when preparing the JS, the proposer should provide a general rationale for each principle, prior to answering the questions associated with the principle.

The principles to which the proposer of the MSS should pay due attention when preparing the justification study are:

1. Market relevance
2. Compatibility
3. Topic coverage
4. Flexibility
5. Free trade
6. Applicability of conformity assessment
7. Exclusions

NOTE   No questions directly refer to the principle 8 ("Ease of use"), but it should guide the development of the deliverable.

Basic information on the MSS proposal

1	What is the proposed purpose and scope of the MSS? Is the document supposed to be a guidance document or a document with requirements?
2	Does the proposed purpose or scope include product (including service) specifications, product test methods, product performance levels, or other forms of guidance or requirements directly related to products produced or provided by the implementing organization?
3	Is there one or more existing committee or non-ISO and non-IEC organization that could logically have responsibility for the proposed MSS? If so, identify.
4	Have relevant reference materials been identified, such as existing guidelines or established practices?
5	Are there technical experts available to support the standardization work? Are the technical experts direct representatives of the affected parties from the different geographical regions?
6	What efforts are anticipated as being necessary to develop the document in terms of experts needed and number/duration of meetings?

7	Is the MSS intended to be a guidance document, contractual specification or regulatory specification for an organization?

Principle 1: Market relevance

8	Have all the affected parties been identified? For example:
	a) organizations (of various types and sizes): the decision-makers within an organization who approve work to implement and achieve conformance to the MSS;
	b) customers/end-users, i.e. individuals or parties that pay for or use a product (including service) from an organization;
	c) supplier organizations, e.g. producer, distributor, retailer or vendor of a product, or a provider of a service or information;
	d) MSS service provider, e.g. MSS certification bodies, accreditation bodies or consultants;
	e) regulatory bodies;
	f) non-governmental organizations.
9	What is the need for this MSS? Does the need exist at a local, national, regional or global level? Does the need apply to developing countries? Does it apply to developed countries? What is the added value of having an ISO or IEC document (e.g. facilitating communication between organizations in different countries)?
10	Does the need exist for a number of sectors and is thus generic? If so, which ones? Does the need exist for small, medium or large organizations?
11	Is the need important? Will the need continue? If yes, will the target date of completion for the proposed MSS satisfy this need? Are viable alternatives identified?
12	Describe how the need and importance were determined. List the affected parties consulted and the major geographical or economical regions in which they are located.
13	Is there known or expected support for the proposed MSS? List those bodies that have indicated support. Is there known or expected opposition to the proposed MSS? List those bodies that have indicated opposition.
14	What are the expected benefits and costs to organizations, differentiated for small, medium and large organizations if applicable?
	Describe how the benefits and the costs were determined. Provide available information on geographic or economic focus, industry sector and size of the organization. Provide information on the sources consulted and their basis (e.g. proven practices), premises, assumptions and conditions (e.g. speculative or theoretical), and other pertinent information.
15	What are the expected benefits and costs to other affected parties (including developing countries)?
	Describe how the benefits and the costs were determined. Provide any information regarding the affected parties indicated.
16	What will be the expected value to society?
17	Have any other risks been identified (e.g. timeliness or unintended consequences to a specific business)?

Principle 2: Compatibility

18	Is there potential overlap or conflict with (or what is the added value in relation to) other existing or planned ISO, IEC, non-ISO or non-IEC international standards, or those at the national or regional level? Are there other public or private actions, guidance, requirements and regulations that seek to address the identified need, such as technical papers, proven practices, academic or professional studies, or any other body of knowledge?
19	Is the MSS or the related conformity assessment activities (e.g. audits, certifications) likely to add to, replace all or parts of, harmonize and simplify, duplicate or repeat, conflict with, or detract from the existing activities identified above? What steps are being considered to ensure compatibility, resolve conflict or avoid duplication?
20	Is the proposed MSS likely to promote or stem proliferation of MSS at the national or regional level, or by industry sectors?

Principle 3: Topic coverage

21	Is the MSS for a single specific sector?
22	Will the MSS reference or incorporate an existing, non-industry-specific MSS (e.g. from the ISO 9000 series of quality management standards)? If yes, will the development of the MSS conform to the ISO/IEC Sector Policy (see ISO/IEC Directives, Part 2), and any other relevant policy and guidance procedures (e.g. those that may be made available by a relevant committee)?
23	What steps have been taken to remove or minimize the need for particular sector-specific deviations from a generic MSS?

Principle 4: Flexibility

24	Will the MSS allow an organization competitively to add to, differentiate or encourage innovation of its management system beyond the standard?

Principle 5: Free trade

25	How would the MSS facilitate or impact global trade? Could the MSS create or prevent a technical barrier to trade?
26	Could the MSS create or prevent a technical barrier to trade for small, medium or large organizations?
27	Could the MSS create or prevent a technical barrier to trade for developing or developed countries?
28	If the proposed MSS is intended to be used in government regulations, is it likely to add to, duplicate, replace, enhance or support existing governmental regulations?

Principle 6: Applicability of conformity

29	If the intended use is for contractual or regulatory purposes, what are the potential methods to demonstrate conformance (e.g. first party, second party or third party)? Does the MSS enable organizations to be flexible in choosing the method of demonstrating conformance, and to accommodate for changes in its operations, management, physical locations and equipment?

30	If third-party registration/certification is a potential option, what are the anticipated benefits and costs to the organization? Will the MSS facilitate joint audits with other MSS or promote parallel assessments?

Principle 7: Exclusions

31	Does the proposed purpose or scope include product (including service) specifications, product test methods, product performance levels, or other forms of guidance or requirements directly related to products produced or provided by the implementing organization?

Appendix 2

(normative)

High level structure, identical core text, common terms and core definitions

NOTE   In the identical text proposals, XXX = an MSS discipline specific qualifier (e.g. energy, road traffic safety, IT security, food safety, societal security, environment, quality) that needs to be inserted. Blue italicized text is given as advisory notes to standards drafters.

Introduction

DRAFTING INSTRUCTION   Specific to the discipline.

This text has been prepared using the "high-level structure" (i.e. clause sequence, identical core text and common terms and core definitions) provided in Annex L, Appendix 2 of the ISO/IEC Directives, Part 1. This is intended to enhance alignment among ISO and IEC management system standards, and to facilitate their implementation for organizations that need to meet the requirements of two or more such standards.

HLS is highlighted in the text (clauses 1 to 10) by the use of blue font. Black represents the ISO or IEC specific discipline text. Strikeout is used to show agreed deletions within the HLS text. The use of blue text and strikeout is only to facilitate analysis and will not be incorporated after the Draft International Standard stage of development for this document.

1. Scope

DRAFTING INSTRUCTION   Specific to the discipline.

2. Normative references

DRAFTING INSTRUCTION   Clause Title shall be used. Specific to the discipline.

3. Terms and definitions

DRAFTING INSTRUCTION 1   Clause Title shall be used. Terms and definitions may either be within the standard or in a separate document.

Common terms and core definitions shall be stated as well as others that are discipline specific.

The arrangement of terms and definitions should preferably be listed according to the hierarchy of the concepts (i.e. systematic order). Alphabetical order is the least preferred order.

For the purposes of this document, the following terms and definitions apply.

DRAFTING INSTRUCTION 2   The following terms and definitions constitute an integral part of the "common text" for management systems standards. Additional terms and definitions may be added as needed. Notes may be added or modified to serve the purpose of each standard.

DRAFTING INSTRUCTION 3   Italics type in a definition indicates a cross-reference to another term defined in this clause, and the number reference for the term is given in parentheses.

DRAFTING INSTRUCTION 4   *Where the text "XXX" appears throughout this clause, the appropriate reference should be inserted depending on the context in which these terms and definitions are being applied. For example: "an XXX objective" could be substituted as "an information security objective".*

3.1
organization
person or group of people that has its own functions with responsibilities, authorities and relationships to achieve its *objectives* (3.8)

Note 1 to entry: The concept of organization includes, but is not limited to, sole-trader, company, corporation, firm, enterprise, authority, partnership, charity or institution, or part or combination thereof, whether incorporated or not, public or private.

3.2
interested party (preferred term)
stakeholder (admitted term)
person or *organization* (3.1) that can affect, be affected by, or perceive itself to be affected by a decision or activity

3.3
requirement
need or expectation that is stated, generally implied or obligatory

Note 1 to entry: "Generally implied" means that it is custom or common practice for the organization and interested parties that the need or expectation under consideration is implied.

Note 2 to entry: A specified requirement is one that is stated, e.g. in documented information.

3.4
management system
set of interrelated or interacting elements of an *organization* (3.1) to establish *policies* (3.7) and *objectives* (3.8) and *processes* (3.12) to achieve those objectives

Note 1 to entry: A management system can address a single discipline or several disciplines.

Note 2 to entry: The system elements include the organization's structure, roles and responsibilities, planning and operation.

Note 3 to entry: The scope of a management system can include the whole of the organization, specific and identified functions of the organization, specific and identified sections of the organization, or one or more functions across a group of organizations.

3.5
top management
person or group of people who directs and controls an *organization* (3.1) at the highest level

Note 1 to entry: Top management has the power to delegate authority and provide resources within the organization.

Note 2 to entry: If the scope of the *management system* (3.4) covers only part of an organization, then top management refers to those who direct and control that part of the organization.

3.6
effectiveness
extent to which planned activities are realized and planned results achieved

3.7
policy
intentions and direction of an *organization* (3.1), as formally expressed by its *top management* (3.5)

3.8
objective
result to be achieved

> Note 1 to entry: An objective can be strategic, tactical, or operational.

> Note 2 to entry: Objectives can relate to different disciplines (such as financial, health and safety, and environmental goals) and can apply at different levels (such as strategic, organization-wide, project, product and *process* (3.12)).

> Note 3 to entry: An objective can be expressed in other ways, e.g. as an intended outcome, a purpose, an operational criterion, as an XXX objective, or by the use of other words with similar meaning (e.g. aim, goal, or target).

> Note 4 to entry: In the context of XXX management systems, XXX objectives are set by the organization, consistent with the XXX policy, to achieve specific results.

3.9
risk
effect of uncertainty

> Note 1 to entry: An effect is a deviation from the expected — positive or negative.

> Note 2 to entry: Uncertainty is the state, even partial, of deficiency of information related to, understanding or knowledge of, an event, its consequence, or likelihood.

> Note 3 to entry: Risk is often characterized by reference to potential "events" (as defined in ISO Guide 73) and "consequences" (as defined in ISO Guide 73), or a combination of these.

> Note 4 to entry: Risk is often expressed in terms of a combination of the consequences of an event (including changes in circumstances) and the associated "likelihood" (as defined in ISO Guide 73) of occurrence.

3.10
competence
ability to apply knowledge and skills to achieve intended results

3.11
documented information
information required to be controlled and maintained by an *organization* (3.1) and the medium on which it is contained

> Note 1 to entry: Documented information can be in any format and media, and from any source.

Note 2 to entry: Documented information can refer to:

— the *management system* (3.4), including related *processes* (3.12);

— information created in order for the organization to operate (documentation);

— evidence of results achieved (records).

3.12
process
set of interrelated or interacting activities which transforms inputs into outputs

3.13
performance
measurable result

Note 1 to entry: Performance can relate either to quantitative or qualitative findings.

Note 2 to entry: Performance can relate to managing activities, *processes* (3.12), products (including services), systems or *organizations* (3.1).

3.14
outsource (verb)
make an arrangement where an external *organization* (3.1) performs part of an organization's function or *process* (3.12)

Note 1 to entry: An external organization is outside the scope of the *management system* (3.4), although the outsourced function or process is within the scope.

3.15
monitoring
determining the status of a system, a *process* (3.12) or an activity

Note 1 to entry: To determine the status, there can be a need to check, supervise or critically observe.

3.16
measurement
process (3.12) to determine a value

3.17
audit
systematic, independent and documented *process* (3.12) for obtaining audit evidence and evaluating it objectively to determine the extent to which the audit criteria are fulfilled

Note 1 to entry: An audit can be an internal audit (first party) or an external audit (second party or third party), and it can be a combined audit (combining two or more disciplines).

Note 2 to entry: An internal audit is conducted by the organization itself, or by an external party on its behalf.

Note 3 to entry: "Audit evidence" and "audit criteria" are defined in ISO 19011.

3.18
conformity
fulfilment of a *requirement* (3.3)

3.19
nonconformity
non-fulfilment of a *requirement* (3.3)

3.20
corrective action
action to eliminate the cause(s) of a *nonconformity* (3.19) and to prevent recurrence

3.21
continual improvement
recurring activity to enhance *performance* (3.13)

4. Context of the organization

4.1 Understanding the organization and its context

The organization shall determine external and internal issues that are relevant to its purpose and that affect its ability to achieve the intended outcome(s) of its XXX management system.

4.2 Understanding the needs and expectations of interested parties

The organization shall determine:

— the interested parties that are relevant to the XXX management system;

— the relevant requirements of these interested parties.

4.3 Determining the scope of the XXX management system

The organization shall determine the boundaries and applicability of the XXX management system to establish its scope.

When determining this scope, the organization shall consider:

— the external and internal issues referred to in 4.1;

— the requirements referred to in 4.2.

The scope shall be available as documented information.

4.4 XXX management system

The organization shall establish, implement, maintain and continually improve an XXX management system, including the processes needed and their interactions, in accordance with the requirements of this document.

5. Leadership

5.1 Leadership and commitment

Top management shall demonstrate leadership and commitment with respect to the XXX management system by:

— ensuring that the XXX policy and XXX objectives are established and are compatible with the strategic direction of the organization;

— ensuring the integration of the XXX management system requirements into the organization's business processes;

— ensuring that the resources needed for the XXX management system are available;

— communicating the importance of effective XXX management and of conforming to the XXX management system requirements;

— ensuring that the XXX management system achieves its intended outcome(s);

— directing and supporting persons to contribute to the effectiveness of the XXX management system;

— promoting continual improvement;

— supporting other relevant managerial roles to demonstrate their leadership as it applies to their areas of responsibility.

NOTE   Reference to "business" in this document can be interpreted broadly to mean those activities that are core to the purposes of the organization's existence.

5.2 Policy

Top management shall establish a XXX policy that:

a) is appropriate to the purpose of the organization;

b) provides a framework for setting XXX objectives;

c) includes a commitment to satisfy applicable requirements;

d) includes a commitment to continual improvement of the XXX management system.

The XXX policy shall:

— be available as documented information;

— be communicated within the organization;

— be available to interested parties, as appropriate.

5.3 Roles, responsibilities and authorities

Top management shall ensure that the responsibilities and authorities for relevant roles are assigned and communicated within the organization.

Top management shall assign the responsibility and authority for:

a) ensuring that the XXX management system conforms to the requirements of this document;

b) reporting on the performance of the XXX management system to top management.

6. Planning

6.1 Actions to address risks and opportunities

When planning for the XXX management system, the organization shall consider the issues referred to in 4.1 and the requirements referred to in 4.2 and determine the risks and opportunities that need to be addressed to:

— give assurance that the XXX management system can achieve its intended outcome(s);

— prevent, or reduce, undesired effects;

— achieve continual improvement.

The organization shall plan:

a) actions to address these risks and opportunities;

b) how to:

— — integrate and implement the actions into its XXX management system processes;

— — evaluate the effectiveness of these actions.

6.2 XXX objectives and planning to achieve them

The organization shall establish XXX objectives at relevant functions and levels.

The XXX objectives shall:

a) be consistent with the XXX policy;

b) be measurable (if practicable);

c) take into account applicable requirements;

d) be monitored;

e) be communicated;

f) be updated as appropriate.

The organization shall retain documented information on the XXX objectives.

When planning how to achieve its XXX objectives, the organization shall determine:

— what will be done;

- what resources will be required;
- who will be responsible;
- when it will be completed;
- how the results will be evaluated.

7. Support

7.1 Resources

The organization shall determine and provide the resources needed for the establishment, implementation, maintenance and continual improvement of the XXX management system.

7.2 Competence

The organization shall:

- determine the necessary competence of person(s) doing work under its control that affects its XXX performance;
- ensure that these persons are competent on the basis of appropriate education, training, or experience;
- where applicable, take actions to acquire the necessary competence, and evaluate the effectiveness of the actions taken;
- retain appropriate documented information as evidence of competence.

NOTE   Applicable actions can include, for example, the provision of training to, the mentoring of, or the re-assignment of currently employed persons; or the hiring or contracting of competent persons.

7.3 Awareness

Persons doing work under the organization's control shall be aware of:

- the XXX policy;
- their contribution to the effectiveness of the XXX management system, including the benefits of improved XXX performance;
- the implications of not conforming with the XXX management system requirements.

7.4 Communication

The organization shall determine the internal and external communications relevant to the XXX management system, including:

- on what it will communicate;
- when to communicate;
- with whom to communicate;

— how to communicate.

7.5 Documented information

7.5.1 General

The organization's XXX management system shall include:

a) documented information required by this document;

b) documented information determined by the organization as being necessary for the effectiveness of the XXX management system.

NOTE   The extent of documented information for a XXX management system can differ from one organization to another due to:

— the size of organization and its type of activities, processes, products and services;

— the complexity of processes and their interactions;

— the competence of persons.

7.5.2 Creating and updating

When creating and updating documented information the organization shall ensure appropriate:

— identification and description (e.g. a title, date, author, or reference number);

— format (e.g. language, software version, graphics) and media (e.g. paper, electronic);

— review and approval for suitability and adequacy.

7.5.3 Control of documented information

Documented information required by the XXX management system and by this document shall be controlled to ensure:

a) it is available and suitable for use, where and when it is needed;

b) it is adequately protected (e.g. from loss of confidentiality, improper use, or loss of integrity).

For the control of documented information, the organization shall address the following activities, as applicable:

— distribution, access, retrieval and use;

— storage and preservation, including preservation of legibility;

— control of changes (e.g. version control);

— retention and disposition.

Documented information of external origin determined by the organization to be necessary for the planning and operation of the XXX management system shall be identified, as appropriate, and controlled.

NOTE   Access can imply a decision regarding the permission to view the documented information only, or the permission and authority to view and change the documented information.

8. Operation

8.1 Operational planning and control

DRAFTING INSTRUCTION   This subclause heading will be deleted if no additional subclauses are added to Clause 8.

The organization shall plan, implement and control the processes needed to meet requirements, and to implement the actions determined in 6.1, by:

— establishing criteria for the processes;

— implementing control of the processes in accordance with the criteria;

— keeping documented information to the extent necessary to have confidence that the processes have been carried out as planned.

The organization shall control planned changes and review the consequences of unintended changes, taking action to mitigate any adverse effects, as necessary.

The organization shall ensure that outsourced processes are controlled.

9. Performance evaluation

9.1 Monitoring, measurement, analysis and evaluation

The organization shall determine:

— what needs to be monitored and measured;

— the methods for monitoring, measurement, analysis and evaluation, as applicable, to ensure valid results;

— when the monitoring and measuring shall be performed;

— when the results from monitoring and measurement shall be analysed and evaluated.

The organization shall retain appropriate documented information as evidence of the results.

The organization shall evaluate the XXX performance and the effectiveness of the XXX management system.

9.2 Internal audit

9.2.1 The organization shall conduct internal audits at planned intervals to provide information on whether the XXX management system:

a) conforms to:

　— — the organization's own requirements for its XXX management system;

　— — the requirements of this document;

b) is effectively implemented and maintained.

9.2.2 The organization shall:

a) plan, establish, implement and maintain an audit programme(s) including the frequency, methods, responsibilities, planning requirements and reporting, which shall take into consideration the importance of the processes concerned and the results of previous audits;

b) define the audit criteria and scope for each audit;

c) select auditors and conduct audits to ensure objectivity and the impartiality of the audit process;

d) ensure that the results of the audits are reported to relevant managers;

e) retain documented information as evidence of the implementation of the audit programme(s) and the audit results.

9.3 Management review

Top management shall review the organization's XXX management system, at planned intervals, to ensure its continuing suitability, adequacy and effectiveness.

The management review shall include consideration of:

a) the status of actions from previous management reviews;

b) changes in external and internal issues that are relevant to the XXX management system;

c) information on the XXX performance, including trends in:

— — nonconformities and corrective actions;

— — monitoring and measurement results;

— — audit results;

d) opportunities for continual improvement.

The outputs of the management review shall include decisions related to continual improvement opportunities and any need for changes to the XXX management system.

The organization shall retain documented information as evidence of the results of management reviews.

10. Improvement

10.1 Nonconformity and corrective action

When a nonconformity occurs, the organization shall:

a) react to the nonconformity and, as applicable:

— — take action to control and correct it;

— — deal with the consequences;

b) evaluate the need for action to eliminate the cause(s) of the nonconformity, in order that it does not recur or occur elsewhere, by:

— — reviewing the nonconformity;

— — determining the causes of the nonconformity;

— — determining if similar nonconformities exist, or can potentially occur;

c) implement any action needed;

d) review the effectiveness of any corrective action taken;

e) make changes to the XXX management system, if necessary.

Corrective actions shall be appropriate to the effects of the nonconformities encountered.

The organization shall retain documented information as evidence of:

— the nature of the nonconformities and any subsequent actions taken;

— the results of any corrective action.

10.2 Continual improvement

The organization shall continually improve the suitability, adequacy and effectiveness of the XXX management system.

Appendix 3

(informative)

Guidance on high level structure, identical core text, common terms and core definitions

Guidance on the high level structure, identical core text, common terms and core definitions is provided at the following URL:

Annex L Guidance documents (http://isotc.iso.org/livelink/livelink?func=ll&objId=16347818&objAction=browse&viewType=1

Annex M
(normative)

Policy for the development of sector-specific management standards and sector-specific management system standards (MSS)

M.1 General

Any technical committee or subcommittee, project committee (SyC in IEC) or International Workshop that proposes development of a sector-specific management standard (M.2.2) or a sector-specific management system standard (MSS) (M.2.4) shall follow the directions specified in this annex. It includes, as applicable, committee specific policies (M.5) which may not be limited to sector-specific management standards or sector-specific management system standards.

M.2 Terms and definitions

M.2.1
generic management standard
management standard designed to be widely applicable across economic sectors, various types and sizes of organizations and diverse geographical, cultural and social conditions

M.2.2
sector-specific management standard
management standard that provides additional requirements or guidance for the application of a *generic management standard* (M.2.1) to a specific economic or business sector

M.2.3
generic management system standard
generic MSS
MSS designed to be widely applicable across economic sectors, various types and sizes of organizations and diverse geographical, cultural and social conditions

M.2.4
sector-specific management system standard (MSS)
sector-specific MSS
MSS that provides additional requirements or guidance for the application of a *generic MSS* (M.2.3) to a specific economic or business sector

M.3 Sector-specific management standards and sector-specific management system standards

Any new proposal for a sector-specific management standard (M.2.2) or sector-specific MSS (M.2.4) shall:

— clearly demonstrate its market relevance and alignment through the completion of appropriate ISO or IEC project approval procedures by means of ISO Form 4, *New Work Item Proposal*, or IEC Form NP,

— [in the case of the development of a sector-specific MSS (M.2.4)] clearly demonstrate that all the rules and principles in Annex L have been followed, including the approval of the justification study (see Annex L), and

— clearly demonstrate that the liaison with the committee responsible for the generic management standard or generic MSS concerned is effective,

— if applicable, conform with the committee specific policies set out below.

M.4 Drafting rules

Sector-specific management standards (M.2.2) and sector-specific MSS (M.2.4) shall respect the following rules:

a) Normative reference shall be made to the generic management standard (M.2.1) or generic MSS (M.2.3). Alternatively, the clauses and subclauses may be reproduced verbatim.

b) If text from the generic management standard (M.2.1) or generic MSS (M.2.3) is reproduced in the sector-specific standard, it shall be distinguished from the other elements of the sector-specific standard.

c) Terms and definitions specified in the generic management standard (M.2.1) or generic MSS (M.2.3) shall be referred to in a normative manner or reproduced verbatim.

M.5 Committee specific policies

M.5.1 General

Sector-specific management standards (M.2.2) and sector-specific MSS (M.2.4) shall not interpret, change, or subtract from the requirements of the generic management standard or generic MSS.

M.5.2.1 Terms and definitions

The following terms and definitions are applicable to environmental policy:

M.5.2.1.1
sector-specific environmental management standard
standard that provides additional requirements or guidance for the application of a generic environmental management standard to a specific economic or business sector

EXAMPLE   The application of an environmental management system (ISO 14001) or life-cycle assessment (ISO 14044) to agri-food or energy sectors.

M.5.2.1.2
aspect-specific environmental management standard
standard that provides additional requirements or guidance for the application of a generic environmental management standard for a specific environmental aspect or aspects within its scope

EXAMPLE   The application of an environmental management system (ISO 14001) for greenhouse gas (aspect) management or life-cycle assessment (ISO 14044) for the water (aspect) footprint of products.

M.5.2.1.3
element-specific environmental management standard
standard that provides additional requirements or guidance for the application of a generic environmental management standard for a specific element or elements within its scope

EXAMPLE   Communications or emergency management (elements) within an environmental management system (ISO 14001) or data collection or critical review (elements) within a life-cycle assessment (ISO 14044).

M.5.2.2 General

Any technical committee, subcommittee, project committee (SyC in IEC) or International Workshop that proposes development of a sector-, aspect- or element-specific environmental management standard shall clearly demonstrate its market relevance and alignment through the completion of appropriate project approval procedures, including:

— ISO Form 4, *New Work Item Proposal* for sector-, aspect- or element-specific specific application of generic environmental management system standards, environmental labeling, life-cycle assessment and greenhouse gas management standards, and

— Annex L *Proposals for management system standards (MSS)* for sector-, aspect- or element-specific specific application of generic environmental MSS.

Approval documentation should include specific justification as to why the relevant generic ISO 14000 series standard(s) insufficiently address sector-, aspect- or element-specific needs and how the proposed new standard would effectively resolve identified issues. Proposers should critically assess whether additional sector-, aspect- or element-specific requirements are needed as opposed to the provision of additional guidance to the generic environmental management standard(s).

M.5.2.3 Any technical committee, subcommittee, project committee (SyC in IEC) or International Workshop that proposes development of a sector-, aspect- or element-specific environmental management standard should consider and reflect the needs of developing countries, economies in transition, small- and medium- enterprises and organizations operating across a variety of sectors.

M.5.2.4 ISO/TC 207 will cooperate in or, where appropriate and as decided by the Technical Management Board, lead joint projects with technical committee, subcommittee, project committee (SyC in IEC) or International Workshop developing sector-, aspect- or element-specific environmental management standards to avoid duplication of effort and promote consistency and alignment. There is no intention to restrict the development of market relevant standards in committees outside of ISO/TC 207.

M.5.2.5 Technical committee, subcommittee, project committee (SyC in IEC) or International Workshop developing sector-, aspect- or element-specific environmental management standards shall:

— include the normative reference of the appropriate generic ISO 14000 series environmental management systems, environmental auditing, environmental labeling, life-cycle assessment and greenhouse gas management standards;

— include the normative reference of the appropriate generic ISO 14050 terms and definitions;

— distinguish ISO 14000 series text if it is reproduced; and

— not interpret, change, or subtract from the requirements of the generic ISO 14000 series environmental management systems, environmental auditing, environmental labeling, life-cycle assessment and greenhouse gas management standards.

M.5.2.6 Any requests for guidance on this sector-, aspect- or element-specific policy or for interpretation of generic ISO 14000 series standards or ISO 14050 terms and definitions or for guidance on a sector-, aspect- or element-specific document shall be submitted to the ISO Central Secretariat as well as the relevant TC 207 subcommittee.

M.5.3 Quality

When an technical committee, subcommittee, project committee (SyC in IEC) or International Workshop wishes to develop quality management system requirements or guidance for a particular product or industry/economic sector it shall respect the following rules.

a) Normative reference shall be made to ISO 9001 in its entirety. Alternatively, the clauses and subclauses may be reproduced verbatim.

b) If text from ISO 9001 is reproduced in the sector document, it shall be distinguished from the other elements of the sector document [see d)].

c) Terms and definitions specified in ISO 9000 shall be referred to in a normative manner or reproduced verbatim.

d) The guidance and criteria provided in Quality management systems – Guidance and criteria for the development of documents to meet needs of specific product and industry/economic sectors, approved by ISO/TC 176, shall be considered not only when determining the need for a sector-specific requirements or guidance document but also in the document development process.

Any requests for guidance on this sector policy or for interpretation of ISO 9000 terms and definitions, ISO 9001 or ISO 9004 shall be submitted to the secretariat of ISO/TC 176.

M.5.4 Asset management

When an technical committee, subcommittee, project committee (SyC in IEC) or International Workshop wishes to develop asset management system requirements or guidance for a particular product or industry/economic sector it shall respect the following rules:

a) Normative reference shall be made to ISO 55001 in its entirety. Alternatively, the clauses and subclauses may be reproduced verbatim.

b) If text from ISO 55001 is reproduced in the sector document, it shall be distinguished from the other elements of the sector document.

c) Terms and definitions specified in ISO 55000 shall be referred to in a normative manner or reproduced verbatim.

Any requests for guidance on a sector-specific document or for interpretation of ISO 55000 terms and definitions or ISO 55001 shall be submitted to the secretariat of ISO/TC 251.

M.5.5 Risk

When a technical committee, subcommittee, project committee (SyC in IEC) or International Workshop wishes to develop risk management requirements or guidance for a particular product or industry/economic sector it shall respect the following rules:

a) Reference shall be made to ISO 31000 in its entirety. Alternatively, the clauses and subclauses may be reproduced verbatim.

b) If text from ISO 31000 is reproduced in the sector document, it shall be distinguished from the other elements of the sector document.

c) Terms and definitions specified in ISO 31000 shall be referred to in a normative manner or reproduced verbatim.

Any requests for guidance on a sector-specific document or for interpretation of ISO 31000 terms and definitions shall be submitted to the secretariat of ISO/TC 262.

M.5.6 Social responsibility

When a technical committee, subcommittee, project committee (SyC in IEC) or International Workshop wishes to develop social responsibility requirements or guidance for a particular product or industry/economic sector it shall respect the following rules:

a) Reference shall be made to ISO 26000 in its entirety. Alternatively, the clauses and subclauses may be reproduced verbatim.

b) If text from ISO 26000 is reproduced in the sector document, it shall be distinguished from the other elements of the sector document.

c) Terms and definitions specified in ISO 26000 shall be referred to in a normative manner or reproduced verbatim.

Annex JA
(normative)
Voting

JA.1 General

JA.1.1 Discussion during ballot period

~~When a document is out for ballot at Committee Stage or any later stage, National Bodies / Liaison Organizations are free to circulate their comments to other National Bodies/Liaison Organizations provided they do not use the formal subcommittee or JTC 1 documentation distribution system. Formal distribution is prohibited because it could create confusion as to the status of the ballot. Documents out for ballot at Committee Stage or any later stage shall not be subject to formal discussion at any working level of JTC 1 during the balloting period. Therefore, National Body positions on a document under ballot are not to be formally discussed at any working level.~~

JA.1.1 Meetings

Votes by P-members in attendance may be cast only by the head of that delegation or an individual designated by the Head of Delegation (HoD).

The chairman has no vote and questions on which the vote is equally divided shall be subject to further discussion.

In a meeting, except as otherwise specified in this *Consolidated JTC 1 Supplement* or in JTC 1 Standing Documents, questions are decided by a majority of the votes cast at the meeting by P-members which are present expressing either approval, disapproval, or declared abstention.

If the meeting is to be conducted by teleconference or using electronic means, see Standing Document 19 on "Meetings" clauses 3 and 4 for additional requirements.

JA.1.2 Letter Ballots

For votes by correspondence (letter ballots) in JTC 1 and its subcommittees, except as specified elsewhere in this *Consolidated JTC 1 Supplement* or in JTC 1 Standing Documents, questions are decided by a majority of the votes cast by P-members expressing either approval or disapproval. Letter ballots are cast by web based balloting.

JTC 1 instructs its secretariats to close all letter ballots on the declared closure date. Late votes and comments shall not be accepted. JTC 1 allows actions to be taken between JTC 1 plenary meetings by 8 week letter ballots within JTC 1; such actions for approval may be proposed by the JTC 1 chairman, JTC 1 subcommittees, or JTC 1 ~~special working groups~~ advisory groups. Otherwise, no letter ballot period shall close in less than 12 weeks from the date of notification of issue.

JA.1.3 Default Ballots

In certain cases, consensus may be confirmed for questions which are expected to contain no controversial issues and for which agreement of the committee is foreseen in advance. Such questions will be distributed for a period of 4 weeks. If no objection is received during this period, the question is considered to be approved. If any JTC 1 P-member objects to the question during this period, the ~~Secretary~~ JTC 1 committee manager shall withdraw the ballot immediately and resolve the question by vote, either at a meeting or by letter ballot. JTC 1 P-members wishing to raise an objection to a Default Ballot are requested to notify the responsible Secretariat as soon as possible to prevent undue delays.

Questions for which this may be used are:

- appointment/change of a registration authority;
- establishment or cancellation of a Category C liaison;
- proposal for stabilization/withdrawal of a standard;
- request for availability free of charge of an ISO/IEC publication which meets the established criteria;
- modification of a subcommittee's program of work, including the establishment of a collaborative interchange or collaborative team with ITU-T, as defined in SD 3 (see 1.17.7.2 for details);
- others as approved by JTC 1.

JA.2 Proposal stage – Votes on new work item proposals

A JTC 1 P-member, the committee secretariat, another technical committee or subcommittee, an organization in liaison, the technical management board or one of its advisory groups and the Chief Executive Officer may submit a new work item proposal either to a subcommittee or to JTC 1. Each new work item proposal shall be voted on by letter ballot ~~(see the new work item proposal letter ballot form in the Templates folder at www.jtc1.org)~~, even if it has appeared on the agenda of a meeting. If the proposal includes the establishment of a collaborative interchange or collaborative team with ITU-T, see 1.17.7.2 for a list of considerations which shall be evaluated to determine the rationale, as well as what is needed in the NP documentation. The acceptance criteria are as specified in 2.3.5. The normal ballot period for a new work item proposal shall be 12 weeks from the date of notification of issue (see 2.3.4).

JA.2.1 Votes on NPs at the SC level

A new work item proposal should be balloted only once within a subcommittee.

It should be noted that if a new work item proposal is submitted for ballot without prior consultation of the subcommittee, there is a risk that the ballot may fail because the necessary consensus and support are absent. A subcommittee chair or secretariat may schedule a newly submitted new work item proposal for discussion at a plenary or working group meeting before issuing a ballot, as long as unreasonable delay is not introduced.

For new work item proposals voted at the subcommittee level, a copy of the subcommittee-level ballot shall be forwarded by the subcommittee secretariat to the JTC 1 secretariat for information in parallel with circulation of the new work item proposal ballot (see Figure JA.1). The JTC 1 secretariat shall circulate this copy of the subcommittee-level ballot to JTC 1 P-members and JTC 1 subcommittee Chairs and Secretariats for concurrent review.

Within 4 weeks of the issuance of the subcommittee-level NP ballot for JTC 1 concurrent review, a JTC 1 P-member may request the JTC 1 Secretariat to initiate a parallel JTC 1-level ballot. The JTC 1 P-member shall provide a rationale for the request and such rationale shall focus on the appropriate placement of the new work item, if approved. Rationale that focuses solely on technical aspects of the new work item proposal is not acceptable. If two or more JTC 1 P-members request such a parallel JTC 1-level ballot, the JTC 1 secretariat shall issue a ballot. The JTC 1-level ballot shall be identical to and with the same closing date as the subcommittee-level ballot. The rationales submitted with the requests shall accompany the JTC 1-level ballot and shall also be sent to the relevant subcommittee. The approval criteria for the JTC 1-level ballot shall be identical to the subcommittee-level NP ballot but the participation commitment requirement shall not apply (see 2.3.5(a)).

Upon completion of the JTC 1-level ballot, the NP is approved only when both the subcommittee-level ballot and the JTC 1 ballot pass.

When approved, the JTC 1 secretariat shall inform all JTC 1 National Bodies and the subcommittee ~~Secretary~~ committee manager of the result (together with the project number assigned).

If the subcommittee-level ballot failed and no JTC 1 ballot had been issued, no further action is taken.

If the subcommittee-level ballot passed and no JTC 1 ballot had been issued, then the new work item proposal is approved.

NOTE If, during the JTC 1 concurrent review, any JTC 1 P-member not participating in the SC has a comment on the NP but does not request a JTC 1 ballot, the JTC 1 P-member may submit such a comment to the SC secretariat conducting the subcommittee-level ballot.

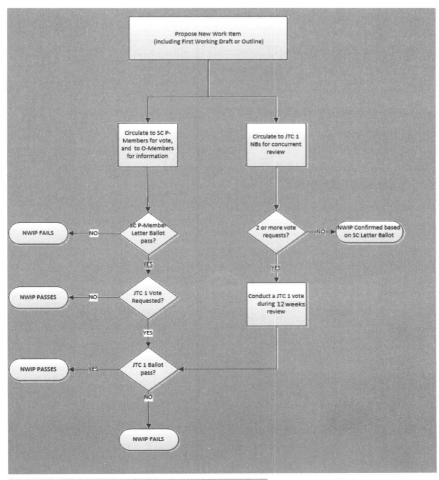

Figure JA.1: Flowchart of NP Ballot Process at SC level

JA.2.2 Votes on new work item proposals at the JTC 1 level

JTC 1 should consider a new work item proposal:

- for a work item originating from a working group which reports directly to JTC 1; or
- in exceptional circumstances, such as a new work item proposal which is not within the scope of an existing subcommittee.

In all other cases, the appropriate subcommittee should ballot the new work item proposal. Each new work item proposal shall be voted on by letter ballot (see the new work item proposal letter ballot form in the Templates folder at www.jtc1.org), even if it has appeared on the agenda of a meeting. The acceptance criteria are as specified in 2.3.5.

JA.3 Preparatory Stage

No votes are foreseen at this stage.

JA.4 Committee Stage - Votes on CDs/~~P~~DAMs/~~P~~DTSs/~~P~~DTRs

If the consideration of committee drafts/~~proposed~~ committee draft amendments/ draft technical specifications/ draft technical reports (CDs/~~P~~DAMs/DTSs/DTRs) is dealt with by correspondence, P-members and technical committees and organizations in liaison are asked to submit their comments (and P-members their votes) by a specified date ~~(see the committee draft letter ballot form in the Templates folder at:~~

~~http://isotc.iso.org/livelink/livelink?func=ll&objId=8913214&objAction=browse&sort=name).~~

In the case of committee drafts/~~proposed~~ committee draft amendments / draft technical specifications/ draft technical reports, this date should be 8, 12, or 16 weeks from the date of notification of issue.

The default for CD/~~P~~DAM/DTS/DTR circulation is 8 weeks.

Abstention by a P-member on committee drafts/~~proposed~~ committee draft amendments/ draft technical specifications/ draft technical reports ballots does not bar the P-member from voting on subsequent versions of the document at the same or later stages.

Consideration of successive committee drafts/~~proposed~~ committee draft amendments/ draft technical specifications/ draft technical reports shall continue until the substantial support of the P-members of the committee has been obtained or a decision to abandon or defer the project has been reached.

Committee drafts/~~proposed~~ committee draft amendments/ draft technical specifications/ draft technical reports produced by a joint working group should be balloted by all P-members of all subcommittees formally involved in the joint work. Each P-member shall have only one vote.

JA.5 Votes on Draft Technical Corrigenda (DCORs)

Consideration of a draft technical corrigendum (DCOR) is dealt with by correspondence ~~(see the draft technical corrigendum ballot form in the Templates folder at www.jtc1.org)~~. SC P-members and organizations in liaison are asked to submit their comments (and SC P-members their votes) by a specified date that should be no less than 12 weeks from the date of notification of issue.

JA.6 Overview of Ballot Periods in ISO/IEC JTC 1

The following table gives an overview of ballot periods that apply in ISO/IEC JTC 1.

TYPE OF VOTE	DURATION	CROSS REFERENCE
New work item proposal – JTC 1 or SC ballot	12 weeks normally	JA.2
New work item proposal from a subcommittee: JTC 1 confirmation	8 weeks	JA.2.1

Committee Draft	8, 12 or 16 weeks (the default is 8 weeks)	2.5.4, JA.4
Draft Technical Specification / Draft Technical Report	8, 12 or 16 weeks (the default is 8 weeks)	JA.4; 2.5.4
Committee Draft Amendment	8, 12 or 16 weeks (the default is 8 weeks)	JA.4; 2.5.4; 2.10.3
Draft International Standard	12 weeks (with 8 weeks translation period)	2.6.1
Fast-Track Draft International Standard	12 weeks (with 8 weeks translation period)	2.6.1; F.4.3
JTC 1 Publicly Available Specification Draft International Standard	12 weeks (with 8 weeks translation period)	2.6.1; F.4.3
Draft Amendment	12 weeks (with 8 weeks translation period)	2.6.1
Final Draft International Standard	8 weeks	2.7.1
Fast-Track Final Draft International Standard	8 weeks	2.7.1, F.4.8
JTC 1 Publicly Available Specification Final Draft International Standard	8 weeks	F.4.9(b)
Final Draft Amendment	8 weeks	2.10.3
Draft Technical Corrigendum	Minimum 12 weeks	JA.6
Withdrawal proposal of Stabilized Standard	4 weeks	JA 1.3
Reinstatement of withdrawn standard	12 weeks (with 8 week translation period), or 8 weeks.	2.9.6
JTC 1 Publicly Available Specification Submitter recognition	12 weeks	F.3.4.1
JTC 1 Publicly Available Specification Submitter reaffirmation	12 weeks	F.3.4.1

JTC 1 - other letter ballot periods	Minimum 12 weeks	JA.1.3
JTC 1 – default letter ballot	4 weeks	JA 1.3
JTC 1 - action between plenary meetings	8 weeks	JA.1.2
Subcommittee Programme of Work Change	4 weeks	JA.1.3
2nd and further Draft International Standard	8 weeks Maximum 12 weeks	2.6.4
2nd and further Draft Amendment	8 weeks Maximum 12 weeks	2.6.4
Withdrawal of standard	4 weeks	2.9.4; JA.1.3
Stabilization of standard	20 weeks (normal systematic review ballot)	2.9.5
Systematic Review	20 weeks	2.9.2

Annex JB
(normative)
ITU-T and ISO/IEC JTC 1 Cooperation

1. The Guide for ITU-T and ISO/IEC JTC 1 cooperation has been drafted by ISO/IEC JTC 1 and ITU-T and approved by ISO/TMB, IEC/-SMB and ITU-T. The text in Standing Document 3: Guide for ITU-T and ISO/IEC JTC 1 cooperation, is identical to the text in Annex A of ITU-T Recommendation A.23.

2. It continues a long-standing agreement among the same organizations concerning collaboration methods by which ITU-T Recommendations and ISO/IEC International Standards developed in ISO/IEC JTC 1 have common texts or identical technical contents.

3. In addition to the normal liaison arrangements already in use by the three organizations and when desirable to reach common text or identical technical content in a particular area of work, ITU-T and ISO/IEC JTC 1 shall use one of two modes of closer cooperation: *collaborative interchange or a collaborative team*.

4. Collaborative interchange involves progressing the technical work on a single text in successive meetings of both the organizations involved, with synchronization of the respective commenting and approval procedures. It shall be used where the work is relatively straightforward and non-controversial, and where common participation in the meetings of the two organizations is sufficient for the interchange to be highly effective. Terms of reference for the work to be accomplished shall be agreed.

5. A single collaborative team shall be set up to progress any work requiring extended dialogue to develop solutions and reach consensus. Terms of reference for the team shall be agreed, and shall include the scope of the effort and the parent body in each organization to which the team reports. Once consensus is achieved, synchronized use is made of the approval procedures in ITU-T, IEC and ISO to achieve publication. The procedures to be followed by collaborative teams may be found in clause 8 of the JTC 1 Standing Document 3: ITU-T and ISO/IEC JTC 1 Cooperation.

6. In either collaboration mode, the approved deliverables may be published as common text (an ITU-T Recommendation and an International Standard using the presentation style specified in Appendix II of the Guide), or as twin text (an ITU-T Recommendation and an International Standard whose texts are technically aligned but not identical), in which case the approval processes do not require exact timing synchronization.

7. The ITU-T Study Group and the ISO/IEC JTC 1 Subcommittee shall agree whether no contact is needed, or liaison, collaborative interchange or a collaborative team will be used in each area of work. The mode may change during a project, again by agreement.

8. In the unusual event that either organization feels that collaboration for a given area of work should be terminated, this situation shall be immediately discussed with the other organization. If satisfactory resolution cannot be obtained, either ITU-T or ISO/IEC JTC 1 may unilaterally terminate collaboration on a project, or decide that no common text should be published. If termination should occur, both organizations can make use of the prior collaborative work. Any work accomplished up to that point may be used by each organization.

Annex JC
(normative)
Proposals for management system standards

JC.1 General

Whenever a proposal is made to prepare a new management system standard (MSS), including sector-specific MSS, a justification study (JS) shall be carried out in accordance with Appendix 1 to this Annex JC.

NOTE No JS is needed for the revision of an existing MSS whose development has already been approved and provided the scope is confirmed (unless it was not provided during its first development).

To the extent possible, the proposer shall endeavour to identify the full range of deliverables which will constitute the new or revised MSS family, and a JS shall be prepared for each of the deliverables.

JC.2 Terms and definitions

For the purposes of this Annex JC, the following terms and definitions apply.

JC.2.1
management system
See definition contained in Appendix 2 (clause 3.4) of this Annex JC.

JC.2.2
Management System Standard
MSS
Standard for management systems (JC 2.1)

Note to entry: For the purposes of this document, this definition also applies to other ISO deliverables (e.g. TS)

JC.2.3
generic MMS
MSS designed to be widely applicable across economic sectors, various types and sizes of organizations and diverse geographical, cultural and social conditions.

JC.2.4
sector-specific MSS
MSS that provides additional requirements or guidance for the application of a generic MSS to a specific economic or business sector.

JC.2.5
Type A MSS
MSS providing requirements

EXAMPLES

— Management system requirements standards (specifications)

― Management system sector-specific requirements standards

JC.2.6
Type B MSS
MSS providing guidelines

EXAMPLES

― Guidance on the use of management system requirements standards

― Guidance of the establishment of a management system

― Guidance on the improvement/enhancement of a management system

JC.2.7
High Level Structure
HLS
outcome of the work of the ISO/TMB/JTCG "Joint technical Coordination Group on MSS" which refers to high level structure (HLS), identical subclause titles, identical text and common terms and core definitions. See Appendix 2 to this Annex JC.

JC.3 Obligation to submit a JS

All MSS (including sector-specific MSS (JC 2.4) see ISO Annex SP) proposals and their JS shall be identified by the relevant TC/SC/PC leadership and their JS shall be sent to the ISO/TMB (or its MSS task force) for evaluation and approval before the NWI ballot takes place. It is the responsibility of the relevant TC/SC/PC secretariat to identify all MSS proposals, without exception, so that there will be no MSS proposals which fail (with knowledge or without knowledge) to carry out the JS or which fail to be sent to the ISO/TMB for evaluation.

No JS is required for a Type B MSS providing guidance on a specific Type A MSS for which a JS has already been submitted and approved.

EXAMPLE, ISO/IEC 27003:2010 (Information technology — Security techniques — Information security management system implementation guidance) does not need to have JS submitted as ISO/IEC 27001:2013 (Information technology — Security techniques — Information security management systems — Requirements) has already had a JS submitted and approved.

JC.4 Cases where no JS have been submitted

MSS proposals which have not been submitted for ISO/TMB evaluation before the NWI ballot will be sent to the ISO/TMB for evaluation and no new ballot should take place before the ISO/TMB decision (project on hold). It is considered good practice that the TC/SC/PC members endorse the JS prior it is sent to the ISO/TMB.

NOTE   Already published MSS which did not have a JS submitted will be treated as new MSS at the time of revision, i.e. a JS is to be presented and approved before any work can begin.

JC.5 Applicability of Annex JC

The above procedures apply to all ISO deliverables including IWAs.

JC.6 General principles

All projects for new MSS (or for MSS which are already published but for which no JS was completed) shall undergo a JS (see JC.1 and Note to JC.3). The following general principles provide guidance to assess the market relevance of proposed MSS and for the preparation of a JS. The justification criteria questions in Appendix 1 to this Annex JC are based on these principles. The answers to the questions will form part of the JS. An MSS should be initiated, developed and maintained only when all of the following principles are observed.

1) Market relevance - Any MSS should meet the needs of, and add value for, the primary users and other affected parties.

2) Compatibility - Compatibility between various MSS and within an MSS family should be maintained.

3) Topic coverage - A generic MSS (JC 2.3) should have sufficient application coverage to eliminate or minimize the need for sector-specific variances.

4) Flexibility - An MSS should be applicable to organizations in all relevant sectors and cultures and of every size. An MSS should not prevent organizations from competitively adding to or differentiating from others, or enhancing their management systems beyond the standard.

5) Free trade - An MSS should permit the free trade of goods and services in line with the principles included in the WTO Agreement on Technical Barriers to Trade.

6) Applicability of conformity assessment - The market need for first-, second- or third-party conformity assessment, or any combination thereof, should be assessed. The resulting MSS should clearly address the suitability of use for conformity assessment in its scope. An MSS should facilitate joint audits.

7) Exclusions - An MSS should not include directly related product (including services) specifications, test methods, performance levels (i.e. setting of limits) or other forms of standardization for products produced by the implementing organization.

8) Ease of use - It should be ensured that the user can easily implement one or more MSS. An MSS should be easily understood, unambiguous, free from cultural bias, easily translatable, and applicable to businesses in general.

JC.7 Justification study process and criteria

JC.7.1 General

This clause describes the justification study (JS) process for justifying and evaluating the market relevance of proposals for an MSS. Appendix 1 to this Annex JC provides a set of questions to be addressed in the justification study.

JC.7.2 Justification study process

The JS process applies to any MSS project and consists of the following:

a) the development of the JS by (or on behalf of) the proposer of an MSS project;

b) an approval of the JS by the ISO/TMB (or ISO/TMB MSS task force).

The JS process is followed by the normal ISO balloting procedure for new work item approval as appropriate.

JC.7.3　Justification study criteria

Based on Annex C of the ISO/IEC Directives, Part 1, 2012, and the general principles stated above, a set of questions (see Appendix 1 to this Annex JC) shall be used as criteria for justifying and assessing a proposed MSS project and shall be answered by the proposer. This list of questions is not exhaustive and any additional information that is relevant to the case should be provided. The JS should demonstrate that all questions have been considered. If it is decided that they are not relevant or appropriate to a particular situation, then the reasons for this decision should be clearly stated. The unique aspect of a particular MSS may require consideration of additional questions in order to assess objectively its market relevance.

JC.8　Guidance on the development process and structure of an MSS

JC.8.1　General

The development of an MSS will have effects in relation to:

— the far-reaching impact of these standards on business practice;

— the importance of worldwide support for the standards;

— the practical possibility for involvement by many, if not all, ISO Member Bodies; and

— the market need for compatible and aligned MSS.

This clause provides guidance in addition to the procedures laid down in the ISO/IEC Directives, in order to take these effects into account.

All MSS (whether they are Type A or Type B MSS, generic or sector-specific) shall, in principle, use consistent structure, common text and terminology so that they are easy to use and compatible with each other. The guidance and structure given in Appendix 2 to this Annex JC shall, in principle, also be followed (based on ISO/TMB Resolution 18/2012).

A Type B MSS which provides guidance on another MSS of the same MSS family should follow the same structure (i.e. clauses numbering). Where MSS providing guidance (Type B MSS) are involved, it is important that their functions be clearly defined together with their relationship with the MSS providing requirements (Type A MSS), for example:

— guidance on the use of the requirements standard;

— guidance on the establishment/implementation of the management system;

— guidance on improvement/enhancement of the management system.

Where the proposed MSS is sector specific:

— it should be compatible and aligned with the generic MSS;

— rules and principles specified in ISO Annex SP shall be followed;

— the relevant committee responsible for the generic MSS may have additional requirements to be met or procedures to be followed (see ISO Annex SP);

— other committees may need to be consulted, as well as CASCO on conformity assessment issues.

In the case of sector specific documents, their function and relationship with the generic MSS should be clearly defined (e.g. additional sector-specific requirements; elucidation; or both as appropriate).

Sector-specific documents should always show clearly (e.g. by using different typographical styles) the kind of sector-specific information being provided.

NOTE 1　Where the identical text or any of the requirements cannot be applied in a specific MSS, due to special circumstances, this should be reported to the ISO/TMB through the TMB Secretary at tmb@iso.org (see JC.9.3).

JC.8.2　MSS development process

JC.8.2.1　General

In addition to the JS, the development of an MSS should follow the same requirements as other ISO deliverables (ISO/IEC Directives, Part 1, Clause 2).

JC.8.2.2　Design specification

To ensure that the intention of the standard, as demonstrated by the justification study, will be maintained, a design specification may be developed before a working draft is prepared.

The responsible committee will decide whether the design specification is needed and in case it is felt necessary, it will decide upon its format and content that is appropriate for the MSS and should set up the necessary organization to carry out the task.

The design specification should typically address the following.

- User needs – The identification of the users of the standard and their associated needs, together with the costs and benefits for these users.
- Scope – The scope and purpose of the standard, the title and the field of application.
- Compatibility – How compatibility within this and with other MSS families will be achieved, including identification of the common elements with similar standards, and how these will be included in the recommended structure (see Appendix 2 to this Annex JC).
- Consistency – Consistency with other documents (to be) developed within the MSS family.

NOTE　Most, if not all of the information on user needs and scope will be available from the justification study.

The design specification should ensure that

a) the outputs of the justification study are translated correctly into requirements for the MSS;

b) the issues of compatibility and alignment with other MSS are identified and addressed;

c) a basis for verification of the final MSS exists at appropriate stages during the development process;

d) the approval of the design specification provides a basis for ownership throughout the project by the members of the TC/SC(s);

e) account is taken of comments received through the NWI ballot phase; and

f) any constraints are taken into account.

The Committee developing the MSS should monitor the development of the MSS against the design specification in order to ensure that no deviations happen in the course of the project.

JC.8.2.3 Producing the deliverables

JC.8.2.3.1 Monitoring output

In the drafting process, the output should be monitored for compatibility and ease of use with other MSS, by covering issues such as:

— the high level structure (HLS), identical subclause titles, identical text and common terms and core definitions the need for clarity (both in language and presentation); and

— avoiding overlap and contradiction.

JC.8.2.4 Transparency of the MSS development process

MSS have a broader scope than most other types of standard. They cover a large field of human endeavour and have an impact on a wide range of user interests.

Committees preparing MSS should accordingly adopt a highly transparent approach to the development of the standards, ensuring that:

— possibilities for participation in the process of developing standards are clearly identified; and

— the development processes being used are understood by all parties.

Committees should provide information on progress throughout the life cycle of the project, including:

— the status of the project to date (including items under discussion);

— contact points for further information;

— communiqués and press releases on plenary meetings; and

— regular listings of frequently asked questions and answers.

In doing this, account needs to be taken of the distribution facilities available in the participating countries.

Where it may be expected that users of a Type A MSS are likely to demonstrate conformity to it, the MSS shall be so written that conformity can be assessed by a manufacturer or supplier (first party, or self-declaration), a user or purchaser (second party) or an independent body (third party, also known as certification or registration).

Maximum use should be made of the resources of the ISO Central Secretariat to facilitate the transparency of the project and the committee should, in addition, consider the establishment of a dedicated open-access website.

Committees should involve the national member bodies to build up a national awareness of the MSS project, providing drafts as appropriate for different interested and affected parties, including accreditation bodies, certification bodies, enterprises and the user community, together with additional specific information as needed.

The committee should ensure that technical information on the content of the MSS under development is readily available to participating members, especially those in developing countries.

JC.8.2.5 Process for interpretation of a standard

The committee may establish a process to handle interpretation questions related to their standards from the users, and may make the resulting interpretations available to others in an expedient manner. Such a mechanism can effectively address possible misconceptions at an early stage and identify issues that may require improved wording of the standard during the next revision cycle.

JC.9 High level structure, identical core text and common terms and core definitions for use in Management Systems Standards

JC.9.1 Introduction

The aim of this document is to enhance the consistency and alignment of ISO MSS by providing a unifying and agreed upon high level structure, identical core text and common terms and core definitions. The aim being that all ISO Type A MSS (and B where appropriate) are aligned and the compatibility of these standards is enhanced. It is envisaged that individual MSS will add additional "discipline-specific" requirements as required.

NOTE   In Annex JC. 9.1 and Annex JC. 9.4 "discipline-specific" is used to indicate specific subject(s) to which a management system standard refers, e.g. energy, quality, records, environment etc.

The intended audience for this document is ISO Technical Committees (TC), Subcommittees (SC) and Project Committees (PC) and others that are involved in the development of MSS.

This common approach to new MSS and future revisions of existing standards will increase the value of such standards to users. It will be particularly useful for those organizations that choose to operate a single (sometimes called "integrated") management system that can meet the requirements of two or more MSS simultaneously.

Appendix 2 to this Annex JC sets out the high level structure, identical core text and common terms and core definitions that form the nucleus of future and revised ISO Type A MSS and Type B MSS when possible.

Appendix 3 to this Annex JC sets out guidance to the use of Appendix 2 to this Annex JC.

JC.9.2 Use

ISO MSS include the high level structure and identical core text as found in Appendix 2 to this Annex JC. The common terms and core definitions are either included or normatively reference an international standard where they are included.

NOTE The high level structure includes the main clauses (1 to 10) and their titles, in a fixed sequence. The identical core text includes numbered sub-clauses (and their titles) as well as text within the sub-clauses.

JC.9.3 Non applicability

If due to exceptional circumstances the high level structure or any of the identical core text, common terms and core definitions cannot be applied in the management system standard then the TC/PC/SC needs to explain their rationale for review by:

a) providing an initial deviation report to ISO/CS with the DIS submission;

b) providing a final deviation report to ISO/TMB (through the ISO/TMB Secretary at tmb@iso.org) upon submission of the final text of the standard for publication.

TC/PC/SC shall use the ISO commenting template to provide their deviation reports.

NOTE 1 The final deviation report can be an updated version of the initial deviation report.

NOTE 2 TC/PC/SC strives to avoid any non-applicability of the high level structure or any of the identical core text, common terms and core definitions.

JC.9.4 Using Annex JC Appendix 2

Discipline-specific text additions to Annex JC Appendix 2 are managed as follows.

1. Discipline-specific additions are made by the individual ISO/TC, PC, SC or other group that is developing the specific ISO management system standard.

2. Discipline-specific text does not affect harmonization or contradict or undermine the intent of the high level structure, identical core text, common terms and core definitions.

3. Insert additional sub-clauses, or sub-sub-clauses (etc.) either ahead of an identical text sub-clause (or sub-sub-clause etc.), or after such a sub-clause (etc.) and renumbered accordingly.

NOTE 1 Hanging paragraphs are not permitted – see ISO/IEC Directives, Part 2, 22.2.3.

NOTE 2 Attention is drawn to the need to check cross referencing.

4. Add or insert discipline-specific text within Appendix 2 to this Annex JC. Examples of additions include:

   a) new bullet points;

   b) discipline-specific explanatory text (e.g. Notes or Examples), in order to clarify requirements;

   c) discipline-specific new paragraphs to sub-clauses (etc.) within the identical text;

d) adding text that enhances the existing requirements in Appendix 2 to this Annex JC.

5. Avoid repeating requirements between identical core text and discipline-specific text by adding text to the identical core text taking account of point 4.2 above.

6. Distinguish between discipline-specific text and identical core text from the start of the drafting process. This aids identification of the different types of text during the development and balloting stages.

NOTE 1　Distinguishing options include by colour, font, font size, italics, or by being boxed separately etc.

NOTE 2　Identification of distinguishing text is not necessarily carried into the published version.

7. Understanding of the concept of "risk" may be more specific than that given in the definition under 3.09 of Appendix 2 to this Annex JC. In this case a discipline-specific definition may be needed. The discipline-specific terms and definitions are differentiated from the core definition, e.g. (XXX) risk.

NOTE　The above can also apply to a number of other definitions.

8. Common terms and core definitions will be integrated into the listing of terms and definitions in the discipline-specific management system standard consistent with the concept system of that standard.

JC.9.5　Implementation

Follow the sequence, high level structure, identical core text, common terms and core definitions for any new management system standard and for any revisions to existing management system standard.

JC.9.6　Guidance

Find supporting guidance in Appendix 3 to this Annex JC.

Appendix 1
(normative)
Justification criteria questions

1. General

The list of questions to be addressed in the justification study is in line with the principles listed in JC.6. This list is not exhaustive. Additional information not covered by the questions should be provided if it is relevant to the case.

Each general principle should be given due consideration and ideally when preparing the JS, the proposer should provide a general rationale for each principle, prior to answering the questions associated with the principle.

The principles the proposer of the MSS should pay due attention to when preparing the justification study are:

1. Market relevance
2. Compatibility
3. Topic coverage
4. Flexibility
5. Free trade
6. Applicability of conformity assessment
7. Exclusions

NOTE No questions directly refer to the principle 8 "ease of use", but it should guide the development of the deliverable

Basic information on the MSS proposal

1	What is the proposed purpose and scope of the MSS? Is the document supposed to be a guidance document or a document with requirements?
2	Does the proposed purpose or scope include product (including service) specifications, product test methods, product performance levels, or other forms of guidance or requirements directly related to products produced or provided by the implementing organization?
3	Is there one or more existing ISO committee or non-ISO organization that could logically have responsibility for the proposed MSS? If so, identify.
4	Have relevant reference materials been identified, such as existing guidelines or established practices?
5	Are there technical experts available to support the standardization work? Are the technical experts direct representatives of the affected parties from the different geographical regions?
6	What efforts are anticipated as being necessary to develop the document in terms of experts needed and number/duration of meetings?
7	Is the MSS intended to be a guidance document, contractual specification or regulatory specification for an organization?

Principle 1: market relevance

8	Have all the affected parties been identified? For example: a) organizations (of various types and sizes): the decision-makers within an organization who approve work to implement and achieve conformance to the MSS; b) customers/end-users, i.e. individuals or parties that pay for or use a product (including service) from an organization; c) supplier organizations, e.g. producer, distributor, retailer or vendor of a product, or a provider of a service or information; d) MSS service provider, e.g. MSS certification bodies, accreditation bodies or consultants; e) regulatory bodies; f) non-governmental organizations.
9	What is the need for this MSS? Does the need exist at a local, national, regional or global level? Does the need apply to developing countries? Does it apply to developed countries? What is the added value of having an ISO document (e.g. facilitating communication between organizations in different countries)?
10	Does the need exist for a number of sectors and is thus generic? If so, which ones? Does the need exist for small, medium or large organizations?

11	Is the need important? Will the need continue? If yes, will the target date of completion for the proposed MSS satisfy this need? Are viable alternatives identified?
12	Describe how the need and importance were determined. List the affected parties consulted and the major geographical or economical regions in which they are located.
13	Is there known or expected support for the proposed MSS? List those bodies that have indicated support. Is there known or expected opposition to the proposed MSS? List those bodies that have indicated opposition.
14	What are the expected benefits and costs to organizations, differentiated for small, medium and large organizations if applicable? Describe how the benefits and the costs were determined. Provide available information on geographic or economic focus, industry sector and size of the organization. Provide information on the sources consulted and their basis (e.g. proven practices), premises, assumptions and conditions (e.g. speculative or theoretical), and other pertinent information.
15	What are the expected benefits and costs to other affected parties (including developing countries)? Describe how the benefits and the costs were determined. Provide any information regarding the affected parties indicated.
16	What will be the expected value to society?
17	Have any other risks been identified (e.g. timeliness or unintended consequences to a specific business)?

Principle 2: compatibility

18	Is there potential overlap or conflict with (or what is the added value in relation to) other existing or planned ISO or non-ISO international standards, or those at the national or regional level? Are there other public or private actions, guidance, requirements and regulations that seek to address the identified need, such as technical papers, proven practices, academic or professional studies, or any other body of knowledge?
19	Is the MSS or the related conformity assessment activities (e.g. audits, certifications) likely to add to, replace all or parts of, harmonize and simplify, duplicate or repeat, conflict with, or detract from the existing activities identified above? What steps are being considered to ensure compatibility, resolve conflict or avoid duplication?
20	Is the proposed MSS likely to promote or stem proliferation of MSS at the national or regional level, or by industry sectors?

Principle 3: topic coverage

21	Is the MSS for a single specific sector?

22	Will the MSS reference or incorporate an existing, non-industry-specific ISO MSS (e.g. from the ISO 9000 series of quality management standards)? If yes, will the development of the MSS conform to the ISO/IEC Sector Policy (see 6.8.2 of ISO/IEC Directives, Part 2), and any other relevant policy and guidance procedures (e.g. those that may be made available by a relevant ISO committee)?
23	What steps have been taken to remove or minimize the need for particular sector-specific deviations from a generic MSS?

Principle 4: flexibility

24	Will the MSS allow an organization competitively to add to, differentiate or encourage innovation of its management system beyond the standard?

Principle 5: free trade

25	How would the MSS facilitate or impact global trade? Could the MSS create or prevent a technical barrier to trade?
26	Could the MSS create or prevent a technical barrier to trade for small, medium or large organizations?
27	Could the MSS create or prevent a technical barrier to trade for developing or developed countries?
28	If the proposed MSS is intended to be used in government regulations, is it likely to add to, duplicate, replace, enhance or support existing governmental regulations?

Principle 6: applicability of conformity

29	If the intended use is for contractual or regulatory purposes, what are the potential methods to demonstrate conformance (e.g. first party, second party or third party)? Does the MSS enable organizations to be flexible in choosing the method of demonstrating conformance, and to accommodate for changes in its operations, management, physical locations and equipment?
30	If third-party registration/certification is a potential option, what are the anticipated benefits and costs to the organization? Will the MSS facilitate joint audits with other MSS or promote parallel assessments?

Principle 7: exclusions

31	Does the proposed purpose or scope include product (including service) specifications, product test methods, product performance levels, or other forms of guidance or requirements directly related to products produced or provided by the implementing organization?

Appendix 2
(normative)

High level structure, identical core text, common terms and core definitions

NOTE In the Identical text proposals, XXX = an MSS discipline specific qualifier (e.g. energy, road traffic safety, IT security, food safety, societal security, environment, quality) that needs to be inserted. Blue italicized text is given as advisory notes to standards drafters.

Introduction

DRAFTING INSTRUCTION Specific to the discipline.

1. Scope

DRAFTING INSTRUCTION   Specific to the discipline.

2. Normative references

DRAFTING INSTRUCTION   Clause Title shall be used. Specific to the discipline.

3. Terms and definitions

DRAFTING INSTRUCTION   1   Clause Title shall be used. Terms and definitions may either be within the standard or in a separate document. To reference Common terms and Core definitions + discipline specific ones.

For the purposes of this document, the following terms and definitions apply.

DRAFTING INSTRUCTION 2   The following terms and definitions constitute an integral part of the "common text" for management systems standards. Additional terms and definitions may be added as needed. Notes may be added or modified to serve the purpose of each standard.
DRAFTING INSTRUCTION 3   Italics type in a definition indicates a cross-reference to another term defined in this clause, and the number reference for the term is given in parentheses.
DRAFTING INSTRUCTION 4   Where the text "XXX" appears throughout this clause, the appropriate reference should be inserted depending on the context in which these terms and definitions are being applied. For example: "an XXX objective" could be substituted as "an information security objective".

3.1
organization
person or group of people that has its own functions with responsibilities, authorities and relationships to achieve its **objectives** (3.8)
Note 1 to entry: The concept of organization includes, but is not limited to sole-trader, company, corporation, firm, enterprise, authority, partnership, charity or institution, or part or combination thereof, whether incorporated or not, public or private.

3.2
interested party (preferred term)
stakeholder (admitted term)
person or **organization** (3.1) that can affect, be affected by, or perceive itself to be affected by a decision or activity

3.3
requirement
need or expectation that is stated, generally implied or obligatory

NOTE 1 to entry: "Generally implied" means that it is custom or common practice for the organization and interested parties that the need or expectation under consideration is implied.

NOTE 2 to entry: A specified requirement is one that is stated, for example in documented information.

3.4
management system
set of interrelated or interacting elements of an **organization** (3.1) to establish **policies** (3.7) and **objectives** (3.8) and **processes** (3.12) to achieve those objectives
NOTE 1 to entry: A management system can address a single discipline or several disciplines.
NOTE 2 to entry: The system elements include the organization's structure, roles and responsibilities, planning, and operation.
NOTE 3 to entry: The scope of a management system may include the whole of the organization, specific and identified functions of the organization, specific and identified sections of the organization, or one or more functions across a group of organizations.

3.5
top management
person or group of people who directs and controls an **organization** (3.1) at the highest level
NOTE 1 to entry: Top management has the power to delegate authority and provide resources within the organization.
NOTE 2 to entry: If the scope of the **management system** (3.4) covers only part of an organization then top management refers to those who direct and control that part of the organization.

3.6
effectiveness
extent to which planned activities are realized and planned results achieved

3.7
policy
intentions and direction of an **organization** (3.1) as formally expressed by its **top management** (3.5)

3.8
objective
result to be achieved

NOTE 1 to entry: An objective can be strategic, tactical, or operational.

NOTE 2 to entry: Objectives can relate to different disciplines (such as financial, health and safety, and environmental goals) and can apply at different levels (such as strategic, organization-wide, project, product and **process** (3.12)).

NOTE 3 to entry: An objective can be expressed in other ways, e.g. as an intended outcome, a purpose, an operational criterion, as an XXX objective or by the use of other words with similar meaning (e.g. aim, goal, or target).

NOTE 4 to entry: In the context of XXX management systems XXX objectives are set by the organization, consistent with the XXX policy, to achieve specific results.

3.9
risk
effect of uncertainty

NOTE 1 to entry: An effect is a deviation from the expected — positive or negative.

NOTE 2 to entry: Uncertainty is the state, even partial, of deficiency of information related to, understanding or knowledge of, an event, its consequence, or likelihood.

NOTE 3 to entry: Risk is often characterized by reference to potential "**events**" (as defined in ISO Guide 73:2009, 3.5.1.3) and "**consequences**" (as defined in ISO Guide 73:2009, 3.6.1.3), or a combination of these.

NOTE 4 to entry: Risk is often expressed in terms of a combination of the consequences of an event (including changes in circumstances) and the associated **likelihood** (ISO Guide 73, 3.6.1.1) of occurrence.

3.10
competence
ability to apply knowledge and skills to achieve intended results

3.11
documented information
information required to be controlled and maintained by an **organization** (3.1) and the medium on which it is contained

NOTE 1 to entry: Documented information can be in any format and media and from any source.

NOTE 2 to entry: Documented information can refer to
— the **management system** (3.4), including related **processes** (3.12);
— information created in order for the organization to operate (documentation);
— evidence of results achieved (records).

3.12
process
set of interrelated or interacting activities which transforms inputs into outputs

3.13
performance
measurable result

NOTE 1 to entry: Performance can relate either to quantitative or qualitative findings.

NOTE 2 to entry: Performance can relate to the management of activities, **processes** (3.12), products (including services), systems or **organizations** (3.1).

3.14
outsource (verb)
make an arrangement where an external **organization** (3.1) performs part of an organization's function or **process** (3.12)

NOTE 1 to entry: An external organization is outside the scope of the **management system** (3.04), although the outsourced function or process is within the scope.

3.15
monitoring
determining the status of a system, a **process** (3.12) or an activity

NOTE 1 to entry: To determine the status there may be a need to check, supervise or critically observe.

3.16
measurement
process (3.12) to determine a value

3.17
audit
systematic, independent and documented **process** (3.12) for obtaining audit evidence and evaluating it objectively to determine the extent to which the audit criteria are fulfilled

NOTE 1 to entry: An audit can be an internal audit (first party) or an external audit (second party or third party), and it can be a combined audit (combining two or more disciplines).

NOTE 2 to entry: An internal audit is conducted by the organization itself, or by an external party on its behalf

NOTE 3 to entry: "Audit evidence" and "audit criteria" are defined in ISO 19011.

3.18
conformity
fulfilment of a **requirement** (3.3)

3.19
nonconformity
non-fulfilment of a **requirement** (3.3)

3.20
corrective action
action to eliminate the cause of a **nonconformity** (3.19) and to prevent recurrence

3.21
continual improvement
recurring activity to enhance **performance** (3.13)

4. Context of the organization

4.1 Understanding the organization and its context

The organization shall determine external and internal issues that are relevant to its purpose and that affect its ability to achieve the intended outcome(s) of its XXX management system.

4.2 Understanding the needs and expectations of interested parties

The organization shall determine
— the interested parties that are relevant to the XXX management system;
— the relevant requirements of these interested parties.

4.3 Determining the scope of the XXX management system

The organization shall determine the boundaries and applicability of the XXX management system to establish its scope.

When determining this scope, the organization shall consider

— the external and internal issues referred to in 4.1,
— the requirements referred to in 4.2.
The scope shall be available as documented information.

4.4 XXX management system
The organization shall establish, implement, maintain and continually improve an XXX management system, including the processes needed and their interactions, in accordance with the requirements of this International Standard/this part of ISO XXXX/this Technical Specification

5. Leadership

5.1 Leadership and commitment

Top management shall demonstrate leadership and commitment with respect to the XXX management system by

— ensuring that the XXX policy and XXX objectives are established and are compatible with the strategic direction of the organization
— ensuring the integration of the XXX management system requirements into the organization's business processes
— ensuring that the resources needed for the XXX management system are available
— communicating the importance of effective XXX management and of conforming to the XXX management system requirements
— ensuring that the XXX management system achieves its intended outcome(s)
— directing and supporting persons to contribute to the effectiveness of the XXX management system
— promoting continual improvement
— supporting other relevant management roles to demonstrate their leadership as it applies to their areas of responsibility.
NOTE Reference to "business" in this International Standard/this part of ISO XXXX/this Technical Specification can should be interpreted broadly to mean those activities that are core to the purposes of the organization's existence.

5.2 Policy

Top management shall establish a XXX policy that

a) is appropriate to the purpose of the organization;
b) provides a framework for setting XXX objectives;
c) includes a commitment to satisfy applicable requirements;
d) includes a commitment to continual improvement of the XXX management system.

The XXX policy shall

— be available as documented information
— be communicated within the organization
— be available to interested parties, as appropriate.

5.3 Organizational roles, responsibilities and authorities

Top management shall ensure that the responsibilities and authorities for relevant roles are assigned and communicated within the organization.

Top management shall assign the responsibility and authority for:

a) ensuring that the XXX management system conforms to the requirements of this International Standard/this part of ISO XXXX/this Technical Specification; and
b) reporting on the performance of the XXX management system to top management.

6. Planning

6.1 Actions to address risks and opportunities

When planning for the XXX management system, the organization shall consider the issues referred to in 4.1 and the requirements referred to in 4.2 and determine the risks and opportunities that need to be addressed to

— give assurance that the XXX management system can achieve its intended outcome(s)
— prevent, or reduce, undesired effects
— achieve continual improvement.

The organization shall plan:

a) actions to address these risks and opportunities,

b) how to

— integrate and implement the actions into its XXX management system processes
— evaluate the effectiveness of these actions.

6.2 XXX objectives and planning to achieve them

The organization shall establish XXX objectives at relevant functions and levels.
The XXX objectives shall

— a) be consistent with the XXX policy
— b) be measurable (if practicable)
— c) take into account applicable requirements
— d) be monitored
— e) be communicated, and
— f) be updated as appropriate.

The organization shall retain documented information on the XXX objectives.

When planning how to achieve its XXX objectives, the organization shall determine

— what will be done
— what resources will be required
— who will be responsible
— when it will be completed
— how the results will be evaluated.

7. Support

7.1 Resources

The organization shall determine and provide the resources needed for the establishment, implementation, maintenance and continual improvement of the XXX management system.

7.2 Competence

The organization shall

— determine the necessary competence of person(s) doing work under its control that affects its XXX performance;
— ensure that these persons are competent on the basis of appropriate education, training, or experience;
— where applicable, take actions to acquire the necessary competence, and evaluate the effectiveness of the actions taken;
— retain appropriate documented information as evidence of competence.

NOTE Applicable actions can include, for example, the provision of training to, the mentoring of, or the re-assignment of currently employed persons; or the hiring or contracting of competent persons.

7.3 Awareness

Persons doing work under the organization's control shall be aware of

— the XXX policy
— their contribution to the effectiveness of the XXX management system, including the benefits of improved XXX performance
— the implications of not conforming with the XXX management system requirements.

7.4 Communication

The organization shall determine the internal and external communications relevant to the XXX management system including

— on what it will communicate;
— when to communicate;
— with whom to communicate;
— how to communicate.

7.5 Documented information

7.5.1   General

The organization's XXX management system shall include

a) documented information required by this International Standard/this part of ISO XXXX/this Technical Specification;

~~b) documented information determined by the organization as being necessary for the effectiveness of the XXX management system.~~

~~NOTE The extent of documented information for a XXX management system can differ from one organization to another due to~~

- ~~the size of organization and its type of activities, processes, products and services;~~
- ~~the complexity of processes and their interactions;~~
- ~~the competence of persons.~~

~~**7.5.2   Creating and updating**~~

~~When creating and updating documented information the organization shall ensure appropriate~~

- ~~identification and description (e.g. a title, date, author, or reference number)~~
- ~~format (e.g. language, software version, graphics) and media (e.g. paper, electronic)~~
- ~~review and approval for suitability and adequacy.~~

~~**7.5.3   Control of documented information**~~

~~Documented information required by the XXX management system and by this International Standard/this part of ISO XXXX/this Technical Specification shall be controlled to ensure~~

- ~~a) it is available and suitable for use, where and when it is needed~~
- ~~b) it is adequately protected (e.g. from loss of confidentiality, improper use, or loss of integrity).~~

~~For the control of documented information, the organization shall address the following activities, as applicable~~

- ~~distribution, access, retrieval and use,~~
- ~~storage and preservation, including preservation of legibility~~
- ~~control of changes (e.g. version control)~~
- ~~retention and disposition~~

~~Documented information of external origin determined by the organization to be necessary for the planning and operation of the XXX management system shall be identified as appropriate, and controlled.~~

~~NOTE Access can imply a decision regarding the permission to view the documented information only, or the permission and authority to view and change the documented information.~~

~~**8. Operation**~~

~~**8.1 Operational planning and control**~~

~~The organization shall plan, implement and control the processes needed to meet requirements, and to implement the actions determined in 6.1, by~~

- ~~establishing criteria for the processes~~
- ~~implementing control of the processes in accordance with the criteria~~
- ~~keeping documented information to the extent necessary to have confidence that the processes have been carried out as planned.~~

~~The organization shall control planned changes and review the consequences of unintended changes, taking action to mitigate any adverse effects, as necessary.~~

~~The organization shall ensure that outsourced processes are controlled.~~

9. Performance evaluation

9.1 Monitoring, measurement, analysis and evaluation

The organization shall determine

— what needs to be monitored and measured
— the methods for monitoring, measurement, analysis and evaluation, as applicable, to ensure valid results
— when the monitoring and measuring shall be performed
— when the results from monitoring and measurement shall be analysed and evaluated.

The organization shall retain appropriate documented information as evidence of the results.

The organization shall evaluate the XXX performance and the effectiveness of the XXX management system.

9.2 Internal audit

9.2.1 The organization shall conduct internal audits at planned intervals to provide information on whether the XXX management system:

a) conforms to
— the organization's own requirements for its XXX management system
— the requirements of this International Standard/this part of ISO XXXX/this Technical Specification;
b) is effectively implemented and maintained.

9.2.2 The organization shall:

a) plan, establish, implement and maintain an audit programme(s), including the frequency, methods, responsibilities, planning requirements and reporting, shall take into consideration the importance of the processes concerned and the results of previous audits;
b) define the audit criteria and scope for each audit;
c) select auditors and conduct audits to ensure objectivity and the impartiality of the audit process;
d) ensure that the results of the audits are reported to relevant management;
e) retain documented information as evidence of the implementation of the audit programme and the audit results.

9.3 Management review

Top management shall review the organization's XXX management system, at planned intervals, to ensure its continuing suitability, adequacy and effectiveness.

The management review shall include consideration of:

a) the status of actions from previous management reviews;
b) changes in external and internal issues that are relevant to the XXX management system;
c) information on the XXX performance, including trends in:
— nonconformities and corrective actions;
— monitoring and measurement results;
— audit results;
d) opportunities for continual improvement.

The outputs of the management review shall include decisions related to continual improvement opportunities and any need for changes to the XXX management system.

The organization shall retain documented information as evidence of the results of management reviews.

10. Improvement

10.1 Nonconformity and corrective action

When nonconformity occurs, the organization shall:

a) react to the nonconformity, and as applicable

— take action to control and correct it;
— deal with the consequences;

b) evaluate the need for action to eliminate the causes of the nonconformity, in order that it does not recur or occur elsewhere, by

— reviewing the nonconformity
— determining the causes of the nonconformity;
— determining if similar nonconformities exist, or could potentially occur;

c) implement any action needed;
d) review the effectiveness of any corrective action taken;
e) make changes to the XXX management system, if necessary.

Corrective actions shall be appropriate to the effects of the nonconformities encountered.

The organization shall retain documented information as evidence of

— the nature of the nonconformities and any subsequent actions taken, and
— the results of any corrective action.

10.2 Continual improvement

The organization shall continually improve the suitability, adequacy and effectiveness of the XXX management system.

Appendix 3
(informative)
Guidance on high level structure, identical core text, common terms and core definitions

Guidance on the high level structure, identical core text, common terms and core definitions is provided at the following URL: Annex SL Guidance documents (http://isotc.iso.org/livelink/livelink?func=ll&objId=16347818&objAction=browse&viewType=1).

Annex JD
(normative)

Matrix Presentation of Stage Codes

JD.1 Introduction to the Harmonized Stage Code

The standardization process has a number of definite steps or stages which can be used both to describe the process and to indicate where in the process any one item has reached. In general terms the methods used to develop and publish standards via the formal standardization process operated by international, regional and national standards bodies are very similar no matter which body is overseeing the process. Thus, at a high level, it is possible to have a common view of the standardization process and with it a common set of stages. There are differences between the processes of individual bodies, however, and this has led to the development of different stage systems for each body.

This Harmonized Stage Code (HSC) system is used in ISO's databases for tracking standards development projects. Its purpose is to provide a common framework for the transfer of core data. The system allows tracking of the development of a given project in the same way in databases being used at international, regional and national levels and the matrix is so constructed that it can easily be adapted to new requirements.

JD.2 Design of the stage code matrix

A series of "stages" representing procedural sequences common to different organizations has been established. These represent the main stages of standards development.

A series of "sub-stages" has been established within each stage, using a consistent logical system of concepts. The terms "stage" and "sub-stage" are hence used to designate the respective axes of the resulting matrix.

Principal stages and sub-stages are each coded by a two-digit number from 00 to 90, in increments of 10. Individual cells within the generic matrix are coded by a four-digit number made up of its stage and sub-stage coordinates. For visual presentation (although not necessarily for the purposes of database operations), the pair of coordinates are separated by a point (e.g. 10.20 for stage 10, sub-stage 20).

All unused stage codes are reserved for future use, to allow for interpolation of additional phases that might be identified, e.g. stage codes 10, 30, 40, 50 and 80.

JD.3 Basic guidelines for using the system

— Other information concerning, for example, document source or document type, should be recorded in separate database fields and should not be reflected in stage codes.

— There is no sub-code to indicate that a project is dormant at any particular stage. It is recommended to use another database field to address this issue.

— The HSC system allows for the cyclical nature of the standards process and for the repeating of either the current phase or an earlier phase. Events that may be repeated in the life of a project are recordable by repetition of the same stage codes.

— Freezing a project at any point is possible by using the code the project has reached. Projects that have been suspended should have this information recorded in a separate database field.

— The HSC system is not concerned with recording either target or actual dates for achieving stages.

Matrix presentation of project stages

STAGE	SUB-STAGE						
	00 Registration	20 Start of main action	60 Completion of main action	90 Decision			
				92 Repeat an earlier phase	93 Repeat current phase	98 Abandon	99 Proceed
00 Preliminary stage	00.00 Proposal for new project received	00.20 Proposal for new project under review	00.60 Close of review			00.98 Proposal for new project abandoned	00.99 Approval to ballot proposal for new project
10 Proposal stage	10.00 Proposal for new project registered	10.20 New project ballot initiated	10.60 Close of voting	10.92 Proposal returned to submitter for further definition		10.98 New project rejected	10.99 Approval to New project approved
20 Preparatory stage	20.00 New project registered in TC/SC work programme	20.20 Working draft (WD) study initiated	20.60 Close of comment period			20.98 Project deleted	20.99 WD approved for registration as CD
30 Committee stage	30.00 Committee draft (CD) registered	30.20 CD study/ballot initiated	30.60 Close of voting/ comment period	30.92 CD referred back to Working Group		30.98 Project deleted	30.99 CD approved for registration as DIS
40 Enquiry stage	40.00 DIS registered	40.20 DIS ballot initiated: 5 months	40.60 Close of voting	40.92 Full report circulated: DIS referred back to TC or SC	40.93 Full report circulated: decision for new DIS ballot	40.98 Project deleted	40.99 Full report circulated: DIS approved for registration as FDIS
50 Approval stage	50.00 FDIS registered for formal approval	50.20 FDIS ballot initiated: 8 weeks. Proof sent to secretariat	50.60 Close of voting. Proof returned by secretariat	50.92 FDIS referred back to TC or SC		50.98 Project deleted	50.99 FDIS approved for publication
60 Publication stage	60.00 International Standard under publication		60.60 International Standard published				
90 Review stage		90.20 International Standard under systematic review	90.60 Close of review	90.92 International Standard to be revised	90.93 International Standard confirmed		90.99 Withdrawal of International Standard proposed by TC or SC

| | 95 Withdrawal stage | | 95.20 | 95.60 | 95.92 | | | 95.99 |

Annex JE
(normative)

Procedures for the standardization of graphical symbols

JE.1 Introduction

This annex describes the procedures to be adopted in the submission and subsequent approval and registration, when appropriate, of all graphical symbols appearing in ISO documents.

Within ISO the responsibility for the coordination of the development of graphical symbols has been subdivided into two principal areas, allocated to two ISO technical committees:

— ISO/TC 145 – all graphical symbols (except those for use in technical product documentation) (see ISO/TC 145 website);

— ISO/TC 10 – graphical symbols for technical product documentation (tpd) (see ISO/TC 10 website).

In addition, there is coordination with IEC/TC 3 (Information structures, documentation and graphical symbols) and with IEC/TC 3/SC 3C (Graphical symbols for use on equipment).

The basic objectives of the standardization of graphical symbols are to:

— meet the needs of users;

— ensure that the interests of all concerned ISO committees are taken into account;

— ensure that graphical symbols are unambiguous and conform to consistent sets of design criteria;

— ensure that there is no duplication or unnecessary proliferation of graphical symbols.

The basic steps in the standardization of a new graphical symbol are:

— identification of need;

— elaboration;

— evaluation;

— approval, when appropriate;

— registration;

— publication.

All steps should be carried out by electronic means.

— Proposals for new or revised graphical symbols may be submitted by an ISO committee, a liaison member of an ISO committee or any ISO member organization (hereafter jointly called the "proposer").

— Each approved graphical symbol will be allocated a unique number to facilitate its management and identification through a register that provides information that can be retrieved in an electronic format.

— Conflicts with the relevant requirements and guidelines for graphical symbols shall be resolved by liaison and dialogue between ISO/TC 145 or ISO/TC 10 and the product committee concerned at the earliest possible stage.

JE.2 All graphical symbols except those for use in technical product documentation

JE.2.1 General

ISO/TC 145 is responsible within ISO for the overall coordination of standardization in the field of graphical symbols (except for tpd). This responsibility includes:

— standardization in the field of graphical symbols as well as of colours and shapes, whenever these elements form part of the message that a symbol is intended to convey, e.g. a safety sign;

— establishing principles for preparation, coordination and application of graphical symbols: general responsibility for the review and the coordination of those already existing, those under study, and those to be established.

The standardization of letters, numerals, punctuation marks, mathematical signs and symbols, and symbols for quantities and units is excluded. However, such elements may be used as components of a graphical symbol.

The review and co-ordination role of ISO/TC 145 applies to all committees that undertake the responsibility for creation and standardization of graphical symbols within their own particular fields.

ISO/TC 145 has allocated these responsibilities as follows:

— ISO/TC 145/SC 1: Public information symbols;

— ISO/TC 145/SC 2: Safety identification, signs, shapes, symbols and colours;

— ISO/TC 145/SC 3: Graphical symbols for use on equipment.

There is also liaison with ISO/TC 10 and with IEC, in particular with IEC/SC 3C, Graphical symbols for use on equipment.

Table JE.1 shows the categories of graphical symbols covered by each coordinating committee.

Table JE.1 — Categories of graphical symbols

	Basic message	Location	Target audience	Design principles	Overview	Responsible committee
Public information symbols	Location of service or facility	In public areas	General public	ISO 22727	ISO 7001	ISO/TC 145/SC 1
Safety signs (symbols)	Related to safety and health of persons	In workplaces and public areas	a) General public or b) authorized and trained persons	ISO 3864-1 ISO 3864-3	ISO 7010	ISO/TC 145/SC 2
Product safety labels	Related to safety and health of persons	On products	c) General public or d) authorized and trained persons	ISO 3864-2 ISO 3864-3	—	ISO/TC 145/SC 2
Graphical symbols for use on equipment	Related to equipment	On equipment	e) General public or f) authorized and trained persons	IEC 80416-1 ISO 80416-2 IEC 80416-3	ISO 7000 IEC 60417	ISO/TC 145/SC 3 IEC/TC 3/SC 3C
tpd symbols	(Product representation)	Technical product documentation (drawings, diagrams, etc.)	Trained persons	ISO 81714-1	ISO 14617 IEC 60617	ISO/TC 10/SC 10 IEC/TC 3

Table JE.2 — Examples of different types of graphical symbols shown in their context of use

Public information symbols	Telephone ISO 7001 – 008	Aircraft ISO 7001 – 022	Sporting activities ISO 7001 – 029	Gasoline station ISO 7001 – 009	Direction ISO 7001 – 001
Safety signs (symbols)	Means of escape and emergency equipment signs: E001 – Emergency exit (left hand)	Fire safety signs: F001 – Fire extinguisher	Mandatory action signs: M001 – General mandatory action sign	Prohibition signs: P002 – No smoking	Warning signs: W002 – Warning; Explosive material
Product safety labels	Supplementary safety information (text or symbol)	Signal Word — Supplementary safety information (text or symbol)			
Graphical symbols for use on equipment	Ventilating fan: Air-circulating fan ISO 7000 – 0089	Parking Brake ISO 7000 – 0238	Weight ISO 7000 – 0430	Lamp; lighting; illumination IEC 60417 – 5012	Brightness / Contrast IEC 60417 – 5435
tpd symbols	Two-way valve ISO 14617-8 – 2101	Surface texture with special characteristics ISO 1302, Figure 4			

JE.2.2 Submission of proposals

Proposers shall submit their proposals on the relevant application form as soon as possible to the secretariat of the appropriate ISO/TC 145 subcommittee in order to allow for timely review and comment. It is strongly recommended that this submission be made by proposers at the CD stage, but it shall be no later than the first enquiry stage (i.e. DIS or DAM) in the case of an International Standard.

Prior to submitting a graphical symbol proposal, the proposer should:

— be able to demonstrate the need for the proposed graphical symbol;

— have reviewed the relevant ISO and/or IEC standards of graphical symbols, in order to avoid ambiguity and/or overlap with existing standardized graphical symbols, and to check for consistency with any related graphical symbol or family of graphical symbols already standardized;

— create the proposed graphical symbol in accordance with the relevant standards and instructions; these include design principles and criteria of acceptance.

JE.2.3 Standardization procedure for proposed graphical symbols

Upon receipt of a proposal, the ISO/TC 145 sub-committee concerned shall review the application form, within 8 weeks, to check whether it has been correctly completed and the relevant graphics file(s) has been correctly provided. If necessary, the proposer will be invited to modify the application, and to re-submit it.

Upon receipt of a correctly completed application form, a formal review process shall be commenced to review the proposal for consistency with standardized graphical symbols, the relevant design principles and criteria of acceptance.

When this formal review process has been completed, the results shall be transmitted to the proposer, together with any recommendations. The proposer will, where appropriate, be invited to modify the proposal, and to re-submit it for a further review.

The procedures outlined on the relevant ISO/TC 145 sub-committee website shall be followed:

— ISO/TC 145/SC 1: Public information symbols (www.iso.org/tc145/sc1);

— ISO/TC 145/SC 2: Safety identification, signs, shapes, symbols and colours (www.iso.org/tc145/sc2);

— ISO/TC 145/SC 3: Graphical symbols for use on equipment (www.iso.org/tc145/sc3).

These websites also provide application forms for the submission of proposals.

Graphical symbols approved by ISO/TC 145 shall be assigned a definitive registration number and included in the relevant ISO/TC 145 standard.

NOTE   In exceptional cases, unregistered symbols may be included in ISO standards subject to TMB approval.

JE.3 Graphical symbols for use in technical product documentation (tpd) (ISO/TC 10)

ISO/TC 10 is responsible for the overall responsibility for standardization in the field of graphical symbols for technical product documentation (tpd). This responsibility includes

— maintenance of ISO 81714-1: Design of graphical symbols for use in the technical documentation of products – Part 1: Basic rules, in co-operation with IEC;

— standardization of graphical symbols to be used in technical product documentation, co-ordinated with IEC;

— establishing and maintaining a database for graphical symbols including management of registration numbers.

Included is the standardization of symbols for use in diagrams and pictorial drawings.

ISO/TC 10 has allocated these responsibilities to ISO/TC 10/SC 10. The Secretariat of ISO/TC 10/SC 10 is supported by a maintenance group.

Any committee identifying the need for new or revised graphical symbols for tpd shall as soon as possible submit their proposal to the secretariat of ISO/TC 10/SC 10 for review and — once approved — allocation of a registration number.

Annex JF
(normative)

Registration Authority ("RA") Policy

JF.1 Scope

A number of International Standards developed by ISO technical committees require the assignment of unique registration elements, and describe the methodology for the assignment of these elements. The elements themselves are not part of the standard but are assigned by an appointed RA, who also maintains an accurate register of the elements that have been assigned. The RA is a competent body with the requisite infrastructure for ensuring the effective allocation of these registration elements. These bodies are designated by ISO to serve as the unique RA for particular standards, which creates a de facto monopoly situation.

This Policy is mandatory and shall be read in conjunction with Annex H of Part 1 of the ISO/IEC Directives.

Where ISO/CS becomes aware of a RA Standard that has not followed this Policy, it shall stop the publication process to allow time to implement this Policy before the RA Standard is published. For this reason committees are encouraged to make the ISO Technical Program Manager (TPM) aware of a project requiring an RA as early in the development process as possible to avoid delays in publication.

JF.2 Definitions

JF.2.1 RA Standard: The standard for which the RA is providing the Registration Services.

JF.2.2 Registration Services: Services provided by the RA in the implementation of the RA Standard and which shall be described in the RA Standard.

JF.2.3 Registration Authority ("RA"): Entity appointed by ISO to fulfill the Registration Services in a RA Standard.

JF.2.4 Registration Agencies: Third parties (e.g. national or regional sub-entities) to which the RA delegates part of the Registration Services. Even when delegated to Registration Agencies, the Registration Services remain under the responsibility of the RA.

JF.2.5 Registration Authority Agreement ("RAA"): Agreement based on the RAA template signed by the RA and the ISO Secretary-General on behalf of ISO, which details the functions, roles and legal obligations of the parties involved. A RAA shall be signed before a RA Standard is published (including revisions).

JF.2.6 Registration Elements: Unique elements, the methodology for which is described in the standard but which themselves are not part of the standard.

JF.2.7 Technical Programme Managers (TPM): Individual within ISO/CS assigned to work with a given committee.

JF.3 Procedure

JF.3.1 Chronology

This Policy addresses the various aspects of a RA in the order of the life cycle of a typical RA noting that some stages may be done in parallel. Each stage is addressed as follows:

— Identifying the need for a RA (JF.3.2)

— Drafting a RA Standard (JF.3.3)

— Selecting a RA (JF.3.4)

— Appointing a RA (JF.3.5)

— Signing the RAA (JF.3.6)

— Terminating a RA (JF.3.7)

— Implementing a RA Standard (JF.3.8)

    — — The Role of the RA (JF.3.8.1)

    — — The Role of the Committee (JF.3.8.2)

    — — The Role of the ISO Central Secretariat (JF.3.8.3)

JF.3.2 Identifying the need for RA

The committee confirms its decision that a standard needs a RA for its implementation by way of a resolution.

The committee secretariat completes the RA Confirmation ("RAC") Form (See www.iso.org/forms.Annex SJ) and submits it to the TPM as soon as possible, but no later than the beginning of the preparatory (WD) stage.

JF.3.3 Drafting a RA Standard

The following shall be included in all RA Standards:

— A description of the identification scheme or the mechanism for generating unique registration elements.

— A description of the Registration Services, and the responsibility of the RA for providing the Registration Services.

— The link to the page on iso.org where users can find out the name and contact information of the RA for a given standard. This should also provide links to the RA's website which contains more information on the Registration Services.

The following shall not be included in RA Standards:

— In accordance with clause 4 of the ISO/IEC Directives, Part 2, contractual or other legal elements shall not be included in the RA Standard.

— Procedures concerning the provision of the Registration Services (e.g. a Handbook made available by the RA) shall not be included in the RA Standard. These are to be publicly provided by the RA (e.g. via a website).

— The name of the RA shall not be mentioned in the RA Standard. Instead, the link to the ISO website shall be provided (see above).

— References to the selection or confirmation process for the RA shall not be included in the RA Standard.

— In case of delegation of Registration Services by the RA to third parties (e.g. national or regional sub-entities referred to as "Registration Agencies") as agreed under the RAA, the RA Standard may mention the fact that some Registration Services have been delegated. However, any further details about Registration Agencies shall not be included in the RA Standard.

The TPM is responsible for coordinating with the committee to ensure that the right elements are included in the RA Standard. Any questions about what should be included in the RA Standard are to be addressed to the TPM.

JF.3.4　Selecting a RA

The selection process of the RA applies to new RA Standards and existing RA Standards in cases where the existing RA ceases to be the RA for any reason.

In the case of revisions, the committee shall decide whether the existing RA should continue or if a selection process should be launched to select additional RA candidates. The decision to launch a selection process should be supported by a rationale. The committee shall confirm its decision by resolution.

The selection process shall be completed before the project reaches the DIS stage. It is important for the DIS to contain details about the nature of the Registration Services needed.

The committee establishes the criteria for the selection of the RA and confirms these by resolution. The minimum elements of the selection process shall be:

— Selection criteria – these shall be clearly explained and with sufficient details for possible RA candidates to assess their ability to meet the criteria and apply on this basis. Included in the selection criteria shall be the requirement that the RA provide the following information in writing:

  — Proof (e.g. Statutes) that it is a legal entity which means that is an organization formed under the laws of a jurisdiction and that it is therefore subject to governance related rules.

  — Expression of willingness to take responsibility for the Registration Services.

  — Confirmation that the RA is technically and financially able to carry out the RA Services described in the RA Standard and the RAA on an international level.

  — Expression of willingness to sign a RAA, the ISO/CS RAA template for which shall be shared with RA candidates.

— Confirmation of whether it intends to delegate part of the Registration Services to Registration Agencies.

— Whether it will charge fees for the services and, if it charges fees, confirmation that any such fees will be on a cost recovery basis.

— Public call for RA candidates – committees shall take the appropriate steps needed to post the call for competent RA applicants to as broad a market as possible, also targeting possible organizations by inviting them to apply.

— Evaluation – RA candidates shall provide their responses in writing. The committee (or a subset thereof) shall decide on objective evaluation criteria based on selection criteria, including the relative weight given to each selection criteria.

— Record-keeping – the committee secretariat shall keep records of all documents in the selection process, including the call for candidates, applications, evaluation, decision, etc.

The committee shall then confirm its recommendation for appointment of the organization selected to be RA via a resolution.

JF.3.5 Appointing a RA

The information that is provided by the committee in the RAC (see JF.3.2 above) is needed to launch the TMB ballot appointing the RA, as well as the ISO Council ballot if the RA intends to charge fees. In the case of revisions, approval from the TMB or ISO Council is not needed if the committee decides that the same RA should continue (see JF.3.4).

The ISO/IEC Directives Part 1 state that RA may charge fees for the Registration Services subject to authorization by the ISO Council, and as long as the basis of charging fees is strictly on a cost recovery basis.

The information shall therefore be fully complete. In the case of JTC 1 RA Standards, a copy of the RAC Form shall also be provided to the IEC since RA appointments shall all be confirmed by the IEC/SMB (and Council Board) where fees are charged.

JF.3.6 Signing a RAA

Only after the TMB (and Council if fees are charged) has appointed the RA (and in the case of JTC 1 RA Standards, involving the IEC) can the RAA be signed. Signing a RAA based on the ISO/CS template is mandatory for all RAs. The RAA shall be signed before publication of a new or revised RA Standard. If a RAA is not signed, the new or revised RA Standard shall not be published.

Because a RA Standard can change throughout the phases of development, the RA shall be provided with the version of the RA Standard that is ready for publication. The list of RA tasks is referenced in the RAA.

To ensure consistency and equality of treatment between the different RAs, any requested deviations from the RAA template which ISO/CS considers to be significant in nature shall be submitted to the TMB for approval.

JF.3.7 Terminating a RA

Termination of RAs could occur when the RA resigns or because the committee and ISO/CS have come to the conclusion (due to withdrawal of the RA Standard, customer complaints, etc.) that the

RA should no longer continue to provide the Registration Services. The committee shall notify the TPM of any issue that it becomes aware of that could lead to the end of the Registration Services by the RA.

The process detailed in JF.3.4 above should be followed in the selection of a replacement RA unless the committee has identified an alternative RA candidate that meets the selection criteria in 3.4 and going through the selection process for additional RA candidates would cause disruption in the RA Services.

The RAA defines a notice period in case of the termination of an RA. This notice period is to be used for the selection process of a new RA so that the transition from one RA to the other is seamless.

JF.3.8 Implementing a RA Standard

JF.3.8.1 Role of the RA

The RA provides the Registration Services by:

— providing the Registration Services described in the RA Standard, and

— respecting the provisions of the RAA.

JF.3.8.2 Role of the committee

Although RAA are signed by the RA and by the ISO Secretary-General, the signature of a RAA by the Secretary-General binds all components in the ISO system, including ISO members and ISO committees. The central role is played by committees. In addition to identifying the need for a RA standard (3.2), drafting the RA Standard (3.3) and selecting a RA (3.4) for both new and revised RA Standards, the committee has the main responsibility for the oversight of the RA as follows:

— Answering questions: The committee shall be available to the RA to answer questions about the RA Standard and clarify any expectations regarding their role in implementing the RA Standard.

— Assessing RA's annual reports: The RAA requires RA to provide annual reports with a specified list of elements. The committee shall ensure that these annual reports are provided and shall read through them.

— Monitoring: In addition to the annual RA report, the committee shall also collect any feedback from industry and users of the RA Standard. Based on all of these elements (RA report and other feedback), the committee shall report to ISO/CS (see below).

— Reporting to ISO/CS: At least once per year and based on the information collected under Monitoring above, the committee shall provide a report to the responsible TPM using the Annual Committee Report to TPM ("ACR") Form (See www.iso.org/forms.Annex S]). The purpose of such reports is to confirm that the RA operates in accordance with the RAA or to raise any concerns (concerns can include: RA Standard not meeting industry or user needs, complaints about the quality of the Registration Services, etc.). Such reports shall be provided at least annually to the responsible TPM or more frequently if the committee deems it necessary. The TPM may also ask for ad hoc reports. If the report identifies concerns, it shall include the planned Corrective measures (see below) needed to address these concerns.

— Corrective measures:

- — By the committee: the committee is responsible for implementing the corrective measures that are within its area of responsibility. These could include: revising the RA Standard, providing advice and guidance to the RA, carrying out audits or recommending the termination of the RAA to ISO/CS in severe cases.

- — By ISO/CS: the corrective measures that fall within the responsibility of ISO/CS (e.g. updating or overseeing the RAA) will be coordinated by the TPM. The TPM may also recommend corrective measures.

- — By the RA: the RA is responsible for any corrective measure that are within its area of responsibility, which would include the Registration Services and the provisions described in the RAA.

— Dispute resolution: In case of a dispute with customers of RA Services, the RA shall act diligently to resolve it. The RA shall bring disputes that it cannot resolve to the attention of the committee (as stated in the RAA) who shall advise the responsible TPM. The role of the committee (and ISO/CS) is limited to supporting the RA address the dispute. The committee shall not take over responsibility for the dispute or become the appellate body for disputes against the RA because this may inadvertently give the impression that ISO is responsible for the Registration Services.

— Maintenance of records: All key communications and documentation from and about the RA shall be archived. The committee secretariat is responsible for ensuring that these are maintained in a separate folder on e-committees.

The committee may create a subgroup [often referred to as a Registration Management Group ("RMG")] in order to help them with the above. Committees (either directly or through the RMG) shall not participate or get involved in providing the Registration Services.

JF.3.8.3 Role of the ISO/CS

The committee's interface with ISO/CS is through the responsible TPM. The role of the TPM includes:

— Identification of RA Standards during the development process if not done by the committee.

— Providing guidance and advice for the drafting of RA Standards.

— Training committees on the RA Policy.

— Coordination with committees to ensure compliance with the RA Policies, quality of Registration Services, appropriate handling of complaints, addressing industry and users' needs, including addressing the concerns raised in the annual reports provided by committees (using the ACR Form) and recommending and assisting in the implementation of any corrective measures (see JF.3.8.2).

— Maintenance of records in relation to his or her involvement.

Reference documents

The following are links to reference documents for the JTC 1 community on a number of important subjects.

- JTC 1's web page (www.jtc1.org)

- ISO/IEC Directives, Parts 1 & 2, Consolidated JTC 1 Supplement (www.iso.org/directives) and (www.jtc1.org)

- JTC 1 Standing Documents (www.jtc1.org)

- How to write standards
(http://www.iso.org/iso/how-to-write-standards.pdf)

- Guidance on twinning in ISO standards development activities
(http://www.iso.org/iso/guidance_twinning_ld.pdf)

- Policy concerning normative references in ISO publications
(https://connect.iso.org/pages/viewpage.action?pageId=27592570)

- Guidelines for the submission of text and drawings to ISO/CS
(http://isotc.iso.org/livelink/livelink/open/18862226)